7411

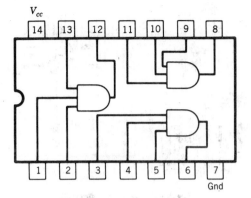

Three-Input AND-gate   7411, 15

7432

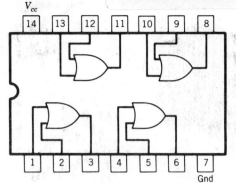

Two-Input OR-gate   7432

7420

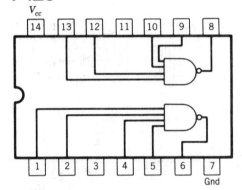

Four-Input NAND-gate   7420, 22, 40

7442

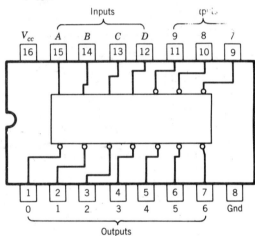

BCD-to-Decimal Decoder   7442

7421

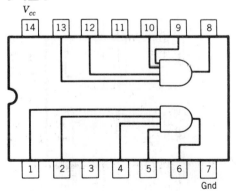

Four-Input AND-gate   7421

7447

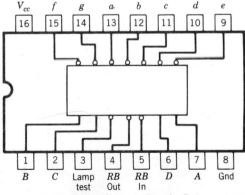

7447   Seven-Segment Decoder/Driver

7451

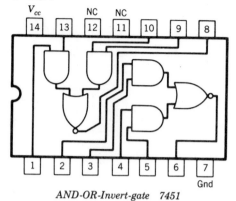

AND-OR-Invert-gate   7451

7474

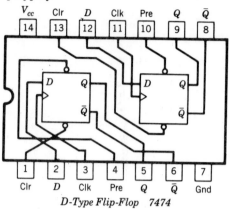

D-Type Flip-Flop   7474

7472

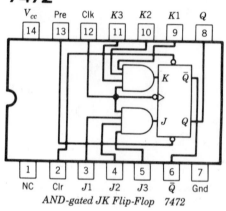

AND-gated JK Flip-Flop   7472

7475

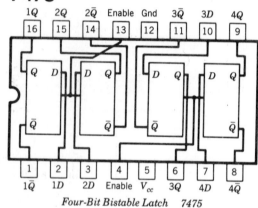

Four-Bit Bistable Latch   7475

7473

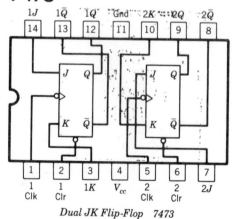

Dual JK Flip-Flop   7473

7476

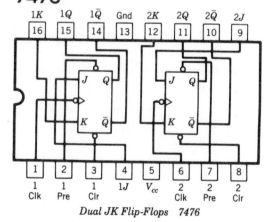

Dual JK Flip-Flops   7476

ESSENTIAL ELECTRONICS

ESSENTIAL ELECTRONICS

Warren Fenton Stubbins
University of Cincinnati

John Wiley & Sons
New York   Chichester   Brisbane   Toronto   Singapore

Copyright © 1986, by John Wiley & Sons, Inc.

All rights reserved. Published simultaneously in Canada.

Reproduction or translation of any part of
this work beyond that permitted by Sections
107 and 108 of the 1976 United States Copyright
Act without the permission of the copyright
owner is unlawful. Requests for permission
or further information should be addressed to
the Permissions Department, John Wiley & Sons.

Library of Congress Cataloging in Publication Data:

Stubbins, Warren Fenton.
  Essential electronics.

  Bibliography: p. 493
  Includes index.
  1. Electronics.   I. Title.
TK7816.S74 1986   621.381   85-9447
ISBN 0-471-88604-1

Printed in the United States of America

10 9 8 7 6 5 4 3 2

This book is dedicated to
Lucille Elizabeth Kehrer Stubbins
a dear and true partner.

PREFACE

Electronics pervades every aspect of modern life. Its importance increases with each advance in technology and with the drive to computerize human tasks and industrial processes. This textbook presents the fundamental concepts on which modern electronics is based and on which the technology will advance. It is written for science and engineering students, and its material was developed in courses we taught over several years.

The nature of electronics is changing rapidly as integrated circuits of unbelievable complexity and capability are developed both in the digital and analog realms and in the marriage between them. But the principles that form the bases of the electronic revolution are the elementary principles of physics and electrical engineering. The aim of this book is to put these basic concepts in perspective and to relate them to the student's desire to understand and apply electronics. We have found that this is the best way to introduce this interesting and challenging subject.

Only a few of the students who were taught the material in this book had earlier experience or instruction in electronics (often consisting of no more than the dc circuits and electricity offered in an introductory physics course), but all were able to learn the material and to do the exercises. The text does not presuppose a more extensive background. The participation of students in the exercises, however, is an essential step in their learning electronics and in making the knowledge useful to them.

The material in the book may be divided into four parts. The first six chapters deal exclusively with digital electronic concepts. In combination with the first 14 laboratory exercises in Appendix B, they may be used as a self-contained text for a course limited to digital electronics. The principles of digital electronics can be taught without exploring or learning the electrical nature of the digital gates because the flow of digital signals is the focus of the work. In the second portion of the text begins the work in analog electronics, in which we must examine the nature of electrical signals rather than the on-off nature of digital signals. Chapters 7 to 10 and Chapter 14 provide the basis for analog electronics. Chapters 11, 12, and 13, which deal with analog electronics and its union with digital electronics, constitute the third portion of the text. The fourth portion includes Appendix A, which discusses computer arithmetic, and Appendix B, the laboratory exercises.

Chapter 1 introduces the basic logic gates and establishes the Boolean algebraic nature of digital signals. Chapter 2 develops Boolean algebra appropriate to the application of logic gates and gives several techniques for simplifying logic-gate circuits. Combinational digital electronics is the subject of Chapter 3, in which combinations of digital gates provide various logic functions. Sequential digital circuits are the subject of Chapter 4, and Boolean algebra is applied to achieve specific sequential patterns. Chapter 5 addresses digital memory elements in various forms. The strictly digital portion of the text concludes in Chapter 6 with an examination of how the digital elements may be brought together to form an elementary microprocessor, in which a simple algorithm is used to solve an arithmetic problem.

In Chapter 7, elementary electrical circuit theory explores dc and ac properties of the circuit elements—resistor, capacitor, and inductance—and their behavior under transient and sine wave signals. This chapter also presents Thevenin's theorem, the rotational operator, and electrical resonance. Chapter 8, which discusses solid-state electronic devices, presents an elementary account of the physics of semiconductors and develops the properties of the *pn* junction. Chapter 8 includes a description and the characteristics of bipolar-junction transistors, field-effect transistors, and metal-oxide-semiconductor field-effect transistors. Chapter 9 examines the nature of electronic amplifiers and the role that feedback plays in changing the amplifying properties. In Chapter 10 a transistor amplifier, the common-emitter amplifier, is used as a design example in preparation for describing the transistor nature of TTL (Transistor-Transistor-Logic) gates. Other material in this chapter pertains to registers and memory elements and their operation. Chapter 14 discusses the power supply, a necessary part of every electronic device, which furnishes regulated voltages to operate the digital and analog electronic elements.

Chapter 11, on the subject of operational amplifiers, explores analog electronics. The emphasis on operational amplifiers reflects our opinion that many analog electronic applications and problems can be satisfied by operational amplifiers and that the use of these elements is the simplest approach for an inexperienced user of electronics. The chapter covers the properties, the limitations, and the arithmetic use of operational amplifiers. Chapter 12 brings analog and digital electronics together and shows how they unite in instrumentation, control, and measuring circuits as analog-to-digital converters and digital-to-analog converters. This chapter also examines the forms of several conversion techniques. Chapter 13 discusses imaginative uses of electronics and suggests further possibilities to the student.

Appendix A concerns the conversions to and from binary, octal, decimal, binary-coded-decimal, and hexadecimal number systems, and gives two algorithms. One shows how division may be performed only by iterative multiplication and addition by a computer to produce the quotient in a fraction of the time required for conventional long division of large numbers. The other is a square root algorithm.

Appendix B is a set of 30 laboratory exercises that illustrate the electronic principles in the text and prepare the student for independent ventures in the applications of electronics. Exercise 1 to 14 are digital exercises, which parallel the presentation of digital electronics in the first portion of the text. The next 11 exercises develop an understanding of analog electronics and the properties of operational amplifiers. Finally, 5 exercises illustrate unusual electronic possibilities and applications to measurements.

These exercises have evolved over several years with the help of student response. Each exercise contains the background and rationale for the exercise and raises points that relate the exercise to the material in the text.

Throughout the chapter discussions, rather than examining particular digital or analog electronic elements, we have restricted the presentation to the concept that the element satisfies. Thus, for example, the property of an AND-gate presented in the text is common to all electronic elements called AND-gates. Only in the exercises do we identify specific elements; there we use the simplest, most reliable, most useful elements, and perhaps the ones whose limitations may be uncovered most easily. From a practical standpoint we discuss in the text and illustrate in the exercises many of the difficulties that may arise when electronic elements are placed together to solve a logic, control, or measurement function.

The instructor's manual contains answers to the chapter questions and to the queries in the exercises and supplies additional technical details about the electronic elements used in the exercises. The manual also lists the instruments required for the laboratory exercises and includes suggestions for using the text and the exercises.

It is my pleasure to acknowledge the encouragement of Ernst Franke and many other colleagues and students and the very competent assistance of Patricia King and Karen Feinberg in the preparation of the text.

Warren Fenton Stubbins

CONTENTS

CHAPTER 1
BASICS OF DIGITAL ELECTRONICS 3

Binary Logic **4**
Electrical Equivalent **4**
Binary Logic
 Elements—Gates **6**
AND-gates **6**
NAND-gates **8**
Inverter **10**

OR-gates **11**
NOR-gates **12**
Exclusive-OR-gates **14**
Equivalence-gate **15**
Summary **16**
Problems **16**

CHAPTER 2
BOOLEAN ALGEBRA FOR DIGITAL SYSTEMS 17

Rules of Boolean Algebra
 for Digital Devices **18**
De Morgan's Theorems **21**

Representation of
 Boolean Algebra **23**
Venn Diagrams **24**

Demonstration of
  De Morgan's Theorems 24
Three-Variable
  Venn Diagrams 26
Equivalent Functions 26
Equivalence by
  Venn Diagrams 27
Equivalence by
  Chart Simplification 27
Simplification
  by Karnaugh Maps 29
Multivariable
  Karnaugh Maps 30
Summary 33
Problems 33

CHAPTER 3
COMBINATIONAL DIGITAL ELECTRONICS 37

Combinational
  Circuits 40
Diodes 40
Logic Gates 43
Fan-Out 43
Open-Collector Gates 44
Logic Functions 45
Half Adder 46
Half Subtractor 48
Binary Subtraction
  by the Complement Method 50
Full Adder 51
Full Adder with
  Two Half Adders 53
Binary and
  Decimal Numbers 54
Decoders and
  Number Systems 55
Binary-to-Octal Decoder 58
Octal-to-Binary Encoder 59
Multiplexer 60
Demultiplexer 61
BCD to Seven-Segment
  Display Decoder 62
Digital Comparators 65
Summary 67
Problems 67

CHAPTER 4
SEQUENTIAL DIGITAL ELECTRONICS 69

RS Flip-Flop 71
Data Flip-Flop 72
Clocked Flip-Flops 74
Clocked *RS* Flip-Flop 74
Characteristic Equation
  for Clocked *RS* Flip-Flop 76
Clocked Data Flip-Flop 77
JK Flip-Flop 78
Toggle Flip-Flop 80
JK Master-Slave Flip-Flop 81
Analysis of
  Sequential Circuits 85
Ripple Counters 90
Divide-by-*N* Counter 91

Design of
 Sequential Circuits **92**
Synchronous
 Sequential Circuits **98**
Synchronous
 Binary Counter **98**
BCD Synchronous
 Counters **99**
Up-Down Counter **100**
Review **100**
Oscillators or Clocks **101**
Monostable Oscillators **102**
Problems **103**

CHAPTER 5
DIGITAL MEMORY 105

Registers **106**
Shift Registers **107**
Serial Adder **108**
Universal
 Shift Registers **110**
Data Registers **112**
High-Density
 Digital Memories **114**
Static RAM Cell **115**
Operative Elemental
 Static RAM Cell **116**
Read Only Memory (ROM) **119**
PROM **120**
EPROM **121**
Summary **122**
Reference **122**

CHAPTER 6
MICROPROCESSOR BASICS 123

Square Root Algorithm **125**
Program for
 a Microprocessor **127**
Operational Code **128**
Multiplication Algorithm **131**
Subroutine for
 Multiplication Algorithm **133**
Program for
 Square Root Algorithm **136**
Microprocessor
 Architecture **139**
Arithmetic Logic Unit **143**
Summary **147**
Problems **148**
References **148**

CHAPTER 7
BASIC ELECTRICAL CIRCUIT THEORY 149

Ohm's Law **150**
Conventions for Using Kirchhoff's Laws **151**
Series Resistors **152**
Parallel Resistors **153**
Kirchhoff's Laws and Conventions **155**
Simplification of Resistor Circuits **156**
Bridge Circuit **160**
Thevenin's Theorem **162**
Time-Dependent Circuits **166**
Capacitance **167**
Inductance **171**
Current-Voltage Relations in Passive Elements **173**
Alternating Currents **174**
Passive Elements and Sine Wave Signals **177**
Rotation Operation j **180**
Reactances **181**
Capacitive Reactance **181**
Inductive Reactance **182**
AC Resistance-Capacitance Circuit **182**
Frequency Dependence of RC Circuit **187**
AC Resistance-Inductance Circuit **189**
Simplification of Circuits with Impedance **190**
Series Resonance Circuit **192**
Parallel Resonance Circuit **195**
Electrical Power **198**
Summary **200**
Problems **200**

CHAPTER 8
SOLID-STATE ELECTRONIC DEVICES 205

Extrinsic Semiconductors **208**
PN Junction **211**
The Diode: Reverse-Biased **212**
The Diode: Forward-Biased **214**
Practical Diode **215**
Zener Diode **216**
The Transistor **217**
Current Relations in a Transistor **219**
Junction Field-Effect Transistor (JFET) **222**
Metal-Oxide-Semiconductor Field-Effect Transistor (MOSFET) **226**
Complementary Metal-Oxide Semiconductor (CMOS) **230**
Summary **231**
Problems **231**
References **233**

CHAPTER 9
AMPLIFIERS AND FEEDBACK 235

Analog Amplifier Characteristics **236**
Decibels **240**
Feedback in Analog Amplifiers **241**
Midfrequency Feedback **243**
Advantages of Negative Feedback **243**
Feedback at all Frequencies **248**
Bode Plot **249**
Stability of Analog Amplifiers with Feedback **250**
Nyquist Criterion for Stability **251**
Bode Criterion for Stability **252**
Summary **253**
Problems **254**

CHAPTER 10
ELEMENTS OF TRANSISTOR CIRCUITS 255

Common-Emitter Amplifier **256**
Design Example **258**
Emitter-Follower Amplifier **261**
Simple Regulating Circuits **262**
Difference Amplifier **264**
Transistor-Transistor-Logic Gates **266**
NAND-gate **269**
NOR-gate **270**
Open-Collector Gates **271**
Tristate Gates **272**
Junction Field-Effect Transistors **273**
MOSFET Amplifiers **276**
MOS and CMOS Logic Gates **278**
Dynamic Shift Registers **281**
Dynamic RAMs **282**
EPROMS **287**
Summary **288**
Problems **289**
References **289**

CHAPTER 11
OPERATIONAL AMPLIFIERS 291

Operational Amplifier **292**
Inverting Amplifier **293**
Noninverting Amplifier **298**
Voltage Follower **299**
Frequency Dependence **300**
Slew Rate **301**
Difference Amplifier **302**
Common-Mode Rejection Ratio **304**

xvi Contents

Arithmetic Uses of
  Operational Amplifiers 305
Summing Amplifier 306
Integrating Amplifier 308
Logarithmic Amplifier 310
Input-Offset Voltage 312

Input-Bias Current 312
Matching Output-Current
  Requirements 313
Summary 314
Problems 314
References 316

CHAPTER 12
ANALOG-DIGITAL CONVERSION 317

Analog-to-Digital Conversion 319
Comparator Ladder 320
Successive-Approximation
  Conversion 322
Subranging 323
Sample-and-Hold Circuits 323
Compression and Expansion 324
Nonlinear Analog-to-Digital
  Conversion 327
Tracking Analog-to-Digital
  Converter 328
Dual-Slope Analog-to-Digital
  Conversion 329
Single-Slope Analog-to-Digital
  Conversion 331

Staircase Analog-to-Digital
  Converter 332
Voltage-Controlled Oscillators 333
Digital Sampling of a Sine Wave
  334
Restoration of the Analog Signal
  337
Digital-to-Analog Conversion 337
Bit-Weighted Summing Amplifier
  338
R-$2R$ Current Ladder 340
Pulse Modulation
  and Demodulation 342
Summary 345
Problems 345

CHAPTER 13
"SMART" ELECTRONICS 347

Suppression of Transients 347
Missing-Pulse Detection 349
Circuits with Hysteresis 350
Impedance Improvement 352
MOSFET Switches 353
Controllable
  Buffer-Inverter 354

Parity Check 355
Error Correction 356
Read Only Memories 358
Programmable Logic Array 358
Summary 361
Problems 362

CHAPTER 14
POWER SUPPLIES 363

Root-Mean-Square Voltage 364
Voltage Amplitude
  Transformation 366
Rectification 367
Capacitance Filter 368
Full-Wave Rectifier 371
Rectifier-Circuit
  Response 372
Example of a
  Power-Supply Design 374
Bridge Rectifier 375
Additional Filtering 376
Voltage Doublers 378

Voltage Regulators 379
Example of a Power-Supply
  Performance 381
Regulation by Operational
  Amplifiers 383
Shifting the
  Regulated Voltage 384
Current Regulation 385
Switching Power Supplies 386
Summary 389
Problems 389

APPENDIX A
COMPUTER ARITHMETIC 391

Binary Codes 391
Conversion of Numbers of
  One Base to Another Base 392
Binary-to-BCD
  Conversion 394
BCD-to-Binary
  Conversion 395
Signed Numbers 396

Decimal Arithmetic 399
Double-Precision
  Arithmetic 400
Wallace Division
  Algorithm 401
Wallace Square Root
  Algorithm 401
Reference 402

APPENDIX B
LABORATORY EXERCISES 403

BIBLIOGRAPHY 493

INDEX 497

ESSENTIAL ELECTRONICS

INTRODUCTION

The nature of electrical signals in electronic circuits readily enables the technology to be divided into classes. One of the classes is *analog* electronics; another is *digital* electronics. Both analog and digital electronics use similar electronic elements, but the manner of use is different, and the technologies appear to be quite distinct. For this reason we shall study them separately until we bring them together as they invariably unite in instrumentation and applications.

Analog electronics pertains to those systems in which the electrical voltage and electrical current are analogous to physical quantities and vary continuously. Electronic circuits that reproduce music must have voltages and currents that are proportional to the sound. A high fidelity amplifying system attempts to keep the analogy as true as possible. Analog electronic circuits are carefully designed to make the electrical voltages and currents follow the input signal. If an input signal doubles in amplitude, the output voltage or current also should double; this is possible because the circuit elements are made to operate within limits that preserve the linearity.

An electrical voltage that is proportional to temperature and changes smoothly as the temperature changes is an analog of temperature. If the temperature range is divided into small increments, however, and a numerical assignment is made to each increment, then the temperature may be indicated by a digital display. As the temperature (voltage) changes smoothly, a decision must be made by an electronic system as to the numerical value to be displayed as the temperature. The circuit making the decision is called an *analog-to-digital converter*, ADC. The inverse process is accomplished by a *digital-to-analog converter*, DAC.

Digital electronic circuits do not require the linearity of analog circuits. Digital circuits act as electronic switches and switch from one state to another. The output state, on or off, is the only signal condition to be examined. In digital circuits the output state is determined by the input signals in as direct a manner as the output voltage of an analog circuit is related to the input signal. In digital circuits the relation between input and output states are expressed as logic equations; the elements of digital electronics are called logic gates. Logic gates switch between states, on or off, very quickly so that they may operate at many megahertz in computers and other applications.

As technical developments continue to provide new and amazing integrated circuits, as they have since the 1960s, both analog and digital systems will be more capable. The designers of electronic systems using integrated circuits will have unlimited possibilities for innovation.

The topics that follow present the foundations of modern electronics. The technology can be comprehended only by emphases on basic electrical and electronic principles. We believe that by learning the basics the student can decipher modern electronics. The topics and the organization of the text were chosen for this purpose.

CHAPTER 1

BASICS OF DIGITAL ELECTRONICS

The approach to digital electronics is straightforward. The details of the involved electronic processes in the transistors and other circuit elements can be put aside, and the character of digital electronics can be revealed by the flow of the electrical signal alone. Taking this simplified approach, we can guide our efforts and illustrate digital technology with binary logic, which considers only two states: on and off.

We can approach the subject by this simplified route because the design of the electronic components and their combination as integrated circuits are the technological miracle of our time. We can compare our approach to digital electronics with that of a machine designer who carefully chooses among the available gears those that will meet his requirements, but does not concern himself with the design of the teeth on the gears. This approach is possible because of the invention of integrated circuits (IC) in 1958, their combination into the more complex medium-scale integrated circuits (MSIC), the still more complex large-scale integrated circuits (LSIC) in the 1970s; finally, the recent very large-scale integrated circuits (VLSIC) has brought this possibility to us.

To prepare ourselves to understand and use integrated circuits, we examine the three basic electronic elements with which all digital circuits are made and the binary logic they follow.

BINARY LOGIC

Binary logic is applied to the study of the state of a system in which the electrical signal is either present or absent. The presence of a signal is designated by 1 and the absence by 0, or by *high* and *low* or *on* and *off*, respectively.

In the actual digital circuits the condition 1 means the presence of a reasonable voltage, for example, 3 to 5 volts (V), while 0 represents a much smaller voltage, for example 0 to 0.5 V. The difference between 1 and 0 must be sufficient to allow no mistake in recognizing the state of the element, that is, 1 or 0. Such a relationship is called positive logic, in which the high voltage is associated with the 1. (Negative logic associates the 1 with the low voltage.) We will use positive logic throughout this work.

The voltages required in systems using integrated circuits are typically ± 12 and $+5$ V, but specific elements may require other voltages. Higher voltages are not generally required except perhaps for output devices. It is extremely important to stay within the rated voltages. For example, a slight overvoltage, such as 5.25 V for 5.0 V ICs, may be acceptable, but greater voltages may degrade the integrated circuits.

To say that an electronic element is in state 1 means that its *output* is near 5.0 V; in state 0, that its *output* is near the zero volt level. These two states are the only ones possible. Later we will consider an element that can be electrically disconnected so that its state is not involved in the binary logic. The element that can be electrically connected or electrically disconnected is called a tristate device: on, off, or electrically disconnected.

All digital circuits, from simple combinations of binary logic circuits to the most sophisticated computers, digital signal processing systems, and artificial intelligence devices are built of three basic binary elements. The three perform the equivalent of multiplication, addition, or negation (inversion). The flow of these voltage states through the binary logic elements constitutes the electronic process in computers and other digital systems.

Our purpose is to understand these three basic binary elements and to learn how they may be combined to serve any control, computational, decision, or other process we wish to implement.

ELECTRICAL EQUIVALENT

Although integrated circuits may contain many electrical circuit components, such as resistors, capacitors, diodes, and transistors, the circuit with binary logic is exactly equivalent to a simple electrical circuit composed of a fixed resistor and a variable resistor in series with the supply voltage across them. The output voltage is taken across the variable resistor. Figure 1-1 shows this simple arrangement.

We designate the fixed resistor by R_L because it is typically called the load resistor. By external control the variable resistor R may have its resistance changed to a value many times greater than R_L or to a value many times smaller than the

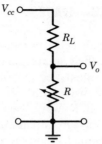

FIGURE 1-1
Electrical Equivalent to a Circuit with Binary Logic
This resistance voltage divider portrays the underlying nature of all electronic devices.

resistance of the fixed resistor R_L. The resistor R in an electronic circuit is an element such as a transistor or even a switch whose conduction (resistance) is determined by the electrical signal on its input. An open switch is equivalent to an extremely large resistance, and a closed switch is the same as a very small resistance. Of course, R may be controlled to have a value near R_L; this is done in the realm of analog electronics, for example, in audio amplifiers. We will present analog electronics after we have examined digital electronics, and then we will show how they work together.

Let us consider the two binary conditions: (a) $R \gg R_L$, and (b) $R \ll R_L$. By simple circuit analysis the voltage V_o is related to the supply voltage V_{cc} by the fraction of the total resistance borne by R.

$$V_o = V_{cc} \frac{R}{R + R_L} \qquad (1\text{-}1)$$

For conditions (a) $R \gg R_L$, the output voltage V_o is approximately V_{cc}. For example, if $R = 20R_L$, then $V_o = \frac{20}{21}V_{cc}$.

For condition (b) $R \ll R_L$, the output voltage V_o is approximately zero. For example, if $R = R_L/20$, then $V_o = \frac{1}{21}V_{cc}$.

In our designation of the binary state the condition (a) is 1 or *on* or *high*, while condition (b) is 0 or *off* or *low*. Since we are examining only the *output* voltage, the logic level is stated independently of the internal operation of the circuit components that make up the element.

In digital circuits the transition from one state to another is made rapidly, and the value of the voltage during the transition is not considered. Thus the smooth time-varying relation between input voltages and output voltages, so important in audio amplifiers and other analog circuits, is not required for digital electronics. The main task is the recognition of the logic state when it has been reached. Digital circuits used in microprocessors and other applications must complete transitions from one state to the other in a few nanoseconds or less.

6 Basics of Digital Electronics

BINARY LOGIC ELEMENTS—GATES

Binary logic elements, commonly called *gates*, are represented by simple symbols that indicate their nature and the binary logic they obey. It is easy to study the flow of electrical signals representing the two binary states by showing only the logic elements and disregarding the connections for electrical power and the grounded reference point. However, *power connections must be made for the gates to perform.*

The logic of signals that are *on* or *off*, the only two states that may exist, has a basis in the branch of formal mathematics called *Boolean algebra.* Boolean algebra is named for George Boole (1815–1864), who wrote *An Investigation of the Laws of Thought, on which are founded the Mathematical Theories of Logic and Probabilities*, published in 1854. Boole was concerned with logical propositions that were true or false. Fortunately his work was available when computer technology needed it; we examine Boolean algebra in the next chapter and use it later to help simplify configurations of logic gates.

AND-GATES

The symbol for the electronic *AND-gate* is shown in Figure 1-2. 1 (5 V) on input *A* and 1 (5 V) on input *B* result in 1 (3 to 5 V) on the output, *C*. Other combinations of input states 1 or 0 result in a 0 (0 to 0.5 V) on the output, *C*. The behavior of the gate is shown in Table 1-1. This tabulation, called a *truth table*, describes the output for any combination of inputs. Each kind of logic gate has its own truth table; Table 1-1 applies to every two-input AND-gate.

The truth table can be summarized by the Boolean algebra expression

$$A \times B = C$$

An AND-gate performs the logical function of multiplication. From Table 1-1 we can confirm that *A* and *B* = output *C*, that is, $0 \times 0 = 0, 0 \times 1 = 0, 1 \times 0 = 0$, and $1 \times 1 = 1$.

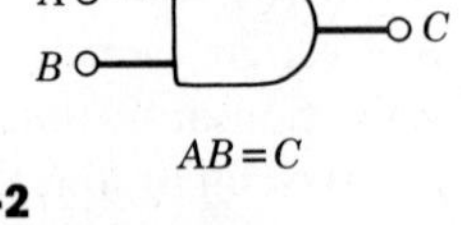

$$AB = C$$

FIGURE 1-2

A Two-Input AND-gate

Input *A* AND input *B* yield the value of output *C* in accord with the Boolean algebra, $A \times B = C$; that is, an AND-gate represents logical multiplication. The connections for electrical power and electrical ground have been omitted. The input and output voltages are measured from reference ground, which is implied but not shown.

TABLE 1-1
Truth Table for Two-Input AND-gates

Input A	Input B	Output C
0	0	0
0	1	0
1	0	0
1	1	1

An AND-gate can have more than two inputs. Figure 1-3 shows an AND-gate with four inputs, and Table 1-2 is the truth table for this figure.

Table 1-2 shows that any combination of the four input states that is not all 1s will give 0 as the output. *A and B and C and D* must have input 1 if the output *W* is to be 1. Table 1-2 is abbreviated because it is not necessary or interesting to show all the $2^4 = 16$ possible combinations of binary inputs. The Boolean algebra is

$$A \times B \times C \times D = W$$

so that if any input is 0, the output is 0. The truth table confirms this formula.

$ABCD = W$

FIGURE 1-3

A Four-Input AND-gate

The output of a four-input AND-gate is determined by the Boolean multiplication, $A \times B \times C \times D = W$.

TABLE 1-2
Truth Table for Four-Input AND-gates

Input A	Input B	Input C	Input D	Output W
0	0	0	0	0
1	0	0	0	0
0	1	0	0	0
·	·	·	·	0
·	·			
1	1	0	1	0
1	1	1	0	0
1	1	1	1	1

8 Basics of Digital Electronics

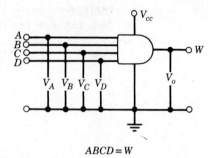

$ABCD = W$

FIGURE 1-4

A Four-Input AND-gate with Reference Ground

This figure shows the electrical power connections and the voltage reference ground from which the inputs and outputs are measured. Circuit lines that cross are not connected. Electrical connections are shown by a dot at the crossings.

Figures 1-2 and 1-3 do not show the power supply connections or the common ground from which the input and output states (voltages) are measured. Figure 1-4 includes the common ground for all the voltages in the system, the measuring points for the four-input AND-gate, and the power connections. Figure 1-4 allows us to illustrate two common features of circuit diagrams involving gates. The small circles at the ends away from the gate on the input and output lines merely indicate connection points, *but this is not the case* when the circle is adjacent to the gate, as we will show below. The second feature is that lines representing conductors crossing each other are *not connected* unless there is a dot at their crossing.

NAND-GATES

Another circuit, which has logic inverse to that of an AND-gate, is called the "Not-AND-gate," or simply the "NAND-gate." We note that in this case the term "Not" does *not* mean the opposite electrical polarity; that is, $+5\text{V}$ does *not* become -5V. It means that the inverse of $+5\text{V}$, which is 0V, pertains; the inverse of presence $(+5\text{V})$ is absence (0V).

A two-input NAND-gate is shown in Figure 1-5. It differs from the AND-gate in Figure 1-2 by a *small circle adjacent to the gate* on the output line. Table 1-3 is the truth table for the two-input NAND-gate.

A comparison of Table 1-1 with Table 1-3 shows that the output state of the NAND-gate $\bar{C}$ is the inverse of the output state of the AND-gate C with the same

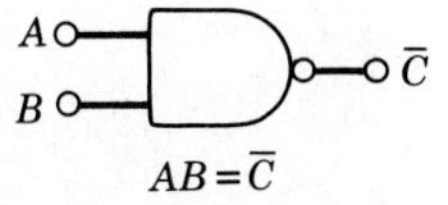

$$AB = \bar{C}$$

FIGURE 1-5

A Two-Input NAND-gate

The output of the NAND-gate is the complement of the output of the AND-gate with identical inputs, as shown in the symbol by the circle at the body of the gate on the output lead. The Boolean algebra is $A \times B = \bar{C}$.

input states. This is designated by $\bar{C}$, a C with a bar over it. In the figures, a small circle adjacent to the gate means that the complement (inverse or "negative") of the state that would otherwise result is provided at that point. The truth table can be summarized by the Boolean algebra expression

$$A \times B = \bar{C}$$

The change of a state from 1 to 0 or from 0 to 1 is termed *taking the complement*. Thus we can say that the output of a NAND-gate is the *complement* of the output of an AND-gate with the same inputs. $\bar{C}$ is the complement of C. The process of taking the complement is discussed more fully in a later chapter.

The simple logic of the NAND-gate is shown by the electrical switch circuit of Figure 1-6. An open switch corresponds to an input of 0, and a closed switch is equivalent to 1 on an input. When both switches A and B are open ($A = 0$, $B = 0$) and when either A or B is open and the other is closed ($A = 0$, $B = 1$ or $A = 1$, $B = 0$), the output is $V_o = V_{cc} = 1$. However, if both switches are closed ($A = 1$, $B = 1$), the output is $V_o = 0$ V (ground), which corresponds to the state 0. The 0 state occurs in the case of $R \ll R_L$, as discussed in conjunction with Figure 1-1, because closed switches (1, 1) are equivalent to extremely small resistance.

TABLE 1-3
Truth Table for Two-Input NAND-gates

Input A	Input B	Output $\bar{C}$
0	0	1
1	0	1
0	1	1
1	1	0

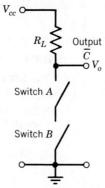

FIGURE 1-6
Electrical Circuit Equivalent of a Two-Input NAND-gate

The switch logic, 0-open, 1-closed, yields an output, V_o, identical with Table 1-3, the NAND-gate truth table.

INVERTER

The second basic logic element needed in forming digital systems is an *inverter*. The inverter is the electronic element that changes its input state 1 to an output state 0, or vice versa. The symbol for an inverter is shown in Figure 1-7. Note the circle adjacent to the element in Figure 1-7. An inverter has a single input and a single output and can be used in any signal line to invert the binary signal. Table 1-4 is the truth table for an inverter.

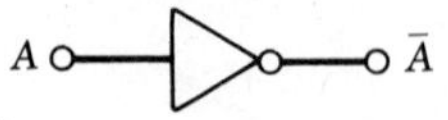

FIGURE 1-7
An Inverter

An inverter complements the input signal by making the output the inverse of the input. The Boolean algebra is $A \leftrightarrow \bar{A}$ so if $A = 0$, $\bar{A} = 1$, and vice versa.

TABLE 1-4
Truth Table for Inverters

Input A	Output $\bar{A}$
0	1
1	0

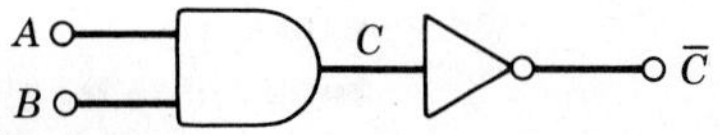

FIGURE 1-8

An AND-gate with an Inverter Becomes a NAND-gate

The relation of C in Table 1-1 to $\bar{C}$ in Table 1-3 verifies the interchange of binary values achieved by the elements of this figure.

A NAND-gate may be made by adding an inverter to the output of an AND-gate to invert its output. Figure 1-8 shows the combination of an AND-gate and an inverter. Note that the complement of the output C of the AND-gate in Table 1-1 is $\bar{C}$, which is the value shown in Table 1-3 for the NAND-gate. Similarly, the addition of an inverter to a NAND-gate yields an AND-gate. The reader will observe that an inverter can be made by tying together the inputs of a NAND-gate, which is equivalent to selecting only the first and last lines in Table 1-3. The electrical-circuit equivalent of an inverter is made by having only one switch in Figure 1-6.

OR-GATES

The third basic logic element needed to form digital systems is an *OR-gate*. The OR-gate provides addition according to Boolean algebra, just as the AND-gate provides multiplication.

The symbol for an OR-gate with three inputs is shown in Figure 1-9. Table 1-5 is the truth table for the three-input OR-gate.

Here we see that the Boolean algebra for OR-gates is addition:

$$A + B + C = Y$$

From Table 1-5 we can confirm that A or B or C = output Y. For example, $1 + 1 + 0 = 1$ illustrates a fundamental property of binary logic: the summation of 1s can only add to 1. This is a statement that binary logic can be true (1) or false (0), but that it cannot be more than true or more than false.

FIGURE 1-9

A Three-Input OR-gate

The Boolean algebra for the three-input OR-gate is given by $A + B + C = Y$.

TABLE 1-5
Truth Table for Three-Input OR-gates

Input A	Input B	Input C	Output Y
0	0	0	0
1	0	0	1
0	1	0	1
0	0	1	1
1	1	0	1
1	0	1	1
0	1	1	1
1	1	1	1

NOR-GATES

The inverse or complement of the output of an OR-gate can be obtained by adding an inverter to its output. Figure 1-10 shows the arrangement of an OR-gate followed with an inverter. The combination is a "Not-OR-gate," or simply a "NOR-gate."

The negative-OR-gate or the not-OR-gate is a common logic element available to the user. The NOR-gate symbol is given in Figure 1-11. It differs from the OR-gate symbol shown in Figure 1-9 by the addition of a circle adjacent to the element

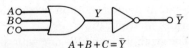

$$A+B+C=\bar{Y}$$

FIGURE 1-10

An OR-Gate with an Inverter Becomes a NOR-gate

The inverter provides the complement of the OR-gate output.

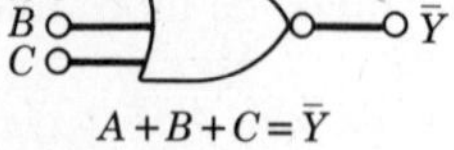

$$A+B+C=\bar{Y}$$

FIGURE 1-11

A Three-Input NOR-gate

The circle on the output lead adjacent to the OR-gate symbol states that the output has been complemented and that it follows the NOR-gate truth table and the Boolean algebra $A + B + C = \bar{Y}$ of addition and complementation.

TABLE 1-6
Truth Table for Three-Input NOR-gates

Input A	Input B	Input C	Output $\bar{Y}$
0	0	0	1
1	0	0	0
0	1	0	0
0	0	1	0
1	1	0	0
1	0	1	0
0	1	1	0
1	1	1	0

at the output. The circle indicates that the OR-gate output is inverted, making it a NOR-gate. Table 1-6 is the truth table for the three-input NOR-gate shown in Figure 1-11; it applies just as well to the combined elements in Figure 1-10.

A comparison of Table 1-5 and Table 1-6 shows that the outputs are the inverse or the complement of each other. Table 1-6 can be summarized by the algebraic equation

$$A + B + C = \bar{Y}$$

We find that the use of NAND-gates and NOR-gates allow us to perform all the logic operations that are possible with AND-gates and OR-gates. Each kind of gate works equally well, but NAND-gates and NOR-gates are more widely used.

The logic element called the NOR-gate is analogous to the arrangement of switches in Figure 1-12. The output $\bar{Y} = 1$ when all switches are open (corresponding to inputs $A = B = C = 0$). $\bar{Y} = 0$ when one or more of the switches are closed

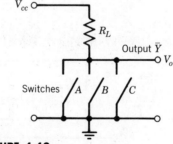

FIGURE 1-12

Electrical Circuit Equivalent of a Three-Input NOR-gate

V_o equals 0 if S_A or S_B or S_C any combination of switches are closed. A binary 1 signifies a closed switch and a binary 0 signifies an open switch.

14 Basics of Digital Electronics

(input = 1). An important point is that the closure of a second or third switch does not give a different output because the output already is grounded (0) by a single switch.

EXCLUSIVE-OR-GATES

A valuable logic element is the exclusive-OR-gate. The exclusive-OR-gate is not considered an elementary gate because it can be made with the elementary gates. However, it is one of many gates available as a unit. It has the property of giving a high output if one input *or* the other is high, but it gives a low output if no input is high *or* if both inputs are simultaneously high. The exclusive-OR-gate is abbreviated as XOR-gate, and the symbol for the XOR-gate is shown in Figure 1-13a. Figure 1-13b is one of several combinations of elementary gates (in this case, NAND-gates) that are equivalent to an XOR-gate.

We emphasize the property of the XOR-gate by generating its truth table, Table 1-7, by applying the truth table for the NAND-gate, Table 1-3, to the elements in Figure 1-13b. The intermediate values, E, F, G, and the output C, are found for each set of values for A and B.

A useful relation for the XOR-gate can be found by writing its truth table in a different manner. Table 1-8 repeats Table 1-7 with $A = 1$ and $\bar{A} = 0$, and the same changes made for B and C.

We see that the second and third lines of Table 1-8 both have C as the output. In these lines we equate C to the terms that produce it (lines 2 and 3) to find

$$C = \bar{A}B + A\bar{B}$$

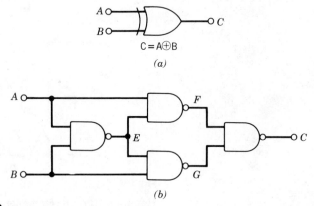

FIGURE 1-13

A Two-Input Exclusive-OR-gate

The exclusive-OR-gate follows the Boolean algebra of $C = \bar{A}B + A\bar{B}$, which is symbolized by $C = A \oplus B$. (*a*) The symbol for an exclusive-OR-gate. (*b*) The combination of elementary logic gates (NAND-gates) forms an XOR-gate.

TABLE 1-7
Truth Table for an XOR-gate equivalent

Line	Input A	Input B	E	F	G	Output C
1	0	0	1	1	1	0
2	0	1	1	1	0	1
3	1	0	1	0	1	1
4	1	1	0	1	1	0

TABLE 1-8
Truth Table for XOR-gates

Line	Input A	Input B	Output C	$D = \bar{C}$
1	$\bar{A}$	$\bar{B}$	$\bar{C}$	D
2	$\bar{A}$	B	C	$\bar{D}$
3	A	$\bar{B}$	C	$\bar{D}$
4	A	B	$\bar{C}$	D

which is the Boolean algebraic relation for the XOR-gate we are seeking. The XOR-gate relation is written as

$$C = A \oplus B$$

which is another way of stating that $C = \bar{A}B + A\bar{B}$, which is read "C equals not-A times B or A times not-B."

EQUIVALENCE-GATE

The addition of an inverter to the output of the XOR-gate creates the equivalence-gate, which provides the complement to the output of the XOR-gate. We designate the complement of C as $\bar{C} = D$. If we examine lines one and four in Table 1-8, we see that the terms in those lines produce the logic relation

$$D = \bar{A}\bar{B} + AB$$

This equation states that $D = 1$ whenever A and B are identical or equivalent; that is, when both equal 0 or both equal 1. This relation is designated by

$$D = A \odot B$$

Another name for the equivalence-gate is the exclusive-NOR-gate or XNOR-gate.

SUMMARY

The three binary logic elements needed to perform all possible digital electronic functions are the AND-gate, the OR-gate, and the inverter. The complementary gates, the NAND-gate and the NOR-gate, along with the inverter, also completely satisfy the logical functions.

We will see that combinations of the three elementary gates yield all the specialized circuits that we will encounter in digital electronics. We will also observe that all the integrated digital circuits, from the simplest to the very largest, are only ingenious combinations of these three basic elements.

PROBLEMS

1-1 Write all the states for the four-input AND-gate shown in Figure 1-4. Repeat for a four-input NAND-gate.

1-2 The range of voltages corresponding to binary states of logic gates made with bipolar junction transistors is 0 to 0.8 V for a *low* (0) and 2.8 to 5.0 V for a high (1). If $R_L = 1 \text{ k}\Omega$, what must be the values of R in Figure 1-1 to ensure these limits when V_{cc} is 5.0 V?

1-3 Use the truth tables for the NAND-gate and NOR-gate to show how each may be made into an inverter.

1-4 Show that by exchanging the positions of R_L and the switches in Figures 1-6 and 1-12, the new truth tables correspond to AND-gate and OR-gate logic, respectively.

CHAPTER 2

BOOLEAN ALGEBRA FOR DIGITAL SYSTEMS

Imagine playing the game called *Twenty Questions*. In Twenty Questions the questioner asks one well-chosen question after another that lead to the identification of a person, object, or place. The person being questioned may answer only "yes" or "no." A record of the answers can be written with ones and zeros corresponding to yeses and nos. After a certain point in the game the score might look like 11001101011..., in which the leftmost character corresponds to the reply to the first question. For example, suppose the person to be identified is George Washington. The first question might be "Is it a person?" The 1 indicates that the answer is "yes." The second question might be "Is it a man?" Again we record a 1. Perhaps the questioner then asks "Is he alive?" and a 0 is recorded. This continues until George Washington emerges as the answer to the accumulated questions.

We can use this game to illustrate several important concepts that are directly related to digital electronics. First, we note how much information can be communicated by a series of binary bits (0 and 1). With twenty bits, $2^{20} = 1,048,576$ different answers can be given. Second, we can imagine a logic gate circuit that uses the bit pattern to point to the specific answer. Third, we perceive that a good question must divide the class of possible answers into two distinct parts, so that a 1 or a 0 can distinguish clearly between two alternatives. If this last condition is satisfied, then the mathematics of binary (two-valued) logic can be used to guide us in the design of our digital circuits.

The mathematics of computers and other digital electronic devices have been developed from the decisive work of George Boole (1815–1864) and many others, who expanded and improved on his work. These include Augustus De Morgan (1806–1871), Friedrich L. G. Frege (1845–1925), Giuseppe Peano (1858–1932), Alfred North Whitehead (1861–1947), Bertrand A. W. Russell (1872–1970), and many later workers. The body of thought that is known collectively as symbolic logic established the principles for deriving mathematical proofs and singularly modified our understanding and the scope of mathematics.

Only a portion of this powerful system is required for our use. Boole and others were interested in developing a systematic means of deciding whether a proposition in logic or mathematics was true or false, but we shall be concerned only with the validity of the output of digital devices. *True* and *false* can be equated with *one* and *zero*, *high* and *low*, or *on* and *off*. These are the only two states of electrical voltage from a digital element. Thus, in this remarkable algebra performed by logic gates, there are only two values, *one* and *zero*; any algebraic combination or manipulation can yield *only these two values*. Zero and one are the only symbols in *binary* arithmetic.

The various logic gates and their interconnections can be made to perform all the essential functions required for computing and decision making. In developing digital systems the easiest procedure is to put together conceptually the gates and connections to perform the assigned task in the most direct way. Boolean algebra is then used to reduce the complexity of the system, if possible, while retaining the same function. The equivalent simplified combination of gates will probably be much less expensive and less difficult to assemble.

Several techniques are used to reduce the complexity of gate arrangements. They include algebraic manipulation, the construction of charts, the use of Karnaugh maps and Venn diagrams, and the use of truth tables. We shall examine these techniques after we have studied the rules of Boolean algebra, and then use them to simplify combinations of logic gates.

RULES OF BOOLEAN ALGEBRA FOR DIGITAL DEVICES

Boolean algebra has three rules of combination, as any algebra must have: the associative, the commutative, and the distributive rules. To show the features of the algebra we use the variables A, B, C, and so on. To write relations between variables, each one of which may take the value 0 or 1, we use $\bar{A}$ to mean "not A," so if $A = 1$, then $\bar{A} = 0$. The complement of every variable is expressed by placing a bar over the variable; the complement of $B = \bar{B} = $ "not B." Two fixed quantities also exist. The first is identity, $I = 1$; the other is null, null $= 0$.

Boolean algebra applies to the arithmetic of three basic types of gates: an OR-gate, an AND-gate and the inverter. The truth tables for the logic gates in Chapter

1 illustrate that the AND-gate corresponds to multiplication, the OR-gate corresponds to addition, and the inverter yields the complement of its input variable.

We have already found that

$$AB = \text{``}A \text{ AND } B\text{''}$$

for the AND-gate and

$$A + B = \text{``}A \text{ OR } B\text{''}$$

for the OR-gate.

The AND, or conjunctive, algebraic form and the OR, or disjunctive, algebraic form must each obey the three rules of algebraic combination. In the equations that follow, the reader may use the two possible values 0 and 1 for the variables A, B, and C to verify the correctness of each expression. Use $A = 0, B = 0, C = 0; A = 1, B = 0, C = 0$; and so on, in each expression.

The associative rules state how variables may be grouped.

For   AND $(AB)C = A(BC) = (AC)B,$

and for   OR $(A + B) + C = A + (B + C) = (A + C) + B$

the rules indicate that different groupings of variables may be used without altering the validity of the algebraic expression.

The commutative rules state the order of variables.

For   AND $AB = BA,$

and for   OR $A + B = B + A$

the rules indicate that either order is allowed. These commutative rules were used in order to write the last terms in the associative rules given above.

The distributive rules state how relations among variables can be expanded.

For   AND $AB + C = (A + C)(B + C)$

and for   OR $(A + B)C = AC + BC$

the rules indicate that the operations can be grouped and expanded as shown.

Before we show the remaining rules of Boolean algebra for digital devices, let us confirm the distributive rule for AND by writing the truth table, Table 2-1. We will discover soon how we knew that we could write $AB + C = (A + C)(B + C)$, which is proved by the truth table to be a proper expansion.

The more complex expression and its simpler form yield identical values. Because binary logic is dominated by an algebra in which a sum of ones equals one, the truth table permits us to identify the equivalence among algebraic expressions. A truth table may be used to find a simpler equivalent to a more complex relation among variables, if such an equivalent exists. We will see shortly how the reduction of complexity may be achieved in a systematic manner with truth tables and other techniques.

TABLE 2-1
Truth Table for the AND-Distributive Rule

A	B	C	AB + C =	Value	(A + C)(B + C) =	Value
0	0	0	0 + 0	0	0 × 0	0
0	0	1	0 + 1	1	1 × 1	1
0	1	0	0 + 0	0	0 × 1	0
0	1	1	0 + 1	1	1 × 1	1
1	0	0	0 + 0	0	1 × 0	0
1	0	1	0 + 1	1	1 × 1	1
1	1	0	1 + 0	1	1 × 1	1
1	1	1	1 + 1	1	1 × 1	1

Some additional relations in the algebra, which use identity and null, are worth noting. Here we illustrate properties of the AND and OR operations that use the distributive rules and the fact that I is always 1 and null is always 0.

AND	$AI = A$	or	$A1 = A$
OR	$A + \text{null} = A$		$A + 0 = A$
AND	$A\bar{A} = \text{null}$		$A\bar{A} = 0$
OR	$A + \bar{A} = I$		$A + \bar{A} = 1$
AND	$A\,\text{null} = \text{null}$		$A0 = 0$
OR	$A + I = I$		$A + 1 = 1$
AND	$AA = A$		
OR	$A + A = A$		

The relation $A + \bar{A} = I$ points out an important fact, that is, that I, the identity, is the universal set. Null is called the empty set.

Before examining the important De Morgan theorems, which connect AND and OR relations, let us point out two more relations in Boolean algebra. As stated above, the complement (opposite or negative) of A is denoted by $\bar{A}$ and is read "not A." Complementing $\bar{A}$ is equivalent to double negation of A, so we note that

$$\bar{\bar{A}} = A.$$

Similarly, $\bar{1} = 0$ and $\bar{0} = 1$.

The two useful relations are the absorption rule and its dual, which state the values that result when a variable (A, B, C, etc.) is common to a product and a sum. The relations are

Absorption rule $A(A + B) = A$

Dual $\bar{A} + \overline{(A + B)} = \bar{A}$

The relation between an algebraic equation and its dual is based upon the properties of Boolean algebra. For two-value Boolean algebra a dual is formed by interchanging OR and AND operators (+ and ·) and replacing variables by their complements.

Note that $\overline{(A + B)}$ is the complement of $(A + B)$ and that it is not the same as $(\bar{A} + \bar{B})$. The term $(A + B)$ is considered a unit. Now we point out how the absorption rule is applied to the relation $AB + C = (A + C)(B + C)$, for which we prepared the truth table (Table 2-1).

We can use absorption rule and its dual to write $A(A + B) = AA + AB$. We note that by making a truth table with the two possible values of A and B, 0 and 1, the expression has the value of A regardless of the value of B.

Now let us examine the relation

$$AB + C = (A + C)(B + C)$$

to see how it was found with Boolean algebra. Multiplying a variable by the identity does not change its value, $IC = C$, and adding a variable to the identity always yields the value $I = 1$, $(I + C = I)$. The algebraic steps make no change in the value, but allow factoring and grouping. Starting with

	$AB + C$	**Algebraic Step**
we get	$AB + CI$	The identity makes no change.
and		
	$AB + C(I + A)$	Adding A to I retains I.
	$AB + CA + C$	Group the terms with factor A.
	$AB + CA + C(C + B)$	Apply the absorption rule.
	$AB + AC + CB + CC$	Group the terms with factor C.
	$A(B + C) + C(B + C)$	Write with common factors.
	$(A + C)(B + C)$	QED

DE MORGAN'S THEOREMS

De Morgan's theorems relate the complements of variables in AND and OR relations. These theorems are useful for manipulating expressions.

De Morgan's theorems state that

$$\overline{A + B} = \bar{A}\bar{B}$$

and

$$\overline{AB} = \bar{A} + \bar{B}$$

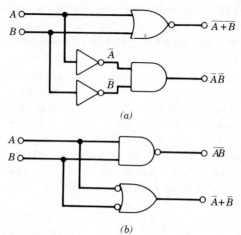

FIGURE 2-1

Gate Equivalents to De Morgan's Theorems

(a) Both gates will give identical outputs for each combination of A and B because $\overline{A+B} = \bar{A}\bar{B}$. (b) These gates are equivalent because $\overline{AB} = \bar{A} + \bar{B}$. Note that the circles adjacent to the OR-gate mean that the input is complemented by the gate before the logic is performed. The same logic output is achieved by placing inverters on both input lines of a two-input OR-gate of the kind shown before. This figure illustrates an option in designating logic inputs and a feature of some gates.

TABLE 2-2
Rules of Boolean Algebra

Rule	Disjunctive (OR)	Conjunctive (AND)
1 Associative	$(A + B) + C = A + B + C$	$(AB)C = A(BC)$
2 Commutative	$A + B = B + A$	$AB = BA$
3 Distributive	$(A + B)C = AC + BC$	$AB + C = (A + C)(B + C)$
4	$A + 0 = A$	$AI = A$
5	$A + \bar{A} = I$	$A\bar{A} = 0$
6	$A + I = I$	$A0 = 0$
7	$A + A = A$	$AA = A$
8 De Morgan's theorem	$\overline{A + B} = \bar{A}\bar{B}$	$\overline{AB} = \bar{A} + \bar{B}$
9 Complement	$\bar{I} = 0;$	$\bar{0} = I$
10 Absorption rule and its dual	$\bar{A} + (\overline{A + B}) = \bar{A}$	$A(A + B) = A$

We read the first as "not the quantity (A plus B) equals not A times not B," and the second as "not the quantity (A times B) equals not A plus not B."

Figure 2-1 shows the arrangements of gates that follow De Morgan's relations. In Figure 2-1b the circles on the input lines adjacent to the OR-gates indicate that the inputs are complemented by the gate before the gate logic is applied. This arrangement is equivalent to the action of the inverters in Figure 2-1a.

The rules of Boolean algebra are summarized in Table 2-2.

REPRESENTATION OF BOOLEAN ALGEBRA

Both algebraic equations and diagrams are used in Boolean algebra to show the relations among the variables. We now turn to Venn diagrams and Karnaugh maps and apply the ideas that they suggest to make systematic diagrams of the binary variables. Venn diagrams illustrate the Boolean logic relations, and Karnaugh maps are an adaptation of Venn diagrams to binary functions.

In Boolean algebra, the identification of a function or a variable (A, B, etc.) with a geometrical pattern allows a graphic representation of the relations that may exist between or among functions. Because $A + \bar{A} = I$, we can consider a square or rectangle to represent I, a universe. Then it can be divided into two parts, one part corresponding to A and the other corresponding to $\bar{A}$ (Figure 2-2a). The size or

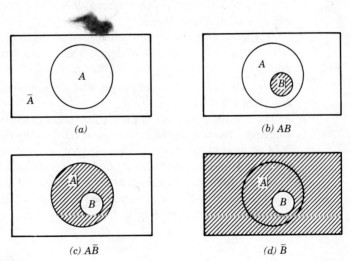

FIGURE 2-2

Venn Diagrams

The rectangle represents the universe. (a) $A + \bar{A} = I$ is the universe. (b) AB is the crosshatched area. B is a subclass of A, a special case. (c) The complement of B included in A is the crosshatched area, $A\bar{B}$. (d) The area outside of B is $\bar{B}$, which includes the portion of A not in B. $B + \bar{B} = I$ is the universe.

shape of the area associated with A or $\bar{A}$ is of no significance in these pictures, but the entire area *must* be filled by A or $\bar{A}$. When several variables are involved, the significance of these patterns is related to the areas in common and the areas that are not in common. An example clarifies the association of variables with areas.

VENN DIAGRAMS

Figure 2-2a shows an area with one portion, the circular area, marked A and the remaining area marked $\bar{A}$. The rectangle represents the universe (that is, all things), while the circular area represents a particular class of things within the universe of interest. For example, the universe could be all living persons, and the class could be all students. If we equate A with students, then $\bar{A}$ is all people who are not students. Recall that $\bar{A}$ means "not A," so $\bar{A}$ means members of the universe (all living persons) who are not specified by A.

In binary logic, where the only two states are 0 and 1, we find the meaning of $\bar{A}$ and A restricted. The universe is equal to $A + \bar{A} = I$ (rule 5 in Table 2-2), which consists of all students and all nonstudents. By being logically careful in our choice of universe and of class, we can make sure that any member of the universe is either in A or in $\bar{A}$, but not in both.

We now identify an additional class of the universe: all science students. Let B represent all science students; then $\bar{B}$ stands for all students who are not science students *and* all other people, because $B + \bar{B} = I$, not A alone. We note that B is a subclass of A, a special case.

In Figure 2-2b the crosshatched area is AB, which represents people who are both students and science students. The crosshatched area represents an AND operation. The remaining area of the universe (all that is not crosshatched) is "not AB" = $\overline{AB}$; that is, all other people who are not students and all students who are not science students.

The crosshatched area in Figure 2-2c represents all students who are not science students. The crosshatched area equals $A\bar{B}$. Finally the crosshatched area in Figure 2-2d represents all people who are not science students. This includes other students as well as nonstudents. $\bar{B}(A + \bar{A}) = \bar{B}I = \bar{B}$ follows from rules 4 and 5 in Table 2-2. It should be clear that the entire area of the rectangle (universe) outside B is identified as $\bar{B}$. The crosshatched portion of Figure 2-2d is identical with the unmarked area of Figure 2-2b.

DEMONSTRATION OF DE MORGAN'S THEOREMS

Universes have classes that need not be inclusive, as A was inclusive of B in the example above. For example, if the universe is all printed matter, then class A may be all books and class B may be all material printed with English. (We say "with"

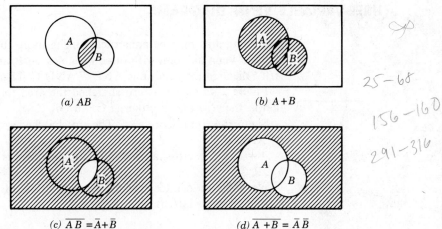

FIGURE 2-3

Demonstrations of De Morgan's Theorems

The complement of (a) is the crosshatched area of (c). The complement of (b) is the crosshatched area of (d). These diagrams represent both De Morgan's theorems.

rather than "in" English to avoid uncertainty about the class containing, for instance, a French-English dictionary. This is an example of the logical care mentioned above.) Figure 2-3a shows the relation of classes A and B to each other and to the universe. The crosshatched area, AB, is all books printed with English. This common area represents things that are both books and printed with English.

The crosshatched area in Figure 2-3b represents members of the universe of printed matter that are either books or all matter printed with English. Thus the crosshatched area represents all books in any language *plus* all newspapers, magazines, pamphlets, and other material printed with English. The crosshatched area is $A + B$. Because magazines or pamphlets, for example, printed in English are not books, it is clear that A does not include all of B in this case.

Figure 2-3c is the complement of Figure 2-3a. It may be written $\overline{AB}$, and is read "not A AND B." The crosshatched area in Figure 2-3c is $\overline{AB}$. An examination of Figure 2-3c will show that the crosshatched area is "not A" OR "not B." Thus we find that $\overline{AB} = \overline{A} + \overline{B}$, which is one of De Morgan's theorems. What is "not B" includes the portion of A that is not common to A and B, and vice versa. We read this De Morgan theorem: that which is "not A AND B" equals that which is "not A" OR "not B." The shaded area is all printed matter except books with English.

Figure 2-3d demonstrates the other De Morgan theorem. The crosshatched area is clearly $\overline{A + B}$, as seen in comparison with Figure 2-3b. It is also the area obtained by taking what is "not A" AND "not B," so that we may equate $\overline{A + B} = \overline{A}\overline{B}$. This is read "not A OR B" equals "not A" AND "not B." The shaded area is all that is *not* books or matter printed with English.

THREE-VARIABLE VENN DIAGRAMS

Venn diagrams may be constructed using more than two variables. Figure 2-4a shows a Venn diagram with three variables: A, B, and C. The crosshatched area is ABC, the area included in A AND B AND C. This is written ABC.

Figure 2-4b shows as crosshatched the area that corresponds to $A + B + C$; that is, the area that is either A OR B OR C.

Figure 2-4c represents $A\bar{B}C$. The crosshatched area in this figure is A AND "not B" AND C.

$A + \bar{B} + C$ is crosshatched in Figure 2-4d, and is read "that which is A OR not B OR C."

We shall return to graphic representation of variables when we examine Karnaugh maps later in this chapter.

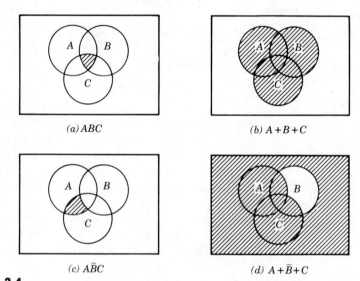

FIGURE 2-4

Venn Diagrams with Three Variables

The three universes $A + \bar{A} = 1$, $B + \bar{B} = 1$, and $C + \bar{C} = 1$ are superimposed as one rectangle. The AND (multiplication) and OR (addition) properties are shown for A, B, and C.

EQUIVALENT FUNCTIONS

An algebraic function of the variables A, B, C, etc. may be reducible to a simpler algebraic function that retains the same logic as the more complicated initial function. When we implement the functions with logic gates, our reduction simplifies the construction and reduces the cost.

We have already demonstrated one method for reducing complexity when we wrote the truth table for $AB + C = (A + C)(B + C)$ (Table 2-1), which demonstrated the equivalence of both functions for all combinations of the values 1 and 0 assigned to the variables A, B, and C. The use of truth tables is not necessarily the most direct way of finding identical functions, but it is a most important way of verifying the equivalence found in another manner.

EQUIVALENCE BY VENN DIAGRAMS

A second method of finding equivalent functions of variables A, B, C, etc. is by constructing Venn diagrams for each term in the function and then identifying common areas among them. The common area will be equivalent to the initial function and may lead to a simpler function.

We demonstrate the use of Venn diagrams with a simple example of the algebraic relation $AB + A\bar{B}$. We see that this can be simplified algebraically as $AB + A\bar{B} = A(B + \bar{B}) = AI = A$. Figures 2-5a and 2-5b show the Venn diagrams of AB and $A\bar{B}$. The superimposition of diagram AB over $A\bar{B}$ shows that the common area is A, which geometrically confirms the relation.

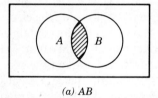
(a) AB

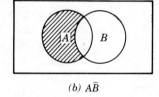
(b) $A\bar{B}$

FIGURE 2-5

Proof of Equivalence with Venn Diagrams

The equivalence relation $A = A(B + \bar{B}) = AI$ is found by superimposing AB on $A\bar{B}$.

EQUIVALENCE BY CHART SIMPLIFICATION

Another technique, chart simplification, uses a truth table written with the variables and their complements. The procedure for simplifying the algebraic relation (if possible) can be programmed for a computer or done by hand. If there is a simpler equivalent function, a function of the variables A, B, C, and so on can be reduced by examining terms in a systematic manner and eliminating terms that do not qualify for retention in the new algebraic relation. The terms that are retained give the equivalent function, and the equivalence can be verified with a truth table. The truth table verification with zeros and ones is an essential check on the reduction process.

TABLE 2-3
Chart of Permutated Variables

Row	Column 1	2	3	4	5	6	7
1	~~A~~	B	~~C~~	AB	AC	BC	~~ABC~~
2	~~$\bar{A}$~~	B	~~C~~	~~$\bar{A}B$~~	~~$\bar{A}C$~~	BC	~~$\bar{A}BC$~~
3	~~A~~	$\bar{B}$	~~C~~	~~$A\bar{B}$~~	AC	~~$\bar{B}C$~~	~~$A\bar{B}C$~~
4	~~A~~	B	$\bar{C}$	AB	~~$A\bar{C}$~~	~~$B\bar{C}$~~	~~$AB\bar{C}$~~
5	~~$\bar{A}$~~	~~$\bar{B}$~~	~~C~~	~~$\bar{A}\bar{B}$~~	~~$\bar{A}C$~~	~~$\bar{B}C$~~	~~$\bar{A}\bar{B}C$~~
6	~~$\bar{A}$~~	~~B~~	~~$\bar{C}$~~	~~$\bar{A}B$~~	~~$\bar{A}\bar{C}$~~	~~$B\bar{C}$~~	~~$\bar{A}B\bar{C}$~~
7	~~A~~	~~$\bar{B}$~~	~~$\bar{C}$~~	~~$A\bar{B}$~~	~~$A\bar{C}$~~	~~$\bar{B}\bar{C}$~~	~~$A\bar{B}\bar{C}$~~
8	~~$\bar{A}$~~	~~$\bar{B}$~~	~~$\bar{C}$~~	~~$\bar{A}\bar{B}$~~	~~$\bar{A}\bar{C}$~~	~~$\bar{B}\bar{C}$~~	~~$\bar{A}\bar{B}\bar{C}$~~

We must construct a truth table or chart with all the permutations of the variables. In our example below, we consider the function $ABC + \bar{A}BC + A\bar{B}C + AB\bar{C}$ and prepare Table 2-3, which contains all of the combinations of the variables and their complements. In using the chart to seek the equivalent simpler function, (a) strike out all rows that do not contain a term in the initial function; (b) column by column, strike out all terms that were eliminated in step (a); (c) eliminate all terms in the row that contain the surviving term as a factor; and (d) collect the remaining terms to form the equivalent function.

To reduce this useful function we first use instruction (a) and strike out rows 5, 6, 7, and 8 because the terms in column 7 of these lines are not found in the initial function. Next, following instruction (b), we strike out all As and $\bar{A}$ in column 1 because they were eliminated in rows 5 to 8. We continue in the same way with all

TABLE 2-4
Truth Table for $ABC + \bar{A}BC + A\bar{B}C + AB\bar{C} = AB + AC + BC$

A	B	C	ABC	+ $\bar{A}BC$	+ $A\bar{B}C$	+ $AB\bar{C}$	= AB	+ AC	+ BC	= Value
0	0	0	0	0	0	0	0	0	0	0
1	0	0	0	0	0	0	0	0	0	0
0	1	0	0	0	0	0	0	0	0	0
0	0	1	0	0	0	0	0	0	0	0
1	1	0	0	0	0	1	1	0	0	1
1	0	1	0	0	1	0	0	1	0	1
0	1	1	0	1	0	0	0	0	1	1
1	1	1	1	0	0	0	1	1	1	1

the other columns. We find that AB survives in column 4, AC survives in column 5, and BC survives in column 6. Now, in accordance with instruction (c), we eliminate from the rows with surviving terms those terms that contain a surviving factor. In row 1 we strike out the term ABC in column 7 because it contains the factor AB (and AC and BC). All members of column 7 are similarly eliminated. Finally, under instruction (d), we collect the surviving terms and equate them to the original relation. We find that $ABC + \bar{A}BC + A\bar{B}C + AB\bar{C} = AB + AC + BC$.

The remaining step is to verify the equivalence by using zeros and ones in the truth table. Table 2-4 verifies the equivalence.

SIMPLIFICATION BY KARNAUGH MAPS

As in Venn diagrams, the functions of the variables A, B, C, and so on, can be simplified by identifying common areas on Karnaugh maps. The Karnaugh map may be considered a graphic form of a truth table; it consists of square or rectangular areas, each of which represents a term in the algebraic expression. Because of their form, Karnaugh maps are easier to use than Venn diagrams for relating areas to functions.

For each variable, A and B, we draw a Karnaugh map, as shown in Figures 2-6a and 2-6b. In Figure 2-6c, A is superimposed on or overlaid with B. The area of the rectangles is the universe representing $A + \bar{A} = I$ and $B + \bar{B} = I$. This is true for all Karnaugh maps because each map contains all the terms and all the complements of terms that fill the universe.

The Karnaugh maps are easier to use if we develop a labeling scheme based on the binary numbering system. To do so, we associate A with 1 and $\bar{A}$ with 0, and do likewise with the other variables. In Figure 2-6c the upper left-hand square can be labeled AB or 11; the upper right-hand square, $\bar{A}B$, becomes 01. The two lower squares, from left to right, are $A\bar{B} = 10$ and $\overline{AB} = 00$.

We use the Karnaugh map in Figure 2-6d to find the equivalence of $AB + A\bar{B}$. The procedure in reducing expressions with Karnaugh maps is to mark the squares corresponding to the terms in the initial function. The reader will observe that adjacent areas corresponding to a variable, A, and its complement, $\bar{A}$, are separated by a line. If a function has the same value in two adjacent squares, it does not depend upon the variable that changes between the squares. Thus, when marked areas are adjacent, they can be combined, and the variable that changes can be eliminated. Squares that are *diagonal* to each other, however, are *not* considered adjacent and *cannot* be combined.

The crosshatched squares in Figure 2-6d represent $AB + A\bar{B}$. They may be grouped, and B and $\bar{B}$ may be eliminated because the areas are adjacent, that is, are on each side of the line between B and $\bar{B}$. Since it is apparent that the crosshatched area covers all of A, we can conclude that $AB + A\bar{B} = A$.

30 Boolean Algebra For Digital Systems

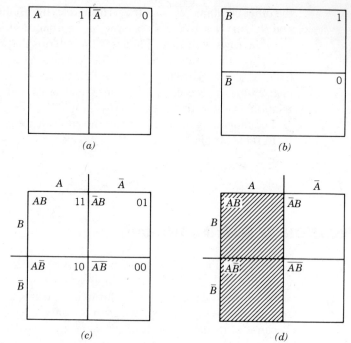

FIGURE 2-6
Two-Variable Karnaugh Maps
The universes $A + \bar{A}$ and $B + \bar{B}$ are superimposed to yield the universe $AB + A\bar{B} + \bar{A}B + \bar{A}\bar{B}$. The crosshatched area in (d) shows that $AB + A\bar{B} = A$, indicating that adjacent areas can be combined to eliminate the variable and its complement, $B + \bar{B}$, which are separated by the line.

This conclusion is easily verified with Boolean algebra: $AB + A\bar{B} = A(B + \bar{B}) = AI = A$.

The reader should also verify that the function $A\bar{B} + \bar{A}B$, which occupies diagonal positions, canot be reduced to a simpler equivalent by the Karnaugh map technique.

MULTIVARIABLE KARNAUGH MAPS

Three- and four-variable Karnaugh maps are prepared by overlaying maps corresponding to the additional variables so as to yield all distinct terms of the function to be reduced. Figure 2-7a shows a possible arrangement of C and $\bar{C}$ to be placed over the maps for A and B. Figure 2-7b is the combined map of the A-map, the B-map, and the C-map. It is necessary to consider that the right and left edges of $\bar{C}$ in Figures 2-7a and 2-7b are adjacent to each other, as if the map were

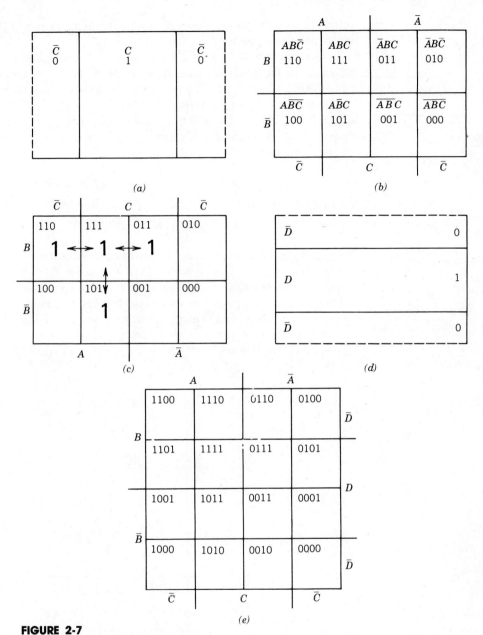

FIGURE 2-7
Multivariable Karnaugh Maps

The universe $C + \bar{C}$ (a) is added to the universe of $AB + A\bar{B} + \bar{A}B + \bar{A}\bar{B}$, (b) to give a three-variable universe (c). A fourth variable, D (d), can be included by adding its universe, $D + \bar{D}$, so as to give distinct regions (e) for all combinations of the variables A, B, C, and D and their complements. The dotted edges of $\bar{C}$ are adjacent, as are the dotted edges of $\bar{D}$, as if they were each wrapped around a cylinder.

wrapped around a cylinder. That is, the squares on the extreme left and the extreme right in the same row are adjacent.

To demonstrate the use of a three-variable Karnaugh map, we will find the equivalent to the function we analyzed above by the chart method. The function $ABC + \bar{A}BC + A\bar{B}C + AB\bar{C}$ is used to select the squares to be marked or shaded. In this case we use an 1 to mark the appropriate squares in Figure 2-7c. We mark squares 111, 011, 101, and 110 to correspond to ABC, $\bar{A}BC$, $A\bar{B}C$, and $AB\bar{C}$. We observe that some of these terms occupy adjacent squares, so the variables may be eliminated as follows.

Squares 101 and 111 unite to give $ABC + A\bar{B}C = AC(B + \bar{B}) = AC$.
Squares 110 and 111 unite to give $AB\bar{C} + ABC = AB(\bar{C} + C) = AB$.
Squares 111 and 011 unite to give $ABC + \bar{A}BC = BC(A + \bar{A}) = BC$.

We sum the reduced expressions to find the equivalent $AB + AC + BC = ABC + \bar{A}BC + A\bar{B}C + AB\bar{C}$. This corresponds to our earlier result.

Figure 2-7d shows a Karnaugh map for a fourth variable, D. Figure 2-7e combines the four maps so that four variables may be considered. Remember that the far edges of the $\bar{C}$ and $\bar{D}$ maps are to be considered adjacent to each other.

A Karnaugh map that has a square of adjacent terms, such as the terms represented by squares 0011, 0111, 1011, and 1111 in Figure 2-8, may be united to

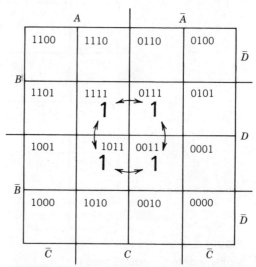

FIGURE 2-8

Karnaugh Map for Reducing $ABCD + A\bar{B}CD + \bar{A}BCD + \bar{A}\bar{B}CD = CD$

A square of marked areas may be combined to eliminate two variables.

eliminate two variables. When we associate the four binary digits with the variables A, B, C, and D, we may write the function and its equivalent. For example, by marking 1111, 1011, 0111, and 0011, we obtain $ABCD + A\bar{B}CD + \bar{A}BCD + \bar{A}\bar{B}CD = CD$.

The complement of a function can be obtained by assembling the terms that are *not* marked in a Karnaugh map for the function.

Notice that adjacent terms may be combined in several ways. Although each combination is a valid reduction of the initial function, one manner of grouping may yield the optimum configuration (perhaps the smallest number) of logic gates. It may be useful to make several attempts to group terms, and it is always advisable to verify the reduction of a function with a truth table. The use of Karnaugh maps is described more completely by Ennes (1978) and Mano (1979).

SUMMARY

In this chapter we have considered several methods of examining logical relations. For the two-value Boolean algebra of digital electronics, the choice of the technique depends upon the nature of the function whose reduction is desired. Some simple functions may be easily reduced by examining their truth table; others require the manipulation of Boolean algebra to reveal the relationship. When we consider the circuit for adding binary numbers, we see that Boolean algebra is required to discover a simplification in that particular application.

The method of chart elimination can be programmed for solution by computer, but for four variables or fewer, it can be done reasonably easily by hand. For most of the work at the level of this text the use of Karnaugh maps is the easiest and probably the quickest technique for finding simpler functions. We will use Karnaugh maps in our study of both combination and sequential gate logic.

In the next chapter we consider the combinations of logic gates for computation and decision making. In reducing our initial configurations to simpler ones, we use the techniques we have discussed here.

PROBLEMS

2-1 Prove De Morgan's theorems by truth tables.

2-2 Write a logical definition of a book so that a piece from the universe of printed matter can be shown to be member of the class "books."

2-3 Design a system of logic gates to determine whether a piece of printed matter is a book.

2-4 Draw the gates and connections to give the logic function $ABC + \bar{A}BC + A\bar{B}C + AB\bar{C}$. Draw the gate arrangement to satisfy $AB + AC + BC$, which was shown to be equivalent to the former function.

2-5 Simplify $A\bar{B}\bar{C} + ABC + \bar{A}BC$ to show that it is equivalent to $A\bar{B}\bar{C} + BC$ by (a) Boolean algebra and (b) chart elimination.

2-6 (a) Write the truth table for the following circuit.

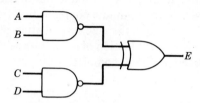

(b) Use the truth table to make a Karnaugh map.
(c) Write the terms found with the Karnaugh map.
(d) Draw the circuit to satisfy the results of (c).

2-7 Show that three NOR-gates may be interconnected to give a two-input AND-gate. *Hint*: Use the De Morgan's theorems and recall that a NOR-gate may also be used as an inverter. Show that three NAND-gates can yield a two-input OR-gate.

2-8 By using truth tables and the dual relation, find the equivalents of the following expressions:

$$\bar{A} + \bar{B}$$
$$\bar{A}\bar{B}$$
$$\overline{AB}$$
$$\overline{A + B}$$

2-9 Draw the Venn diagrams (Figure 2-4) for the following relations: $\bar{A}BC$, $\overline{ABC}$, $\bar{A} + \bar{B}C$, $\bar{A} + \bar{B} + \bar{C}$, $\overline{ABC}$, $\bar{A}B + \bar{C}$, $AB + BC + AC$.

2-10 Use a truth table to verify the absorption rule, $A(A + B) = A$, and its dual, $\bar{A} + (\overline{A + B}) = \bar{A}$.

2-11 Show that $\overline{A + B} \neq \bar{A} + \bar{B}$ by a truth table.

2-12 Show that $\overline{A + B} = \bar{A}\bar{B}$ and that $\overline{AB} = \bar{A} + \bar{B}$ by extending, where necessary, the truth table of problem 1.

2-13 Confirm that $A + \bar{A} = I$ is the dual of $A\bar{A} = 0$ and that $A + I = I$ is the dual of $A0 = 0$.

2-14 Show by a truth table that the dual of $AB + AC + BC$ is $(\bar{A} + \bar{B})(\bar{A} + \bar{C})(\bar{B} + \bar{C}) \neq \overline{(A + B)}\overline{(A + C)}\overline{(B + C)}$.

REFERENCES

1953   Karnaugh, M., "The Map Method for Synthesis of Combinational Logic Circuits," *Trans. AIEE Communications and Electronics*, **72**, 593 (1953).

1956   Newman, James R., (Ed.), *The World of Mathematics*, Part XIII Mathematics and Logic, Simon & Schuster, New York.

1978   Ennes, Harold E., *Boolean Algebra for Computer Logic*, Howard W. Sams, Indiana.

1979   Mano, M. Morris, *Digital Logic and Computer Design*, Prentice-Hall, Englewood Cliffs, N.J.

CHAPTER 3

COMBINATIONAL DIGITAL ELECTRONICS

We have examined the Boolean algebra involving the binary variables and have found that the logical multiplication and addition of variables can be performed by AND-gates and OR-gates. In this chapter we implement some useful algebraic equations with logical gates, and use simplification techniques to reduce the number of gates required. We emphasize important logic-gate applications, such as the addition of binary numbers, the encoding and decoding of binary and octal numbers, and other tasks required in computers and digital communications.

These gate circuits respond immediately to the binary inputs according to their arrangement and interconnections. They are known as *combinational circuits*, in contrast to *sequential circuits*, which is presented in the next chapter. The output of a combinational circuit is established by the inputs, but the output of a sequential circuit depends upon its earlier state as well. For example, a binary counter will indicate the next number when it receives new input; the number depends upon its previous input.

Both combinational and sequential circuits use the three basic logic gates to perform their functions. The basic gates are available as integrated circuits (IC) and several identical gates are supplied to a package. The combinations and sequential logic gate circuits that we will make with the basic gates are available themselves as integrated circuits.

DIP is the acronym for dual in-line package, the most common package of logic gates, inverters, and other elements used in digital electronics and computers. (Analog circuit elements are available in the same form.) DIPs are small units with from 8 to 40 or more pins, which connect to the integrated circuits (ICs) encapsulated within the plastic or ceramic packages. The 40-pin DIPs are microprocessors (μp), peripheral interface adaptors (PIAs), and other elements that perform many functions that are separately available in DIPs with fewer pins. Flat packs and other forms containing encapsulated ICs are available for use when space is restricted, such as in electronically controlled cameras and sensor probes.

The basic logic gates and other elements are usually found in DIPs with 8 to 16 pins. The DIPs contain several identical basic logic gates, a single power pin for the supply voltage, and a common ground pin. Thus, in most cases, only a single power connection is made. The power connections are not shown in most circuit diagrams, but they *must* be made for the ICs to operate.

Integrated circuits are made by several different electronic fabrication techniques. In this chapter, we consider TTL (transistor-transistor logic) devices in particular because they are electrically the most rugged and because they are used in the laboratory exercises. The other widely used types are MOS (metal-oxide-semiconductor) and CMOS (complementary metal-oxide semiconductor) elements. (The circuit details of basic logic gates and the physics of semiconductors are presented in later chapters.) Power drain, speed of switching, compactness in fabrication, and other considerations make one type preferable to another for a specific application, but each kind of logic gate, an AND-gate, for example, performs the same logic regardless of the fabrication technique. The manuals issued by the manufacturers of the integrated circuits provide extensive technical details, such as switching time, allowed and expected voltage, and current ranges required for engineering design of circuits employing ICs.

The TTL integrated circuits are designated by the 7400 DIP series. The 5400 series contains equivalent TTL DIPs with more stringent military specifications, such as the range of temperatures in which they must perform reliably. CMOS integrated circuits are designated as the 4000 and 74C00 series.

The 7400 series DIPs receive power through the pin marked V_{cc}, and are grounded through the pin marked *Gnd*. The 4000 series uses the terminal marked V_{dd} for the power connection and the pin marked V_{ss} to complete the circuit, usually to ground. Both series operate with low voltages. The TTL (7400) series requires $+5.0$ V and the CMOS (4000) series uses from 3 to 15 V. The current drawn is quite small, rarely more than a few milliamperes. The rated voltage maximum for the integrated circuit *must not be exceeded* or it may be damaged.

A dot or groove at one end of the DIP identifies the end from which the pins are numbered. When viewed from the top, the numbers begin at the lower left-hand corner and proceed with the pins pointing away from the viewer in a counterclockwise manner. Figure 3-1 shows the configuration of the 7400 quadruple two-input NAND-gate. The pins are numbered in the figure, but are not numbered on the DIP, and the connections to the gates are not shown on the DIPs. The DIP number (7400 in this example), the manufacturer's code, and other data are printed

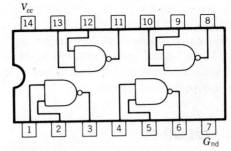

FIGURE 3-1

A 7400 Quadruple Two-Input NAND-gate

The schematic layout of the DIP containing four two-input NAND-gates. The pins are numbered in counterclockwise order, starting at the end with the notch, with the pins pointing away from the observer. The actual DIP does not show the gate layout or have pin numbers. The type number of the DIP and the manufacturer's codes are printed on the top.

TABLE 3-1
Several 7400-Series DIPs

DIP Number	Function
7400	Quadruple two-input NAND-gate
7402	Quadruple two-input NOR-gate
7403	Quadruple two-input NAND-gate (open collector)
7404	Hex inverters
7405	Hex inverters (open collector)
7408	Quadruple two-input AND-gate
7410	Triple three-input NAND-gate
7420	Dual four-input NAND-gate
7427	Triple three-input NOR-gate
7432	Quadruple two-input OR-gate
7486	Quadruple two-input XOR-gate

on the case. Letters may separate parts of the type number; for example, a 7402 NOR-gate may be labeled 74S02, 74LS02, SN7402N, or otherwise. Numbering indicates manufacturer and/or technical differences, but not functional differences.

Table 3-1 lists several of the 7400 series, with the DIP number, the gate, and the number of duplicate gates contained in the DIP. The pinouts for many DIPs are given on the end papers.

COMBINATIONAL CIRCUITS

Digital circuits made with logic gates serve a wide variety of functions. In this chapter we will examine representative examples of the circuits commonly found in computers and digital instruments. Our purpose is to illustrate how the digital function can be implemented by combinations of the basic logic gates and how complicated arrangements can be simplified by using the rules of Boolean algebra and the other methods given in Chapter 2.

Combinational circuits add binary numbers, compare binary numbers to determine their relative magnitudes, allow binary numbers to be displayed as decimal equivalents, route signals from several data paths to one data path or vice versa, and perform many other useful tasks. In this chapter we give examples of combinational circuits that do these and other things, with emphasis on their underlying operating principles. Because combinational circuits are available as integrated circuits packaged in DIPs and other forms, a user will not usually make them with the basic gates. It is important, however, to understand the principles of digital electronics, which allow the user to create circuits to meet special needs.

DIODES

Before we use the ICs we will see that an AND-gate and an OR-gate can be made with the simplest electronic element, a diode. (Diodes are not available as DIPs.) A diode, as the name suggests, is a device with two elements, which conducts electricity in one direction but not in the other. In Chapter 8, we discuss the solid state physics that creates this and other properties of diodes. Here we merely note that the voltage drop across the diode is quite small, a fraction of a volt, in the direction of conduction. A large inverse voltage may be applied without conduction.

Elementary AND-gates and OR-gates may be formed with diodes and resistors. These were the gates initially considered for computers, but they are not easily used, for a reason that we will see presently. The IC gates are more sophisticated, but the principles of the two basic gates are readily illustrated by analyzing gates made with diodes and resistors. They show the actual voltage relations that exist in AND-gates and OR-gates.

The diode symbol seen in Figure 3-2 is a triangle and a bar. The triangle represents the anode (the positive terminal) and the bar represents the cathode (the

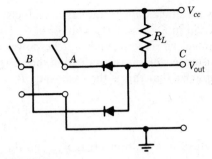

FIGURE 3-2

An AND-gate with Diodes

A two-input AND-gate made with diodes and a resistor. The diode conducts electricity when its cathode (the bar) is at a lower voltage than its anode (the triangle) and does not conduct with the opposite electrical polarity. The output is C. Switches A and B establish inputs high (V_{cc}) or low (ground).

negative terminal). Charge flows from the anode to the cathode when the anode is at a more positive voltage than the cathode. When the opposite is true, the charge cannot flow and the effective resistance of the diode is quite large. When the cathode and the anode are at the same voltage, there is no current through the diode. A dot, a band, or another mark indicates the cathode end of the diode.

The AND-gate in Figure 3-2 operates as follows. In positive logic, you may recall, the AND-gate output C is 1 (high) when inputs A and B are both 1 (high). See the truth table in Table 3-2 for the two-input AND-gate with $AB = C$.

The inputs A and B may either be grounded or placed at V_{cc} by positioning the switches. In Figure 3-2, A is high (1) and B is high (1), corresponding to the last line of Table 3-2. In this configuration the cathodes of both diodes are at the same voltage as the anode; therefore, there is no current through the diodes or through the resistor R_L, and output C is high, $V_{out} = V_{cc}$.

When one of the switches is moved to ground, the charge can flow through R_L, and the diode whose switch is grounded because the cathode of that diode is at a

TABLE 3-2
Truth Table for a Two-Input AND-gate

A	B	C
0	0	0
1	0	0
0	1	0
1	1	1

lower voltage than its anode; it conducts. The voltage drop, $V = IR_L$, caused by the current in R_L lowers the voltage of C. The upper end of R_L attached to the supply voltage, V_{cc}, is kept high by the power supply. Because the diode has very low resistance when it is conducting the voltage at C, V_{out} is near zero. The reader will recognize that this is the case when $R \ll R_L$, as in Figure 1-1; that is,

$$V_{out} = \left(\frac{R_{diode}}{R_{diode} + R_L}\right) V_{cc}$$

where R in Equation 1-1 is R_{diode} in this case.

If the other switch is used instead, the same reasoning applies. These alternatives correspond to the middle lines in Table 3-2.

When both switches are grounded, corresponding to the top line in Table 3-2, both diodes share the conduction, and the IR-drop in R_L lowers the output voltage, V_{out}, at C.

A rearrangement of the diodes and the resistor allows the circuit to become an OR-gate, as shown in Figure 3-3. Note that the polarity of the diodes is reversed from that of Figure 3-2. When either or both switches are closed to connect the diodes to V_{cc}, the current can come through R and the diode because its cathode is lower in voltage than its anode. The voltage drop in R, $V = IR$, raises the upper end of R above ground, and we can nearly measure V_{cc} at C, which is identified as high (1). The condition of one or both switches closed to V_{cc} corresponds to last three lines of Table 3-3. If both switches are closed to ground, however, there is no current through the diodes or through R; thus, C is at ground (0). This condition corresponds to the first line of Table 3-3.

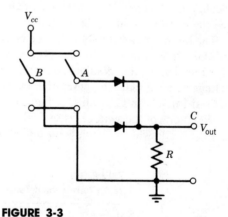

FIGURE 3-3

An OR-gate with Diodes

A two-input OR-gate made with diodes and resistors. The output is C. Note that the diodes are arranged in the opposite direction from Figure 3-2.

TABLE 3-3
Truth Table for a Two-Input OR-gate

A	B	C
0	0	0
0	1	1
1	0	1
1	1	1

We now can explain why AND-gates and OR-gates made with diodes are not easily used and why logic gates made with transistors are preferred. The difficulty arises when several diode-made gates are connected to satisfy a logic function. If a diode in one gate is conducting, it provides a path for the current in the resistor associated with it, but it may also provide a current path for a resistor or diode in another gate in the circuit. If this happens, the other gate may not have the proper output state. In other words, the action of one diode-made gate is not easily isolated from other gates to ensure that each will operate as it should. Transistor-made gates, however, do not have this unfortunate property.

LOGIC GATES

From this point on we study systems formed from basic logic gates made with transistors. Before we consider the combinational circuits made with transistors, however, it is useful to discuss a few features related to their interconnections. Because these clever elements operate by the flow of electrical charges (current) through their internal transistor-resistor networks, the effect of interconnecting them electrically is important to their operation. They are designed to be interconnected, but since they are low power electrical devices, they have limits on the currents they can carry.

FAN-OUT

Logic gates made with transistors are designed so that the output of one gate may be connected to the inputs of several gates. This arrangement is called fan-out. The manufacturer's technical manual specifies the number of gates that each gate can drive. The limit may be stated as the amount of current that the output circuit of the gate can supply and still ensure that the output state (0 or 1) is unambiguous. As the inputs of more gates are tied to the output of a single gate, the increase in current drain or *load* will ultimately lower the output voltage. The loading problem can be solved by adding *buffers* between the gate output and the other gate inputs.

44 Combinational Digital Electronics

Buffers are noninverting amplifiers with good fan-out ability that reproduce their input state at their output.

A different problem arises if the *outputs* of several logic gates are attached *together* because logic gate outputs are designed only to provide digital signals to the inputs of other logic gates. If the outputs of two or more logic gates are tied together, the gates may be damaged. This will happen when one gate is high and another one is low because each gate fights to maintain its own proper logic state on its output line. The large internal currents that arise to maintain the states may damage the gates.

OPEN-COLLECTOR GATES

An *open-collector gate* has a transistor as an output circuit element. One part of a transistor is the collector, through which the main current of the transistor passes. The term "open-collector" simply means an unconnected collector. When the collector is open, the output circuit has been arranged so that the current from the supply voltage, V_{cc}, must come through an external resistor to enter the transistor through its collector. Without the external resistor to the collector, there is no connection to the power supply at the gate output. The details of open-collector gates are given in Chapter 10.

Open-collector (*oc*) gates are designed to allow high currents in their output circuits, but they require an external resistance to complete the electrical path for the current; otherwise the gate will not operate. An open-collector gate may appear to work without the external resistor when its output is tied to the *input* of another gate, but it is an unreliable element. In any case, a *pull-up* resistor to V_{cc} must be connected to each (*oc*) output$_{pin}$ for proper operation. A pull-up resistor is simply a resistor connected to the supply voltage to provide addition current to a gate.

Figure 3-4a shows an OR-gate made with open collector inverters. Because it is made by connecting inverters with resistors, it is called a wired OR-gate. The external resistor, R, connected to V_{cc} lets the charge flow through the collector of the output transistor of either or of both of the inverters when their output state is low, to give the proper OR-gate response, $A + B = C$, to the input states A and B. Note that the *oc* in the symbol for the inverter designates the element as an open-collector device.

Figure 3-4b shows the symbol for the wired OR-gate. Note that the OR-gate shown, with the wires entering it from the sides, does not exist physically; an OR-gate is not present. Its function is represented by the symbol, but the circuit uses only resistors and inverters, as shown in Figure 3-4a. The removal of the inverter on the right will make the circuit a wired NOR-gate.

The reader should distinguish between this circuit, which is made with inverters, and the circuits made with diodes. Because the inverters are transistor-based elements, the wired OR-gate can be connected to the inputs of additional gates without adverse effects.

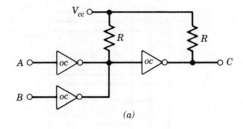

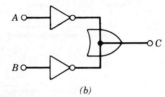

FIGURE 3-4

A wired OR-gate

(*a*) An OR-gate is made with open collector inverters and resistors. (*b*) The symbol for a wired OR-gate.

The integrated circuits that decode and drive displays to show numeric or alphabetic information are open collector devices because of the relatively high currents required to illuminate the displays. A segment in the light-emitting diode, LED, and its current-limiting resistor form the pull-up resistor for the open-collector element.

LOGIC FUNCTIONS

Many logic operations are required to use binary data in computation, control, decoding, display, and other processes. Presently we will discuss the interconnections of basic gates to satisfy several of these applications, but first let us look at an important logic function and then consider others that are widely used.

The logic function simplified in Chapter 2,

$$ABC + \bar{A}BC + A\bar{B}C + AB\bar{C} = X = AB + AC + BC$$

is implemented with basic gates in Figure 3-5a, which represents the left-hand side of the function. Figure 3-5b illustrates the basic gate configuration that satisfies the equivalent function on the right-hand side of the equation.

We emphasize that the arrangement of Figure 3-5a is the straightforward implementation of the logic function $ABC + \bar{A}BC + A\bar{B}C + AB\bar{C}$, using AND-gates, OR-gates, and inverters. The same function could be satisfied with NAND-gates, NOR-gates and inverters.

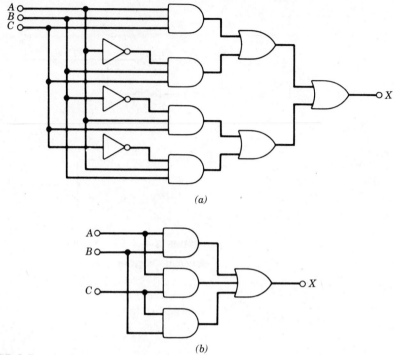

FIGURE 3-5

Implementations of Logic Function X

(a) The straightforward implementation of $ABC + \bar{A}BC + A\bar{B}C + AB\bar{C} = X$. (b) Implementation of reduced form $AB + AC + BC = X$.

HALF ADDER

Addition is the fundamental arithmetic process. All the other arithmetic operations may be accomplished by addition. Subtraction may be performed by adding the negative of the number being subtracted (subtrahend) to the number from which it is subtracted (minuend) to yield the difference. Multiplication may be performed by repeated addition and division by repeated subtraction.

In the addition of 76 and 47, for example,

$$\begin{array}{r} 11 \\ 76 \\ 47 \\ \hline 123 \end{array}$$

we add the rightmost numbers, $6 + 7 = 13$. In adding the two least significant numbers (the rightmost), we call 3 the sum and 1 the carry. The carry 1 is added to the second column, as shown, to obtain the new sum 2, and the carry, 1. We repeat this process until we obtain the total sum.

TABLE 3-4
Binary and Decimal Equivalents

10^0	2^2	2^1	2^0
0	0	0	0
1	0	0	1
2	0	1	0
3	0	1	1
4	1	0	0

The same process takes place when the numbers are the binary numbers formed with 0_2 and 1_2. The numbers 0_2 and 1_2 are subscripted with a 2 to show that they belong to the binary (power of 2) arithmetic system rather than the decimal (power of 10) system. They are called *bits*; their decimal equivalents are shown in Table 3-4. Computer arithmetic is discussed in Appendix A.

A half adder adds the two least significant bits (LSB). Because the LSB are the rightmost bits in the binary numbers, a carry is not to be involved in adding them. The sum may or may not produce a carry bit to the next binary column, that is, the bits in the next higher power of 2 column.

Before examining the gate arrangement let us review the process of adding binary bits. Table 3-5 shows the process of addition in both the decimal and the binary systems of numbers.

In the binary number system the bit or digit in the right-hand column may be zero or one. A bit in this column is called the *least significant bit* (LSB) because of its positional value relative to the bits to its left in positional arithmetic. The sum of two 1_2s in the LSB column is 0_2, which generates a bit 1_2 in the next column to the left, just as $6_{10} + 7_{10} = 13_{10}$ in decimal addition places a one in the next column of

TABLE 3-5
Addition of Binary and Decimal Equivalents

$$\begin{array}{cc} 0_{10} & 0\ 0_2 \\ +\ 0_{10} & +\ 0\ 0_2 \\ \hline 0_{10} & 0\ 0_2 \end{array}$$

$$\begin{array}{cc} 0_{10} & 0\ 0_2 \\ +\ 1_{10} & +\ 0\ 1_2 \\ \hline 1_{10} & 0\ 1_2 \end{array}$$

$$\begin{array}{cc} 1_{10} & 0\ 1 \\ +\ 1_{10} & +\ 0\ 1_2 \\ \hline 2_{10} & 1\ 0_2 \end{array}$$

TABLE 3-6
Truth Table for a Half Adder for $A + B$

A	B	C (Carry)	S (Sum)
0	0	0	0
1	0	0	1
0	1	0	1
1	1	1	0

the powers of 10. The generated bit is called the *carry* bit. The bit in the initial column, which results from addition, is called the *sum* bit.

Because the half adder is the logic-gate arrangement that adds the two LSBs, it does not include the possibility of a carry bit from a previous adder. The full adder, however, adds two binary bits and the carry bit. A full adder must be used except for adding the least significant bits.

We may write the binary part of Table 3-5 in a slightly more general form as a truth table for a half adder, as in Table 3-6.

We can see from the truth table that the sum column is the exclusive OR-gate logic for the inputs A and B. Thus the LSB is

$$A \oplus B = S = A\bar{B} + \bar{A}B$$

The carry column is the logic response of an AND-gate to the inputs A and B. Thus

$$AB = C$$

Figure 3-6 shows an arrangement of basic gates that form a half adder.

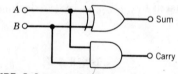

FIGURE 3-6

Implementation of a Half Adder

The gates satisfy the binary logic relations $S = A \oplus B$ and $C = AB$ to find the sum, S, and the carry, C, in adding LSBs.

HALF SUBTRACTOR

Subtracting numbers in any positional number system may be done in several ways. We will implement one way here and discuss other methods in Appendix A, "Computer Arithmetic." The subtraction of a decimal number will be used to illustrate a subtraction process useful for binary numbers.

Half Subtractor

TABLE 3-7
Truth Table for a Half Subtractor for $A - B$

A	B	$\langle B \rangle$ (Borrow)	D (Difference)
0	0	0	0
1	0	0	1
0	1	1	1
1	1	0	0

Consider the subtraction of $7 - 4 = 3$. The difference, 3, can be found by adding the complement of 4 to 7. The complement is the difference between the number and the next power of 10 above it, that is, 15 is the complement of 85 because $15 + 85 = 100$. In our example, since 6 is the complement of 4,

(Minuend)	7	(Minuend)	7
(Subtrahend)	-4	(Complement of subtrahend)	$+6$
(Difference)	3	Answer with overflow	13

but the most significant digit 1_{10} is not part of the numerical answer. It is dropped to yield the difference 3.

Before we discuss binary subtraction by the complement method, let us write the truth table for the subtraction of one binary bit from another binary bit. For $A - B$ the difference is denoted by D and the borrow is designated by $\langle B \rangle$.

In finding $A - B$, a borrow must be allowed if A is less than B. This corresponds to the third row in Table 3-7. An examination of Table 3-7 shows that the difference is the exclusive OR-gate logic corresponding to the inputs A and B. This is the same relation as found for the half adder.

$$D = A \oplus B = A\bar{B} + \bar{A}B$$

The borrow column in the truth table follows the binary logic equation

$$\bar{A}B = \langle B \rangle$$

which yields the proper digit for the borrow.

The arrangement of basic gates to form a half subtractor is shown in Figure 3-7.

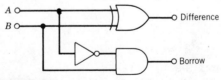

FIGURE 3-7

Implementation of a Half Subtractor

The gates satisfy the binary logic relations $D = A \oplus B$ and $\langle B \rangle = \bar{A}B$ to find the difference, D, and the borrow, $\langle B \rangle$, in subtracting LSBs.

BINARY SUBTRACTION BY THE COMPLEMENT METHOD

In the binary number system, in which the base is 2, any number is written in the manner shown in Table 3-4. In our example, $7 - 4 = 3$ we write the equivalent binary numbers as four-bit numbers. We use the next higher binary position, 1 0000, to form the one's complement because our original number is a four-bit number, just as we had to use 10_{10} with two-decimal digits to find the complement of 4, a signal decimal digit.

The one's complement is found as follows. First, we subtract 0001 from 10000 to get 01111. Then, we subtract $0100_2 = 4_{10}$ to get

$$01111 - 0100 = 01011 = \overline{0100}$$

The one's complement of 0100 is written $\overline{0100} = 1011$.

Now we see why the one's complement is useful in binary arithmetic. To obtain the one's complement of any binary number, we exchange all ones for zeros and vice versa. This step of complementing the binary number is a very simple logic-gate task, performed by inverters.

However, there is another step to facilitate the subtraction by complementation. We add $0001_2 = 1_{10}$ to the one's complement to get the two's complement. The two's complement of 4 is

$$1011 + 0001 = 1100$$

We use this in the subtraction example with decimal and binary numbers, as shown in Table 3-8.

The method demonstrated in Table 3-8 shows that binary numbers can be subtracted by adding circuits when the complement of the number to be subtracted is used. It is not necessary to provide two types of arithmetic circuits. It is only necessary to take complements, to add the 1_2, and to drop the leftmost bit to complete the process.

Subtraction can be more complicated than the example shows. For instance, when the number being subtracted is greater than the number from which it is taken, additional steps are required. For the details of generating and performing arithmetic with signed (positive and negative) numbers, see Appendix A.

TABLE 3-8
Subtraction by the Complement Method

7	=7	=7	0111 =	0111	Minuend
−4	+$\overline{4}$	+6	−0100	+1100	Subtrahend
3		13	0011	10011	
Discard left 1		3		0011	Difference

FULL ADDER

The addition of a pair of LSB binary digits may generate a carry bit. Each pair of more significant bits, when added, must be able to accept as well as generate a carry. A full adder satisfies this requirement. We will use the truth table of a full adder and the techniques of reducing logic functions to implement the gate configuration for full addition.

The truth table of a full adder is given in Table 3-9. Both zeros and ones and the binary variables A, B, and so on will be used in this table. For convenience the rows are arranged in the ascending numerical order of A, B, and Z. This practice is helpful when we discuss decoders, scalers, and other digital circuits. The carry-in is designated by Z and the carry-out is designated by C. Table 3-9 uses both methods of writing truth tables with ones and zeros, along with variables and their complements. The latter are useful in deriving the logic equations contained in the truth tables.

The logic functions for C (carry) and S (sum) can be obtained from the truth table, Table 3-9, by collecting terms that yield the quantity. We find from lines 3, 5, 6, and 7 that

$$C = \bar{A}BZ + A\bar{B}Z + AB\bar{Z} + ABZ$$

and from lines 1, 2, 4, and 7 that

$$S = \bar{A}\bar{B}Z + \bar{A}B\bar{Z} + A\bar{B}\bar{Z} + ABZ$$

where the + sign means OR, literally as well as logically.

The function yielding the carry C is the same one that we reduced in Chapter 2, whose logic-gate arrangement we show with its reduced equivalent in Figure 3-5. We have

$$C = AB + AZ + BZ$$

to implement with logic gates to give the carry.

TABLE 3-9
Truth Table for a Full Adder

Row	A	B	Z	C	S	A	B	Z	C	S
0	0	0	0	0	0	$\bar{A}$	$\bar{B}$	$\bar{Z}$	$\bar{C}$	$\bar{S}$
1	0	0	1	0	1	$\bar{A}$	$\bar{B}$	Z	$\bar{C}$	S
2	0	1	0	0	1	$\bar{A}$	B	$\bar{Z}$	$\bar{C}$	S
3	0	1	1	1	0	$\bar{A}$	B	Z	C	$\bar{S}$
4	1	0	0	0	1	A	$\bar{B}$	$\bar{Z}$	$\bar{C}$	S
5	1	0	1	1	0	A	$\bar{B}$	Z	C	$\bar{S}$
6	1	1	0	1	0	A	B	$\bar{Z}$	C	$\bar{S}$
7	1	1	1	1	1	A	B	Z	C	S

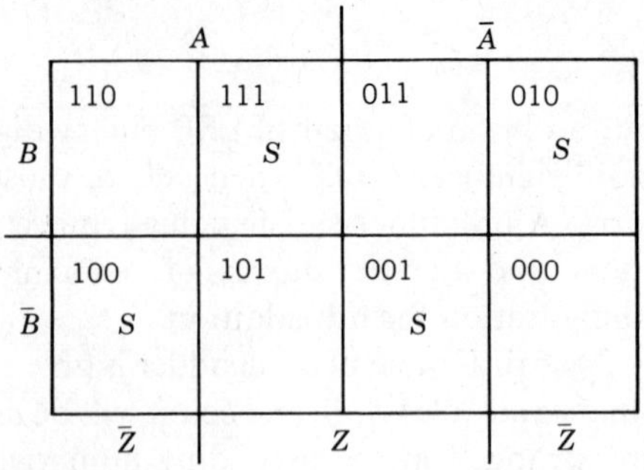

	A		$\bar{A}$	
B	110	111 S	011	010 S
$\bar{B}$	100 S	101	001 S	000
	$\bar{Z}$	Z		$\bar{Z}$

FIGURE 3-8

A Karnaugh Map for $AB Z + A\bar{B}\bar{Z} + \bar{A}B\bar{Z} + \bar{A}\bar{B}Z = S$

This is the Karnaugh map for the equation for the sum, S, in a full adder. Map simplification is not possible.

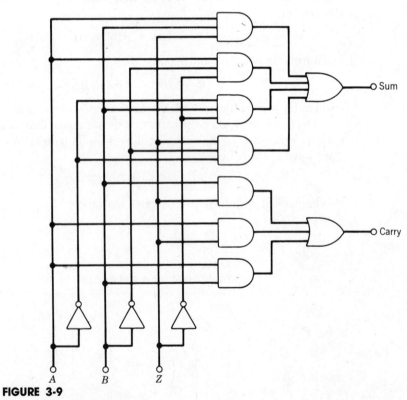

FIGURE 3-9

A Full Adder gate Configuration

The straightforward implementation of the addition of two bits, A and B, and a carry-in, Z, with logic gates.

We shall use a Karnaugh map to examine the possible reduction of the sum S. Figure 3-8 is marked to identify the terms in S. Because Figure 3-8 shows no adjacent terms (terms separated by a single line), however, the function for S is not reducible by the map technique. We must implement

$$S = \bar{A}\bar{B}Z + \bar{A}B\bar{Z} + A\bar{B}\bar{Z} + ABZ$$

Figure 3-9 is the direct implementation of C and S and is a logic-gate configuration of a full adder.

FULL ADDER WITH TWO HALF ADDERS

Two half adders can be used to add two binary bits and the incoming carry bit to yield the sum and the outgoing carry. Although a Karnaugh map failed to reduce the function representing the sum, we can reduce it by using the rules of Boolean algebra, and De Morgan's theorems (Table 2-2) and S becomes a double exclusive-OR function.

To show this result, we recall first the functional relation in XOR-gate logic with two inputs, $X \oplus Y = \bar{X}Y + X\bar{Y}$. Next we write the sum S and factor it.

$$S = \bar{A}\bar{B}Z + \bar{A}B\bar{Z} + A\bar{B}\bar{Z} + ABZ$$
$$= \bar{Z}(\bar{A}B + A\bar{B}) + Z(\bar{A}\bar{B} + AB)$$
$$= \bar{Z}(\bar{A}B + A\bar{B}) + Z(A\bar{A} + \bar{A}\bar{B} + AB + B\bar{B})$$

In the last step we add $0 = A\bar{A} = B\bar{B}$, which does not change the logic equation but permits it to be factored again.

$$S = \bar{Z}(\bar{A}B + A\bar{B}) + Z(A + \bar{B})(\bar{A} + B)$$

By De Morgan's theorem the factors of Z in the rightmost term may be rewritten as

$$S = \bar{Z}(\bar{A}B + A\bar{B}) + Z(\overline{\bar{A}B + A\bar{B}})$$

We apply the exclusive-OR relation here and once again in the final step to get

$$S = \bar{Z}(A \oplus B) + Z\overline{(A \oplus B)}$$
$$= Z \oplus (A \oplus B)$$

The sum in the full adder can be found by an exclusive-OR logic of the carry Z with the term formed by the exclusive-OR logic of the bits to be added. The carry is generated by the AND-gates, and the OR-gate, as we showed in Figure 3-5b. Because a half adder has the exclusive-OR function, two half adders can be combined to form a full adder. Figure 3-10 shows the gate configuration for making a full adder with two half adders.

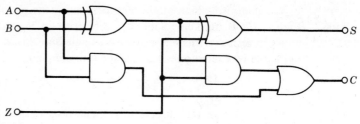

FIGURE 3-10
A Full Adder with Two Half Adders
The addition of two bits, A and B, and a carry-in, Z by combining two half adders and an OR-gate.

BINARY AND DECIMAL NUMBERS

This discussion provides an introduction to the numbers used in computers and is useful throughout our study. Below we present octal and hexadecimal numbers and the means of conversion among number systems. To find the equivalent decimal value of the binary numbers, we recall the meaning of positional numbers, an extremely important invention.

When we write a number such as 123 in base 10 (a decimal number), we know that it means one hundred plus twenty plus three, or $100 + 20 + 3$. Because $100 = 10^2$ and $10 = 10^1$ and $1 = 10^0$, we find that 123 means, "take 1 of the 10^2 values, add to it 2 of the 10^1 values, and then add 3 of the 10^0 values." The relation can be seen easily in Table 3-10, where the columns are labeled with the powers of 10 and the rows represent the number of times the power of 10 in the column is to be used.

TABLE 3-10
Decimal Positional Numbers

10^3	10^2	10^1	10^0	Decimal Value
0	0	0	0	0
0	0	0	1	1
0	0	0	7	7
0	0	4	1	41
0	1	2	3	123
2	0	7	0	2070
9	9	9	9	9999

TABEL 3-11
Binary Positional Numbers

2^3	2^2	2^1	2^0	Binary Number	Decimal Equivalent
0	0	0	0	0_2	0_{10}
0	0	0	1	1_2	1_{10}
0	0	1	0	10_2	2
0	0	1	1	11_2	3
0	1	0	0	100_2	4
0	1	0	1	101_2	5
0	1	1	0	110_2	6
0	1	1	1	111_2	7
1	0	0	0	1000_2	8
1	0	0	1	1001_2	9
1	0	1	0	1010_2	10_{10}
1	0	1	1	1011_2	11_{10}
1	1	0	0	1100_2	12
1	1	0	1	1101_2	13
1	1	1	0	1110_2	14
1	1	1	1	1111_2	15

The binary number system is based on the powers of 2. Only two integers, 0 and 1, may be used, whereas in the decimal system 10 integers, 0 to 9, are used. Recalling that the powers of 2 are $2^0 = 1$, $2^1 = 2$, $2^2 = 4$, $2^3 = 8$, and so on, we can form a positional number system for the binary numbers. Table 3-11 expands Table 3-4 and gives the binary numbers and their decimal equivalent. Wherever it is useful to avoid confusion, we add a subscript to indicate the base to which the number belongs. Any number written with 2 to 9 cannot be a binary number because those digits do not belong to the binary number system.

To write the next higher number in Table 3-10 or in Table 3-11, we need an additional column for the next power of the base. Note in Table 3-11 that four binary digits are required to provide all 10 decimal digits, 0 to 9.

DECODERS AND NUMBER SYSTEMS

In this and the following sections we look at the use of the basic gates for decoding, display, data routing, and other applications. The examples will indicate how designs for special purposes may be made and implemented with basic gates. Circuits for nearly all of the tasks in control, computation, instrumentation, and other functions are available in medium-sized integrated (MSI) circuits packaged in dual in-line packages (DIP).

TABLE 3-12
Several Number Systems

Binary	Octal		Hexadecimal		BCD	Decimal
00000000	000 000	0_8	0000 0000	0_{16}	0000 0000	0
00000001	000 001	1	0000 0001	1	0000 0001	1
00000010	000 010	2	0000 0010	2	0000 0010	2
00000011	000 011	3	0000 0011	3	0000 0011	3
00000100	000 100	4	0000 0100	4	0000 0100	4
00000101	000 101	5	0000 0101	5	0000 0101	5
00000110	000 110	6	0000 0110	6	0000 0110	6
00000111	000 111	7	0000 0111	7	0000 0111	7
00001000	001 000	10	0000 1000	8	0000 1000	8
00001001	001 001	11	0000 1001	9	0000 1001	9
00001010	001 010	12	0000 1010	A	0001 0000	10
00001011	001 011	13	0000 1011	B	0001 0001	11
00001100	001 100	14	0000 1100	C	0001 0010	12
00001101	001 101	15	0000 1101	D	0001 0011	13
00001110	001 110	16	0000 1110	E	0001 0100	14
00001111	001 111	17	0000 1111	F	0001 0101	15
00010000	010 000	20	0001 0000	10	0001 0110	16
00010001	010 001	21	0001 0001	11	0001 0111	17
00111111	111 111	77	0011 1111	3F	0110 0011	63
01000000_2	001 000 000	100_8	0100 0000	40_{16}	0110 0100_{BCD}	64_{10}

Decoders are logic-gate arrangements that accept binary inputs in one form and generate binary outputs in another. The most obvious example is the decoding of decimal numbers to binary form and vice versa. Decoding between other number systems is also required; however, we will discuss first the relations among binary, octal, hexadecimal, decimal, and binary-coded decimal (BCD) numbers. In Table 3-11 we showed the relation between four-bit binary positional numbers and decimal numbers. Here we will expand the relationships and point out the manner of converting from one system to another.

It will be useful to use eight-bit binary numbers in our examples. An eight-bit binary number is called a *byte*. A byte can express $256 = 2^8$ values from 00000000 to 11111111_2 or from 0_{10} to 255.

An octal number is a positional number based on the powers of 8. In each position the number may be 0 to 7, a total of eight digits. Similarly, a hexadecimal number is a positional number based on the powers of 16. The numbers, which may appear in any position, are 0 to 9 and A to F. Table 3-12 lists the numbers in the various systems.

You will observe that binary-coded decimal (BCD) numbers are the binary equivalent of 0 to 9 in each position. Because four binary bits are required to express numbers 0 to 9, each decimal position will correspond to four bits. As we will see, the conversion from binary to BCD or vice versa is a simple process only for the least significant decimal digit. The details of binary-BCD conversion in general are given in Appendix A.

When a binary number is grouped in sets of three bits, starting at the least significant bit (LSB), each set of three binary bits is equivalent to an octal digit. A decoder can transform each three-bit set to the octal digit from 0 to 7. Zeros are added to the end of the binary number to make each set contain three bits if the end set is not complete.

Similarly, the conversion of a binary number to a hexadecimal number requires that the binary number be grouped into sets of four binary bits. Zeros are added if necessary. Each set of four binary bits can be decoded into the hexadecimal digits 0 to 9 and A to F.

Hexadecimal numbers and, to a lesser extent, octal numbers are used in programming microprocessors in their own operational code. The program code usually consists of one, two, or three bytes for each operational step the microprocessor takes. It is extremely difficult to write such numbers in binary bits because the zeros and ones may be transposed and because it is difficult to read the values in a chain of 8 to 24 bits. Hexadecimal numbers are much easier to check in a program than a chain of bits, and they are easily decoded for use by the microprocessor. Each byte of the program code is expressed by two hexadecimal numbers.

A string of binary bits representing a number can be easily transformed to an octal or hexadecimal string by appropriate grouping and encoding. The inverse process can form a binary number from the octal or hexadecimal string. The binary digits correspond exactly to the binary equivalent of these two number systems. The relation among the systems can be seen clearly in Table 3-12.

The advantage of binary-coded decimal (BCD) numbers is that a decoder can display the decimal digit 0 through 9 in the position corresponding to the power of 10 that corresponds to the set of four bits. An examination of Table 3-12 shows that simply grouping the binary number in sets of four bits does not yield the BCD number when the decimal value exceeds nine.

There is no simple way to group the binary number to arrive at the BCD number beyond nine. It is simple, however, to obtain the BCD number between 10_{10} and 19_{10} from the bits of the binary number whose value lies in that range. For a number larger than $9_{10} = 1001_2$ the BCD is obtained by adding $6_{10} = 0110_2$. For example, since $D_{hex} = 13_{10} = 1101_2$ is larger than 9_{10}, we add 0110_2 to get $0001\,0011_{BCD} = 13_{10}$.

MSI (medium-sized integrated) circuits are used to convert large binary numbers to BCD; other MSI circuits convert in the other direction.

Before we study decoders that use binary or other number systems to address memory locations, drive displays, and perform other tasks, we will confirm that the numerical value of each number in the-next-to-last line of Table 3-12 is the same. To do this we multiply the digit with the power of the base of its system corresponding to the position of the digit. We choose the line corresponding to 63_{10} and calculate the equivalent value of each term in the other systems, as shown in Table 3-13.

TABLE 3-13
Equivalent Values in Number Systems

Binary positional		2^7	2^6	2^5	2^4	2^3	2^2	2^1	2^0
Decimal value		128	64	32	16	8	4	2	1_{10}
Binary number		0	0	1	1	1	1	1	1
		0 +	0 +	32 +	16 +	8 +	4 +	2 +	1 = 63_{10}
Octal positional			8^2		8^1			8^0	
Decimal value			64		8			1_{10}	
Octal number					7			7	
			0 +		7 × 8 +			7 × 1 = 63_{10}	
Hexadecimal positional			16^2		16^1			16^0	
Decimal value			256		16			1_{10}	
Hexadecimal number					3			F = 15_{10}	
			0 +		3 × 16 +			15 × 1 = 63_{10}	
Binary-coded decimal			10^2		10^1			10^0	
Decimal value			100		10			1_{10}	
BCD number			0000		0110			0011	
			0 +		6 × 10 +			3 × 1 = 63_{10}	

BINARY-TO-OCTAL DECODER

The binary-to-octal decoder may be made with basic gates arranged to satisfy the logic function found in the truth table. In this task we want an output to correspond to each one of the octal digits from 0 to 7 whenever the binary number of three bits indicates that value. The truth table for this purpose is Table 3-14. The three binary bits are denoted by X, Y, and Z, and the octal lines are designated by A_0 to A_7. In our table $X = 1$ and $\bar{X} = 0$; $Y = 1$ and $\bar{Y} = 0$; and so on.

TABLE 3-14
Truth Table for a Binary to Octal Line Decoder

X	Y	Z	A	Octal Digit
$\bar{X}$	$\bar{Y}$	$\bar{Z}$	A_0	0
$\bar{X}$	$\bar{Y}$	Z	A_1	1
$\bar{X}$	Y	$\bar{Z}$	A_2	2
$\bar{X}$	Y	Z	A_3	3
X	$\bar{Y}$	$\bar{Z}$	A_4	4
X	$\bar{Y}$	Z	A_5	5
X	Y	$\bar{Z}$	A_6	6
X	Y	Z	A_7	7

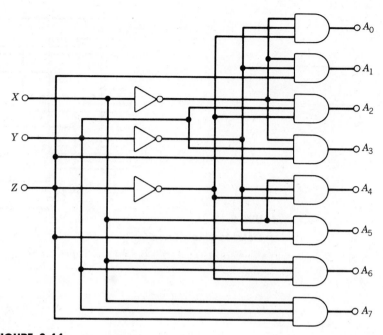

FIGURE 3-11
A Binary-to-Octal Line Decoder
One of eight outputs, A_0 to A_7, is selected by the three-bit input, X, Y, and Z.

Only one line can become high (1) for each independent set of binary bits. Thus each *minterm* (one of the set of three binary bits above) equals one output. For example, $\bar{X}Y\bar{Z} = A_2$ is satisfied by a three-input AND-gate with $\bar{X}$, Y, and $\bar{Z}$ as inputs. Figure 3-11 shows a logic-gate arrangement that accomplishes the task of putting an output of 1 (5 V) on the line designated by the three bits.

OCTAL-TO-BINARY ENCODER

The inverse process of forming a binary number from one of eight octal digits, 0 to 7, is also easily accomplished by basic logic gates. From Table 3-14 we can identify the input; in this case the inputs are A_0 to A_7, which involve X, Y, or Z. For example,

$$A_4 + A_5 + A_6 + A_7 = X$$

X can be generated by an OR-gate with these four inputs. Similarly, $A_2 + A_3 + A_6 + A_7 = Y$ and $A_1 + A_3 + A_5 + A_7 = Z$. Note that A_0 is not required as an input to any of the three OR-gates generating X, Y, or Z. Figure 3-12 shows an octal-to-binary encoder.

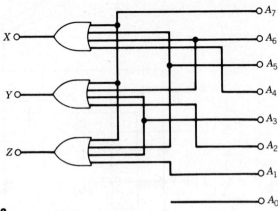

FIGURE 3-12
An Octal-to-Binary Encoder
The three-bit binary equivalent to an octal number is generated by the gates.

MULTIPLEXER

Multiplexing is the technique of selecting and transmitting signals from several data sources to a single data path. In the example here, the signals of the four input lines are connected one at a time to a single output line. The input line selected for transmission over the output line is governed by an address code. The address code must be able to open one path to each possible data source while the other paths are closed. Thus two address signals are required for a four-input multiplexer. The truth table for the four-to-one-line multiplexer is given in Table 3-15. The address lines are A_1 and A_0; the input lines are $I_3, I_2, I_1,$ and I_0; the output line is denoted by W.

The four-to-one multiplexer that satisfies this truth table is shown in Figure 3-13. Multiplexers are used in computers to transmit signals over data buses and to read from and write into memories; therefore, many MSI circuits are available for multiplexing.

TABLE 3-15
Truth Table for a Four-to-One Multiplexer

A_1	A_0	W
0	0	I_0
0	1	I_1
1	0	I_2
1	1	I_3

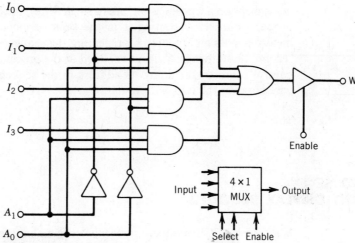

FIGURE 3-13
A Four-to-One Line Multiplier
A two-bit code selects the input line, I, to be routed to the output line, W. The enable signal is required for the data to be available at W.

Many multiplexers have an *enable* line that must be "on" so that the signals routed through the multiplexer can be carried on the output data line. When the enable signal is "off," the multiplexer is disconnected electrically from the output line. When the output is disconnected, the data line is free to transmit signals from another multiplexer or another tristate gate. (We discuss the tristate gate later.) The enable connection is shown in Figure 3-13 as a side input on the buffer amplifier. In our analysis of the gate logic an enable signal is not considered an input variable, but it is understood that an output is sensed only when the enable signal is present.

DEMULTIPLEXER

A demultiplexer serves the complementary function of a multiplexer by directing the incoming data on a single line to one of several data paths. Address signals determine that a particular data path will receive the incoming signal.

Multiplexers and demultiplexers are frequently used at the source end and the use end of data links to reduce the number of transmission lines. For example, consider an array of digital instruments that monitor an industrial process, such as oil refining. The data they generate—temperature, pressure, flow rate, density, and volume—are to be recorded in and to be analyzed to control the process from a central control room. The multiplexer places the data sequentially from each instrument on a transmission link (a single wire); the demultiplexer directs it to the proper recorder by routing it in the same sequence at the control room. The measurements are updated by repeating the sequence perhaps a hundred times a second. If another instrument is added at the processing site, it can be read at the control room by including its address in the sequence without running another wire between the two end points.

BCD TO SEVEN-SEGMENT DISPLAY DECODER

The four-bit BCD number represents a decimal number from zero to nine. A BCD to seven-segment display decoder allows the number to be seen as a decimal digit by lighting the proper segments of the LED (light-emitting diode) display. Similar decoders are used for liquid crystal displays.

A seven-segment display forms the numbers 0 to 9 by connecting the appropriate set of segments to the power supply or to ground through a current-limiting resistor. The decoder for LEDs are open-collector devices, requiring external connections for the displays to operate.

The segments of the display are numbered a, b, c, d, e, f, and g. A decimal point, dp, is usually provided in the display. The arrangement of the segments and the numbers associated with them are shown in Figure 3-14; the segments lighted for each of the hexadecimal digits are also shown. Because many numbers have common segments, different numbers may be displayed merely by extinguishing one or two segments and lighting one or two other segments. Segment c is lighted in all decimal numbers except 2; the decoder extinguishes segment c only when displaying 2. Table 3-16, a truth table, is used to generate the logic functions for the seven-segment display decoder. The logic for one segment is illustrated in the following example.

Table 3-16 lists the decimal number, the BCD number in two forms, and the segments of the seven-segment display to be lighted to display the decimal equivalent to the BCD code.

The segment designated as e is lighted when the numbers 0, 2, 6, and 8 are displayed. We equate

$$e = \overline{W}\overline{X}\overline{Y}\overline{Z} + \overline{W}\overline{X}Y\overline{Z} + \overline{W}XY\overline{Z} + W\overline{X}\overline{Y}\overline{Z}$$

as the logical relation to light segment e. We now simplify this relation by constructing a Karnaugh map (Figure 3-15), in which the squares corresponding to e are marked. The adjacent squares are combined to yield

$$e = \overline{W}X\overline{Z} + \overline{W}\overline{X}\overline{Z} + \overline{X}\overline{Y}\overline{Z}$$

This relation is confirmed by truth table in Table 3-17.

BCD to Seven-Segment Display Decoder

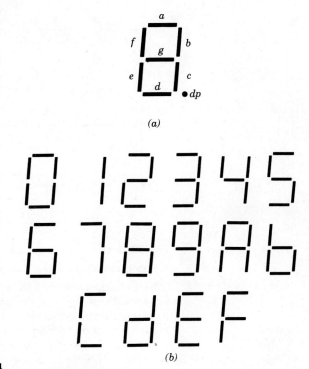

FIGURE 3-14

Seven-Segment LED Displays

(*a*) Segment designation. The segments that may be independently lighted are labeled *a* through *g*. (*b*) Hexadecimal digits. The lighted segments for each hexadecimal number. Segment *a* on 6 and segment *d* on 9 may or may not be used in the display for decimal numbers only.

TABLE 3-16
Truth Table for a Seven-Segment Display

Decimal Number	BCD	Number	Segments to Be Lighted
0	$\overline{W}\overline{X}\overline{Y}\overline{Z}$	0000	a b c d e f
1	$\overline{W}\overline{X}\overline{Y}Z$	0001	b c
2	$\overline{W}\overline{X}Y\overline{Z}$	0010	a b d e g
3	$\overline{W}\overline{X}YZ$	0011	a b c d g
4	$\overline{W}X\overline{Y}\overline{Z}$	0100	b c f g
5	$\overline{W}X\overline{Y}Z$	0101	a c d f g
6	$\overline{W}XY\overline{Z}$	0110	a c d e f g
7	$\overline{W}XYZ$	0111	a b c
8	$W\overline{X}\overline{Y}\overline{Z}$	1000	a b c d e f g
9	$W\overline{X}\overline{Y}Z$	1001	a b c d f g

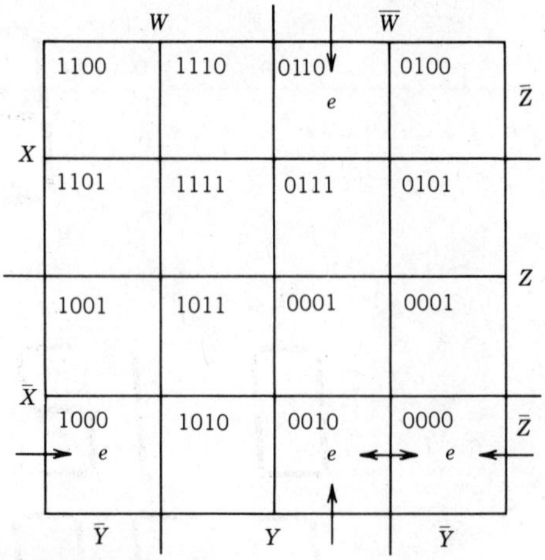

Combinational Digital Electronics

	W		$\overline{W}$		
1100	1110	0110↓ e	0100	$\overline{Z}$
1101	1111	0111	0101	
1001	1011	0001	0001	Z
1000 e →	1010	0010 e ←→	0000 e ←	$\overline{Z}$
$\overline{Y}$	Y		$\overline{Y}$	

FIGURE 3-15

A Karnaugh Map for LED Segment e.
The binary logic equation for segment e in a decimal seven-segment decoder is simplified by the map technique. Arrows indicate the manner of combining blocks.

An examination of Table 3-17 shows an interesting pattern. In the lines for 0 and 2 the term $\overline{W}\overline{X}\overline{Z}$ yields the same value as one of the other terms, but in the lines for 6 and 8, where $e = 1$, it makes no contribution. We conclude that the term is not necessary; we need only to form the logic-gate configurations to satisfy the equation

$$e = \overline{W}Y\overline{Z} + \overline{X}\overline{Y}\overline{Z}$$

TABLE 3-17
Truth Table for $e = \overline{W}Y\overline{Z} + \overline{W}\overline{X}\overline{Z} + \overline{X}\overline{Y}\overline{Z}$

W X Y Z	$\overline{W}Y\overline{Z}$	$\overline{W}\overline{X}\overline{Z}$	$\overline{X}\overline{Y}\overline{Z}$	e
0 0 0 0	0	1	1	1
0 0 0 1	0	0	0	0
0 0 1 0	1	1	0	1
0 0 1 1	0	0	0	0
0 1 0 0	0	0	0	0
0 1 0 1	0	0	0	0
0 1 1 0	0	0	0	0
0 1 1 0	1	0	0	1
0 1 1 1	0	0	0	0
1 0 0 0	0	0	1	1
1 0 0 1	0	0	0	0

The algebraic elimination of the term $\overline{W}\overline{X}\overline{Z}$, by using the rules for Boolean algebra given in Table 2-2, is not straightforward because it is not equal to one or the other term for all cases.

The variables can be identified with the other segments in a similar manner and can be simplified. The manufacturers' specification sheets for BCD-LED decoder drives show the gate arrangements in their medium-sized integrated circuits. An examination will show the suppliers' gate configuration for this task.

Some seven-segment decoders may use all states possible on the four input lines and display the hexadecimal numbers 0 to 9 and A to F. Some commercially available decoders display symbols other than A to F for the hexadecimal equivalents of 10 to 15.

DIGITAL COMPARATORS

Knowledge that one number is equal to, larger than, or smaller than another number is important for computation, data routing, and other logic decisions. Comparators examine binary numbers bit by bit to learn the magnitude relation between them. Truth tables will show the arrangement of logic gates to accomplish the comparison. Figure 3-16 shows a two-bit comparator corresponding to the truth table given in Table 3-18.

An examination of Table 3-18 shows that the magnitude relation is determined by the more significant bits unless they are identical; in that case the relation is resolved by comparing the lower value bits.

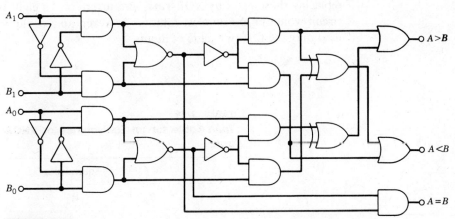

FIGURE 3-16

A Two-Bit Comparator

This logic-gate arrangement determines whether the two-bit number $A_1 A_0$ is equal to, larger than, or smaller than $B_1 B_0$.

TABLE 3-18
Truth Table for a Two-Bit Comparator

A_1	A_0	B_1	B_0	Comparison	Output
0	0	0	0	$A = B$	X
0	0	0	1	$A < B$	Y
0	0	1	0	$A < B$	Y
0	0	1	1	$A < B$	Y
0	1	0	0	$A > B$	Z
0	1	0	1	$A = B$	X
0	1	1	0	$A < B$	Y
0	1	1	1	$A < B$	Y
1	0	0	0	$A > B$	Z
1	0	0	1	$A > B$	Z
1	0	1	0	$A = B$	X
1	0	1	1	$A < B$	Y
1	1	0	0	$A > B$	Z
1	1	0	1	$A > B$	Z
1	1	1	0	$A > B$	Z
1	1	1	1	$A = B$	X

In Figure 3-16 observe that the two AND-gates, the inverters on their inputs, and the NOR-gates that use the inputs A_1, B_1, A_0, and B_0 constitute equivalence circuits whose truth table is Table 3-19.

We see that the input portions of the comparator could be simplified by replacing those gates by XOR-gates and inverters. An eight-bit comparator MSI circuit would be the simplest solution, however, and would be cheaper even if we used only the first two pairs of inputs.

TABLE 3-19
Truth Table for an Equivalence or XNOR-gate

		Equivalence
A_n	B_n	W_n
0	0	1
0	1	0
1	0	0
1	1	1

SUMMARY

Each task illustrated by the arrangements of integrated circuits has been accomplished by combinations of the three basic logic gates. In most applications there would be a large number of gates. For example, to add two bytes we would require eight full adders. With a least significant byte, of course, we could get by with seven full adders and a half adder, but we would probably not want to build a circuit that had a restriction in its use.

If we counted the logic gates to add two bytes, we would find a large number, probably 50 and, if we used DIPs with several gates on each, we would require at least 15 to complete the two-byte adder. The cost of the DIPs, their sockets, and the labor and time needed to wire them together would be quite high. Medium-sized integrated circuits are arrangements of all the basic gates and interconnections on a single silicon chip. MSI four-bit adders are available for less than a dollar each at this writing, and they can be used in tandem to add two bytes. A more sophisticated solution might be to use two four-bit MSI arithmetic logic units (ALUs). These circuits cost less than $2.50 each and can add, subtract, decrement, compare, and perform 12 other arithmetic operations. Each ALU contains the equivalent of 75 gates.

The same remarks apply to the other circuits we discussed. MSI elements exist for decoding, multiplexing, and so on; these are the most practical elements to use. Even so, in many applications a few basic logic gates can provide a control, measurement, or decision function that has not been implemented in a MSI.

Our purpose is to understand how an efficient digital circuit can be made to satisfy any logic equation. The designers of MSI elements have used the same knowledge to create and simplify their circuits. We have learned that regardless of how complicated a digital device might appear, it is made with the basic logic gates, whose action we understand. New techniques in making circuits and their components will be developed as time passes, but binary Boolean logic will endure.

Many other special functions may be implemented with basic gates. In a systematic approach for design, the contents of a truth table are expressed in logic function equations. The logic functions are simplified, if possible, and basic gates are arranged to satisfy the logic functions. Frequently, a single MSI circuit is unavailable, but a number of MSI circuits, rather than many basic gates, may be combined to serve the special purpose.

PROBLEMS

3-1 Without using an exclusive-OR gate, design other basic gate arrangements to form the half adder. There are four simple arrangements.

3-2 Design a two-to-four-line demultiplexer.

3-3 With Table 1-4, the truth table for the inverter, show that the circuit in Figure 3-4a yields the truth table for an OR-gate. Are the wired gates limited to two inputs?

3-4 Rather than using inverters in Figure 3-4a, consider using two open-collector two-input NAND-gates. Find the truth table and identify the circuit equivalence. Repeat for two-input NOR-gates as replacements for the inverters.

3-5 Implement $ABC + \bar{A}BC + A\bar{B}C + AB\bar{C} = X$ (a) with NAND-gates only and (b) with NOR-gates only.

3-6 Subtract 87 from 133 by the complement method illustrated in the text.

3-7 A seven-segment display can be used to give the hexadecimal numbers. After 0 to 9, one option is to display $A = 10_{10}$, $B = 11_{10}$, $C = 12$, $D = 13$, $E = 14$, and $F = 15$. Find the logic expressions to illuminate segments e and a for a hexadecimal display. Note that segment a is used in 6_{10} to avoid confusing $b = 11_{10}$ with 6_{10}.

3-8 Find the logic functions for a two-bit comparator. Draw the logic-gate configuration for the comparator and rewrite Table 3-18 with A_0, A_1, B_0, B_1, and so on.

3-9 Write the value 18_{10} in bits. Convert the binary number to BCD.

3-10 Design a binary-to-decimal decoder.

CHAPTER 4

SEQUENTIAL DIGITAL ELECTRONICS

In the logic-gate circuits that we have considered up to now, the output states are determined by the input states and the interconnections; they are called *combinational circuits*. We now turn to another class of logic-gate circuits, called *sequential digital circuits*, whose output is determined by the inputs and responses at an earlier point. In these circuits the output state is related sequentially to the input states. An obvious example of a sequential digital circuit is a counter or scaler whose count number depends upon the number of pulses received earlier.

Sequential digital electronics are based upon the property by which certain configurations of gates are in one or the other of two states. These configurations of gates are called *bistable elements*. A common name for a class of bistable elements is *flip-flop*, which describes their response to the proper input. Bistable elements, as their name implies, remain in either one state or the other indefinitely or until an input initiates the change in state.

Another type of sequential circuit changes state spontaneously at a regular time interval. Circuits of this type are called *oscillators*; if an oscillator is used to synchronize the change of states of bistable elements, the oscillator is called a *clock*. The term for the sequence of pulses is *clock pulses* (CP).

Still another sequential digital circuit is the *monostable element*. The monostable element responds to the proper input by changing state. After a preset time interval it returns spontaneously to its initial state, where it remains indefinitely or until an input again initiates the temporary change of state.

In using the term "proper input" above, we suggest that an input may or may not initiate a change in state. This will be illustrated below. The digital logic of the gate configuration governs the response of the sequential circuit to inputs and establishes the input states, zeros and ones, to which it can respond. The gate logic also may require an enable signal or a clock pulse or both for the state to change with the proper input.

Bistable elements perform an important function because they can remain in one binary state or the other for an indefinite time. That is, a bistable element remembers its state; thus, a bistable element can be used as a *memory cell*. An array of bistable elements is a *memory array*. (Various forms of digital memories are discussed in the next chapter.) A small array of bistable elements storing binary information may be called a *register*. Registers are used for intermediate storage in arithmetic operations and elsewhere.

In sequential digital electronics it is customary to define the state of the element as that state (0 or 1) which exists at the terminal (pin) marked Q. That is, the state of the element is high (1) if the voltage at Q is high, and the state of the element is low (0) if the voltage at Q is low. In integrated circuits the complement of Q, marked $\bar{Q}$, is also available at a terminal (pin) of the bistable element.

The state after a sequential event, whether or not the state has changed, is called $Q(t+1)$. In the truth tables for sequential elements we list Q and $Q(t+1)$ to show the initial and final states for a set of inputs.

The integrated circuits have terminals (pins) that permit the state, Q, to be established initially. The state of a bistable element is initialized to $Q=1$ by a signal (voltage pulse) on the *set* or *preset* terminal. To make $Q=0$ the signal on a different terminal *resets* or *clears* the element.

We can see the underlying principles of operation of bistable elements by examining the least complicated arrangement of the basic logic gates that has the property of being and remaining in one logic state or the other and that can change states upon an input signal. Two options for making the bistable elements are shown; one uses NOR-gates and the other uses NAND-gates.

The key to the bistable operation lies in the connections between inputs and outputs in the gate arrangement. Some input and output states, however, are logically inconsistent. These are called *indeterminate states* and must be avoided. We will examine the simplest arrangements and the improvements that are made when other logic gates are added to the circuits. Later we will use integrated circuit bistable elements to devise sequential circuits that follow a desired pattern of response to input signals.

Throughout the chapter the circuits will be analyzed with the simplification techniques developed in Chapter 2: truth tables, Boolean algebra, and Karnaugh maps. The examples will show how any desired sequence of states can be designed to occur in a sequential circuit. It will also become apparent that there is no unique configuration of elements or interconnection of elements for any specific sequential device.

RS FLIP-FLOP

Perhaps the simplest of the flip-flops is the set-reset flip-flop, designated the *RS* FF. Figure 4-1 shows *RS* FFs made with basic gates.

The operation of the NOR-gate flip-flop and the NAND-gate flip-flop are similar but because the truth tables for those gates differ, different inputs are required to achieve a particular state Q and to maintain that state. This point is important because truth tables govern the kind of logic gates that may be added to enhance the operation of the simplest flip-flops. Because integrated circuit bistable elements do not have the limitations of these simplest circuits, there is no restriction on the logic gates that may precede or follow the commerically available flip-flops. We will see that as we enhance the simple flip-flops we also will remove the limitations.

At this point in our study the flip-flops in Figure 4-1 are still combinational circuits because the state Q depends only upon the inputs to the R and S terminals. The sequential properties emerge as we examine more sophisticated arrangements of logic gates and flip-flops. In Figure 4-1a a high (1) on the set terminal and a low (0) on the reset terminal will result in the state $Q = 1$. To confirm this finding we write the NOR-gate truth table, using $A + B = \bar{C}$.

For the lower NOR-gate in Figure 4-1a, let $S = 1$. This requires that $\bar{Q} = 0$, the value corresponding to any of the last three lines in Table 4-1. If $R = 0$, then both inputs to the top NOR-gate are 0, corresponding to the first line in Table 4-1; therefore, $Q = 1$ as expected because of the high on the set line. This condition makes both inputs to the lower NOR-gate high, corresponding to the last line in Table 4-1. If the high (1) on the set line is removed—that is, becomes low (0)—the value of $\bar{Q}$ remains 0 as seen in the middle lines of Table 4-1, and there is no change in either Q or $\bar{Q}$. The reader should confirm that the choice of other inputs on S and R will generate the truth table in Table 4-2. The logical contradiction arises when both R and S are made high, because each gate has inputs corresponding to one of the last three lines in Table 4-1, which require that both Q and $\bar{Q}$ be low.

An identical analysis, using the NAND-gate logic equation $AB = \bar{C}$ generates Table 4-3. It shows the allowed and the unallowed inputs for the *RS* FF made with NAND-gates in Figure 4-1b.

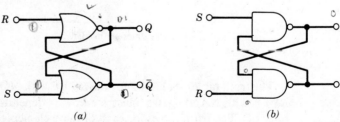

FIGURE 4-1

Set-Reset Flip-Flops

The bistable elements are formed by the basic gates. (*a*) *RS* flip-flop made with NOR-gates. (*b*) *RS* flip-flop made with NAND-gates.

TABLE 4-1
Truth Table for a Two-Input NOR-gate

A	B	$\bar{C}$
0	0	1
0	1	0
1	0	0
1	1	0

TABLE 4-2
Truth Table for a RS Flip-Flop with NOR-gates

S	R	Q	$\bar{Q}$
1	0	1	0
0	0	1	0
0	1	0	1
0	0	0	1
1	1	0[a]	0[a]

[a] These states are invalid.

TABLE 4-3
Truth Table for a RS Flip-Flop with NAND-gates

S	R	Q	$\bar{Q}$
1	0	0	1
1	1	0	1
0	1	1	0
1	1	1	0
0	0	1[a]	1[a]

[a] These states are invalid.

The truth tables, Table 4-2 for the RS FF with NOR-gates and Table 4-3 for the RS FF with NAND-gates, illustrate how Q depends on the values of set (S) and reset (R). The truth tables for each of these elements show that the digital logic is violated when S and R are both high with use of the NOR-gates or both low in case of the NAND-gates. By definition $\bar{Q}$ is the complement of Q, yet this relation is violated with unallowed inputs. In normal operation the possibility of unallowed inputs is avoided.

The NOR-gate *RS* FF is set ($Q = 1$) or reset ($Q = 0$) with a high (1) pulse on the respective inputs. The NAND-gate *RS* FF is set or reset with a low (0) pulse on its inputs. Otherwise the inputs are kept low for the NOR-gate *RS* FF and high for the NAND-gate *RS* FF. These are the *quiescent* inputs. The truth tables show that the quiescent inputs allow the state to persist.

In application, the inputs *S* and *R* in the *RS* flip-flop with NOR-gates are low (0) except when one or the other is raised momentarily to set *Q* to 1 or to reset *Q* to 0.

Similarly, the inputs *S* and *R* in the *RS* flip-flop with NAND-gates are high (1) except when one is lowered momentarily to set (or reset) *Q*.

DATA FLIP-FLOP

The *RS* FF may be modified by adding an inverter to use a single input in order to create a data flip-flop whose state *Q* follows that of the input *D*. The data flip-flop is shown in Figure 4-2, and its truth table in Table 4-4. The importance of the data flip-flop will become apparent. The reader will note that the data flip-flop in Figure 4-2 selects only lines 1 and 3 in Table 4-3 for inputs to the *RS* FF. The invalid state cannot arise because the inverter causes *R* and *S* to be complements.

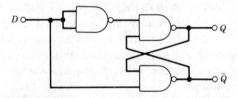

FIGURE 4-2

A Data Flip-Flop

The state of the flip-flop, *Q* is identical to that of the input *D* in this element made with NAND-gates.

TABLE 4-4
Truth Table for a Data Flip-Flop

D	Q	Q̄
0	0	1
1	1	0

CLOCKED FLIP-FLOPS

The usefulness of flip-flops is greatly enhanced when the flip-flop is prevented from responding to input states except in the presence of a signal that will be called a clock pulse. In the sequential circuits the clock pulses may occur regularly, for example, to synchronize the program steps in a computer. We will also designate pulses that are random in time, however, as clock pulses. We chose to use the term "clock pulse" because sometimes other enable signals are in use in sequential circuits. The usefulness of the other kinds of enable signals will be illustrated later.

In the material that follows we assume that a clock pulse is a high (1). The absence of a clock pulse means that a low (0) exists on the clock pulse line.

A clocked flip-flop uses additional gates as well as the clock pulse, enabling the flip-flop to make the logical decision whether or not it should respond to its input states. Consequently we should recognize that clocked flip-flops are sequential circuits, which is consistent with our statement that the response of a sequential circuit depends upon the inputs and on the present state of the circuit. The time dependence of states is shown by designating the initial (present) state as Q and the subsequent state as $Q(t+1)$.

We wish to emphasize that clock pulses and enable signals are not input or output states and thus are not shown as variables in Boolean algebra logic equations. They are simply what their name implies: the presence or absence of a voltage that allows a sequential circuit to respond or prevents it from responding to its inputs in accordance with its present state Q.

Because the RS flip-flop and the D flip-flop discussed above are combinational circuits, like all of those shown in Chapter 3, it is proper to describe their behavior by truth tables, as we did. A sequential circuit, however, is somewhat different; thus, for sequential circuits, we will designate the relation among the inputs, the present state Q, and the subsequent state $Q(t+1)$ as a *characteristic table*. The Boolean algebra relation between the time-dependent states is shown in a *characteristic equation*.

CLOCKED *RS* FLIP-FLOP

The addition of AND-gates to the input lines of the NOR-gate *RS* flip-flop (Figure 4-3) places lows (0) on the flip-flop input lines until a clock pulse (1) is present. Table 4-2 shows that the NOR-gate flip-flop maintains the state when both inputs are low. With the AND-gates on the input lines a change of state can occur when a clock pulse (1) enables the flip-flop. Figure 4-3 shows the clocked *RS* flip-flop, and Table 4-5 is the characteristic table. Note that the table is written with the binary order of increasing magnitude, as if *QSR* were a binary number.

The addition of AND-gates does not prevent the use of unallowed input states, which may give rise to indeterminate states when there is a clock pulse (1). If unallowed inputs exist and the clock pulse changes to zero, the stronger NOR-gate

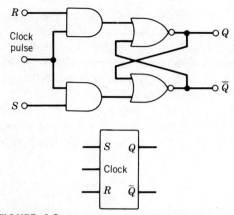

FIGURE 4-3
A Clocked RS Flip-Flop

A clock or enable pulse must occur for the RS flip-flop to respond to its input states.

TABLE 4-5
Characteristic Table for a Clocked RS Flip-Flop

Q	S	R	Q(t + 1)
0	0	0	0
0	0	1	0
0	1	0	1
0	1	1	Indeterminate
1	0	0	1
1	0	1	0
1	1	0	1
1	1	1	Indeterminate

will dominate the behavior of the flip-flop as it returns to an allowed state (i.e., 0 inputs). It is impossible to know which state, $Q(t + 1) = 1$ or $Q(t + 1) = 0$, will result. We conclude that the simultaneous presence of highs on S and R at the time of the clock pulse, as in Figure 4-3, is to be avoided. We have an advantage, however, that did not exist before the AND-gates were added: R and S can have any input states, including the unallowed inputs, as long as there is no clock pulse; whenever a clock pulse (1) is present, however, the flip-flop will respond to the input signals as long as the clock pulse lasts.

In Figure 4-3 the symbol for the clocked *RS* flip-flop does not indicate whether the integrated circuit uses NOR-gates or NAND-gates (or even OR-gates or AND-gates) to form the flip-flop. The characteristic equation is the same for all *RS* FFs; thus a designation in the symbol for the kind of gate employed is meaningless. We show both NOR-gate and NAND-gate flip-flops to highlight the difference between their quiescent states: $R = 0$, $S = 0$ for the NOR-gate FF, and $R = 1$, $S = 1$ for the NAND-gate FF. The absence of a clock pulse on the gating circuits must yield the proper quiescent state. The type of quiescent state dictates the kind of gates to be used to clock the flip-flops.

CHARACTERISTIC EQUATION FOR CLOCKED *RS* FLIP-FLOP

The characteristic equation for the clocked *RS* flip-flop is obtained by making a Karnaugh map of its characteristic table. The information in Table 4-5 is mapped in Figure 4-4. The indeterminate states are marked as *X*s. The ones refer to the $Q(t + 1)$ states; Q is the present state before the clock pulse.

When the adjacent terms are combined, as shown in Figure 4-4, the characteristic equation for the clocked *RS* FF is

$$Q(t + 1) = S + Q\bar{R}$$

We must also state the restriction that $SR = 0$ to prevent the indeterminate states from being selected.

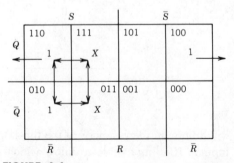

FIGURE 4-4

A Karnaugh Map for Clocked *RS* Flip-Flop

The blocks are labeled as *QSR* and marked with $Q(t + 1) = 1$ to correspond to Table 4-5. The indeterminate states are marked as *X*.

CLOCKED DATA FLIP-FLOP

The clocked data flip-flop shown in Figure 4-5 is made with a NAND-gate flip-flop that follows the logic in Table 4-3. The quiescent condition is maintained in this flip-flop with simultaneous highs on the inputs to the flip-flop. In this case the input NAND-gates ensure the highs when there is no clock pulse. An indeterminate state does not arise because the inverter prevents the simultaneous presence of identical inputs. The characteristic table for the clocked data flip-flop is given in Table 4-6.

The characteristic equation can be written by inspecting Table 4-6. D and $Q(t + 1)$ are always the same, independent of the value of Q. We note this by $Q(t + 1) = D$.

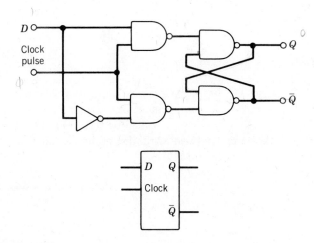

FIGURE 4-5

A Clocked Data Flip-Flop

Without a clock or enable pulse the output state does not respond to the input D. This circuit is called a data latch because it can hold its output state. The symbol for the clocked data flip-flop is shown.

TABLE 4-6
Characteristic Table for a Clocked Data Flip-Flop

Q	D	$Q(t + 1)$
0	0	0
0	1	1
1	0	0
1	1	1

If we look ahead a bit (no pun intended), we see that the clocked data flip-flop is the core of a digital memory cell. The other elements of the memory cell are the gates that are part of the enabling logic allowing a bit to be stored (written) in the cell, or a stored bit to be read from the cell.

The data flip-flop can also be used to store a bit for a period of time so that it may be observed or examined. In a computer that is doing arithmetic the bits on a line may change with each step, but it may be useful to examine the present value until a newer value is computed. When a data flip-flop is gated (i.e., when a single clock pulse occurs), it accepts the bit at its input and retains it until a new clock pulse occurs. Throughout the interval between the clock pulses, the bit to be displayed remains available. The gate or sampling pulse enables the *D* FF to accept a datum, and the bit is "latched in" as long as the gate pulse is absent. The data flip-flop used in this manner is called a *gated D latch*.

JK FLIP-FLOP

The *JK* flip-flop, a clocked flip-flop shown in Figure 4-6, is the next more complicated arrangement of a sequential element. The indeterminate states in the *RS* flip-flop can be avoided by returning the outputs *Q* and $\bar{Q}$ to the inputs, where they are AND-ed with the input signals and the clock pulse. The inputs are renamed *K* and *J*.

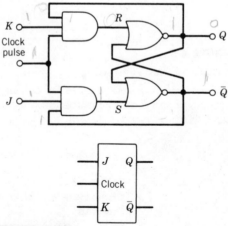

FIGURE 4-6

A *JK* Flip-Flop

The *JK* flip-flop returns its output state *Q* and its complement $\bar{Q}$ to the inputs. A clock pulse is required for the *JK* flip-flop to respond to its input states.

TABLE 4-7
Characteristic Table of a JK Flip-Flop

Q	J	K	Q(t+1)
0	0	0	0
0	0	1	0
0	1	0	1
0	1	1	1
1	0	0	1
1	0	1	0
1	1	0	1
1	1	1	0

The characteristic table for the JK flip-flop is given in Table 4-7 and is formed by selecting all sets of initial states and by examining the response when a clock pulse occurs. The initial states are the values of Q and of the inputs J and K, which govern the change of the flip-flop to the next state $Q(t+1)$. For example, let $Q = 0$, $J = K = 1$ as the initial states (line 4 of Table 4-7) and let there be a positive clock pulse. The upper three-input AND-gate in Figure 4-6 has the inputs $Q = 0$, $K = 1$, and $CP = 1$; therefore, its output is 0. The lower three-input AND-gate has $\bar{Q} = 1$, $K = 1$, and $CP = 1$ as its inputs; thus its output is 1. The NOR-gate RS flip-flop has inputs $R = 0$, $S = 1$, corresponding to the first line in Table 4-2. The new state $Q(t+1) = 1$ is shown in line 4 of Table 4-7. After the clock pulse ($CP = 0$), the quiescent values $R = 0$ and $S = 0$ maintain the new state. Other lines in the characteristic table of the JK flip-flop (Table 4-7) are generated by using other combinations of values of Q, J, and K.

As our example shows, the unallowed inputs of 1 and 1 on the RS flip-flop appear to be allowed in the JK flip-flop. Things are not quite right, however, as we can see by comparing the fourth and the eighth lines of Table 4-7. These lines show that with $J = K = 1$, the output changes to its complement with a high (1) on the clock line. In this situation we say that the output "toggles" between 1 and 0. As long as the clock pulse persists, the JK flip-flop will change states repeatedly when J and K are high. This is called "racing," and when it occurs, one cannot be certain of the final value of $Q(t+1)$ after the clock pulse drops to 0. If the length of a clock pulse is shorter than the time for electronic signal propagation through the gates, then the JK flip-flop responds only once in accordance with the characteristic table, but this is not the usual circumstance.

The solution to the unacceptable toggling is to use two JK flip-flops in tandem and to gate them with the clock pulse and its complement. This arrangement is called the *master-slave flip-flop*, which not only suppresses the racing but also has the desirable property of synchronizing the propagation of binary signals. The master-slave flip-flop is discussed below.

TOGGLE FLIP FLOP

The flip-flop made by tying J and K together is called a *toggle flip-flop*. The output changes state with each clock pulse when the input, T, is high. Table 4-8 is the characteristic table for the toggle flip-flop. As shown in Figure 4-7, the toggle flip-flop will change states on a rising clock pulse with a high on the input. In order to avoid unwanted racing, flip-flops are made with additional gates, shown in Figure 4-7, so that they will trigger on the changing clock pulse. These are called *edge-triggered flip-flops*. With edge triggering, the flip-flop makes a transition only because the clock pulse is rising. It is not enabled (it is disabled, in fact) while the clock pulse is high. This property is provided internally in the integrated circuits

TABLE 4-8
Characteristic Table of a Toggle Flip-Flop

Q	T	$Q(t+1)$
0	0	0
0	1	1
1	0	1
1	1	0

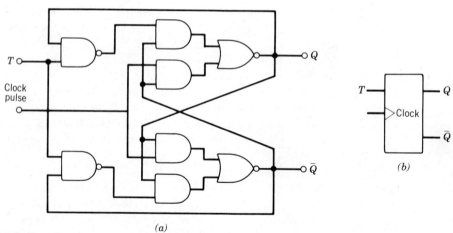

(a)

(b)

FIGURE 4-7

A Toggle Flip-Flop with Edge-Triggering

The inputs are tied together so that with T high (1), the output state Q changes with each clock pulse, but with T low (0), a state is retained. (a) A toggle flip-flop, whose output changes as the clock pulse rises. (b) The symbol with a triangle on the clock input indicates that this kind of toggle flip-flop changes states only on the rising part of the clock pulse. It symbolizes a "dynamic" T FF.

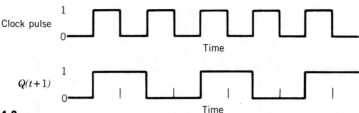

FIGURE 4-8
Binary Scaling by Toggle Flip-Flop
When T = 1 the toggle flip-flop output state Q(t + 1) alternates at half the rate of the clock pulse, and the clock rate is divided by two. The transition is edge-triggered by the positive-going clock pulse.

and is shown by the triangle on the clock pulse line. The triangle designates "dynamic" response, that is, the flip-flop and the clock pulse change together.

Table 4-8 has a familiar look; the relation of $Q(t + 1)$ to Q and T is exactly the same as that of Table 1-7 for an XOR-gate. We can write the characteristic equation for the toggle flip-flop directly as

$$Q(t + 1) = T \oplus Q = Q\bar{T} + \bar{Q}T$$

The toggle flip-flop is the core of scaler or counter circuits. Table 4-8 shows an interesting property of the toggle flip-flop when T is high: two input cycles, CP low-high-low-high, can be made to yield one cycle, $Q(t + 1)$ low-high, as an output. Thus the binary input on the clock line is divided by two. Figure 4-8 shows binary scaling of the clock pulse by an edge-triggered toggle flip-flop when T is high, corresponding to lines 2 and 4 in Table 4-8. Note that the transitions occur only when the clock pulse changes from low to high.

JK MASTER-SLAVE FLIP-FLOP

A more complex bistable element is made by combining two flip-flops and arranging the clock pulses to enable them in sequence rather than simultaneously. This arrangement, a master-slave flip-flop, is much more versatile than simpler elements. Master-slave flip-flops are essential in many applications to ensure the proper sequence of binary logic. Commercially available flip-flops are configured as master-slave elements.

The master-slave JK flip-flop shown in Figure 4-9 returns the outputs Q and $\bar{Q}$ of the slave element to the inputs of the master unit. The NAND-gates in the flip-flop on the left are three-input gates; the set of four NAND-gates on the right is simply the clocked RS flip-flop. An inverter is included in the clock pulse line to the slave flip-flop. Because of the inverter, Q and $\bar{Q}$ do not change until the clock pulse falls. The master element with the inputs J and K responded earlier on the rising clock pulse to establish the values at R and S. The falling clock pulse then causes Q

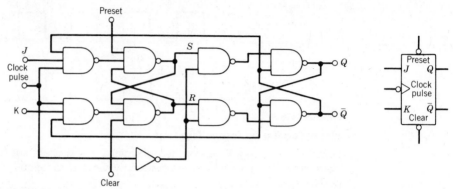

FIGURE 4-9

A Master-Slave *JK* Flip-Flop

The inversion of the clock pulse between the bistable elements causes the output state $Q(t+1)$ to be established after a complete clock pulse, which provides time isolation between input and output states. The small circle at the clock pulse input on the symbol of the flip-flop indicates that the state $Q(t+1)$ is established on the negative-going edge of the clock pulse. The circles on the preset and clear lines indicate that the preset or the clear action occurs if the line is pulsed low (0). These lines are held high (1) for all other operations.

to correspond to the values set on R and S at the slave unit, completing the response of the JK flip-flop. For clarity the additional gates to make the JK FF respond dynamically are not shown in Figure 4-9.

The symbol for the JK FF in Figure 4-9, shows that the transition is dynamic; that is, it responds on the changing clock pulse, but, as noted, the response is completed only on the falling edge of the clock pulse. The small circle adjacent to the small triangle, which denotes the dynamic change, states that the clock pulse transition from high to low is required to bring about $Q(t+1)$. The circle indicates that the complement of the change in the clock pulse pertains, that is, the transition high to low is the complement of the transition low to high occurring in the edged triggered flip-flop whose dynamic response is shown by only a triangle on the clock pulse lines.

The master-slave arrangement not only prevents racing during a clock pulse that exceeds the propagation time for signals through the gates but also allows the sequential response to states in a highly advantageous way. We will see below that a sequence of states can be propagated through a chain of master-slave flip-flops as if they are handed from one stage to the next. The significance of this will become apparent when we discuss scalers and registers.

Figure 4-9 shows two input lines, one labeled "preset" and the other "clear." These must be kept high during the operation of the flip-flop. When the preset line is grounded (making the line low), the state Q becomes high (1) regardless of the other signals or the clock pulse. (Similarly, grounding the clear line makes $Q = 0$.) That is, a low on preset makes S high and R low. This sets the slave RS flip-flop to

TABLE 4-9
Characteristic Table for a Master-Slave *JK* Flip-Flop

Q	J	K	Q(t + 1)
0	0	0	0
0	0	1	0
0	1	0	1
0	1	1	1
1	0	0	1
1	0	1	0
1	1	0	1
1	1	1	0

$Q = 1$ in accordance with the characteristic equation of the RS FF: $Q(t + 1) = S + Q\bar{R}$. Grounding the clear line similarly overrides the element to clear the output, $Q = 0$. These connections are shown on the symbol for the JK FF in Figure 4-9. The small circles state that the preset or the clear occurs when the line is made low.

Table 4-9 is the characteristic table for the master-slave JK flip-flop. Note that it is identical with Table 4-7 for the JK flip-flop in Figure 4-6 even though the JK FF was made with NOR-gates. This confirms that the type of gate used in making the JK FF is not significant. The only difference is the time sequence: the master-slave JK FF will yield $Q(t + 1)$ only after a full or complete clock pulse.

The Karnaugh map of the characteristic table is given in Figure 4-10. The result of combining the adjacent blocks is

$$Q(t + 1) = J\bar{Q} + \bar{K}Q + J\bar{K}$$

A check of the truth table shows that the last term, $J\bar{K}$, adds nothing to the expression. Since the $J\bar{K}$ term has a value of 1 only when one of the other terms is also 1, it is dropped.

The characteristic equation of all JK flip-flops is

$$Q(t + 1) = J\bar{Q} + \bar{K}Q$$

In the master-slave flip-flop the state $Q(t + 1)$ is not present until the clock pulse has returned to zero.

Table 4-9 is a time-dependent tabulation; thus it may be used to find the sequences that will occur when a chain of clock pulses is applied to the master-slave JK flip-flop. The state $Q(t + 1)$, which is established by one complete clock pulse becomes the present state Q for the next clock pulse, and so on for each pulse. State diagrams are made by selecting Q, J, and K and by using Table 4-9 to find $Q(t + 1)$. The state diagram is used to find $Q(t + 2)$, the value after the second clock pulse when $Q(t + 1)$ is considered a present state at the time of the second clock pulse.

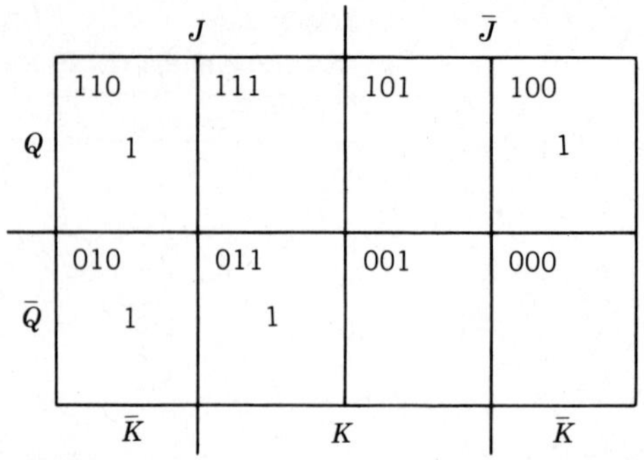

	J		$\bar{J}$	
Q	110 1	111	101	100 1
$\bar{Q}$	010 1	011 1	001	000
	$\bar{K}$	K	K	$\bar{K}$

FIGURE 4-10

A Karnaugh Map of Master-Slave JK Flip-Flop

The blocks are labeled with state values of Q, J, K and marked with $Q(1t+1) = 1$ from Table 4-9.

Figure 4-11 is made by labeling the circles with the values of Q, J, K and connecting one circle to the circle with the next state. For example, line 3 in Table 4-9 has $(Q, J, K) = (010)$, which after the clock pulse is $(Q(t+1), J, K) = (110)$, as in line 7. The next clock pulse makes the $(Q, J, K) = (110)$ become $(Q(t+1), J, K) = (110)$.

After the transition from 010 to 110 occurs, the state 110 continues to reproduce itself; that is, it persists with each clock pulse as long as $J = 1$ and $K = 0$. The third set of circles from the left in Figure 4-11 illustrates this sequence.

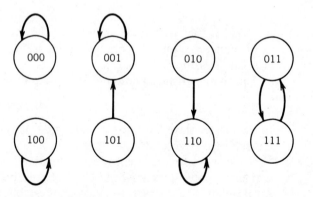

FIGURE 4-11

State Diagrams of Master-Slave JK Flip-Flop

The numbers in the diagrams correspond to Q, J, K; the transitions to $Q(t+1)$, J, K are shown by arrows. States that persist are indicated by looped arrows. The state diagram on the right corresponds to the toggle condition $J = K = 1$.

In Figure 4-11 the states that do not change with each full clock pulse are indicated by an arrow that turns back upon the state from which it started. The diagrams show the initial state and the state following one complete clock pulse, (Q, J, K), $(Q(t + 1), J, K)$, which results with specific inputs on J and K. The diagram at the far right in Figure 4-11 corresponds to the alternating output in the toggle mode when J and K are high. The other six circles illustrate other patterns of transition in the JK master-slave flip-flop.

ANALYSIS OF SEQUENTIAL CIRCUITS

Flip-flops are connected together to make scalers or counters, registers, timers, and a variety of other sequential digital logic circuits. In this section we will discuss and illustrate how the sequential behavior of these circuits may be analyzed. (Later we will design a circuit with a particular sequence.) Underlying the behavior of all sequential circuits are the Boolean algebra of bistable elements given by their characteristic functions, the time relations established by the clock pulses, and the interconnections to the inputs of the bistable elements. We will see how the interconnections can determine the cycle length such as making binary scalers count in a decade (by tens) sequence.

The characteristic tables are the bases for analyzing the sequential circuits, just as truth tables are used to analyze combinational configurations. The characteristic function for the former and the logic function for the latter are found by plotting the tables onto Karnaugh maps. Also, basic gates may be used in the sequential circuits to ensure the sequential logic for the interconnected bistable elements.

The characteristic functions for several of the flip-flops are listed together in Table 4-10.

Sequential circuits may be synchronous, with each element responding at the same time because of a common clock pulse, or asynchronous, in which case the circuit responds when a new input signal occurs. The counter that records random events, such as the signals from a radiation detector, is an asynchronous sequential circuit. By contrast, synchronous elements in computer circuits provide for the orderly flow of signals.

TABLE 4-10
Characteristic Functions for Flip-Flops

Flip-Flop	Inputs	Characteristic Function
RS	SR	$Q(t + 1) = S + \bar{R}Q$ with $SR = 0$
JK	JK	$Q(t + 1) = J\bar{Q} + \bar{K}Q$
D	D	$Q(t + 1) = D$
T	T	$Q(t + 1) = T \oplus Q$

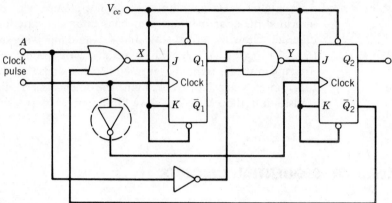

FIGURE 4-12

A Sequential Circuit under Analysis

These interconnected JK flip-flops are clocked in coincidence when the circled inverter is absent from the circuit and in sequence when it is present.

Combinational circuits sometimes ensure sequential logic and often provide the input signals and enable signals to sequential circuits. That is, the basic AND-gates and OR-gates are used to select the sequential operation for computation, data handling, and other digital operations. Even so, the analysis of complicated sequential circuits is made manageable by considering the response to inputs and enable signals step by step as clock pulses occur.

We can illustrate the technique of analyzing sequential circuits by considering the circuit in Figure 4-12. This circuit provides examples of features that may be found in sequential circuits. It also allows us to show how the response depends upon the manner in which clock pulses are provided to the bistable elements. In the first case both bistable elements have the same (coincident) clock pulse, and the circled inverter is presumed absent. In the second case one bistable element has the complementary clock pulse of the other.

Our first step is to collect the logic functions and the characteristic functions for the elements in the circuit. They are written with the notations on the circuit by taking into account the connections in the circuit. Table 4-11 contains these relations, in which Q_1, X, and Y are internal variables.

The K inputs are made high by being attached to V_{cc}. In TTL elements the unattached input is high but unreliable; thus it is advisable to tie inputs that are to be high to V_{cc}. This is also true of preset and clear lines.

Initially we consider the circuit in Figure 4-12 without the inverter (circled) in the clock pulse line; that is, we assume that it is not present. Both JK flip-flops make transitions in coincidence upon the completion of the clock pulse.

Table 4-12 is the characteristic table written with variables A, Q_1, and Q_2 as the independent state. Q_1 and Q_2 can each have either state (0 or 1) as an initial

Table 4-11
Functions of Elements in a Circuit

Element	Function	Function in Circuit Notation
NOR-gate	$C = \overline{A + B} = \bar{A}\bar{B}^a$	$X = \bar{A}Q_2$
NAND-gate	$C = \overline{AB} = \bar{A} + \bar{B}^a$	$Y = \bar{Q}_1 + A$
JK flip-flop	$Q(t + 1) = J\bar{Q} + \bar{K}Q$	$Q_1(t + 1) = X\bar{Q}_1 + \bar{K}_1 Q_1$
		$= X\bar{Q}_1$ with K_1 high
		$Q_2(t + 1) = Y\bar{Q}_2 + \bar{K}_2 Q_2$
		$= Y\bar{Q}_2$ with K_2 high

[a] By De Morgan's theorem.

condition because we assume no preset or clear action in this example. The intermediate variables X and Y shown on Figure 4-12 are used to simplify the analysis. The variable $X = \overline{A + \bar{Q}_2}$, but with De Morgan's theorem this may be rewritten as $X = \bar{A}Q_2$, as shown in Table 4-11. The $Q(t + 1)$ terms shown in Table 4-11 are the relations for the JK FF from Table 4-10 with $K = 1$. The clock pulse is not a variable but an enable signal. In the characteristic table, we identify the states of $Q_1(t + 1)$ and $Q_2(t + 1)$ that arise from each initial state.

Starting with any initial state A, Q_1, and Q_2, Table 4-12 gives the state, $A, Q_1(t + 1), Q_2(t + 1)$, after the clock pulse. For example, if we start with (000), the next state is (001) = $[A, Q_1(t + 1), Q_2(t + 1)]$. Figure 4-13 is the diagram for all the states that arise in the characteristic table with coincident transitions.

The state diagrams show that the two separate sequences depend upon the value of A. When $A = 0$ there are three states around which the transitions occur. We note also that the state (011) finds its way to the three-state loop. In the loop the values of Q_2 and Q_1 follow sequences of one high and two lows for every three clock pulses, with Q_1 one clock pulse displaced from Q_2.

TABLE 4-12
Characteristic Table with Coincident Transitions

A	Q_1	Q_2	$\bar{A}Q_2 = X$	$A + \bar{Q}_1 = Y$	$X\bar{Q}_1 = Q_1(t+1)$	$Y\bar{Q}_2 = Q_2(t+1)$
0	0	0	0	1	0	1
0	0	1	1	1	1	0
0	1	0	0	0	0	0
0	1	1	1	0	0	0
1	0	0	0	1	0	1
1	0	1	0	1	0	0
1	1	0	0	1	0	1
1	1	1	0	1	0	0

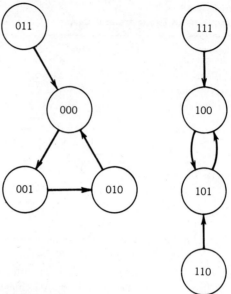

FIGURE 4-13

State Diagrams for Coincidence Transitions

The sequential states of the circuit in Figure 4-12 with both flip-flops clocked in coincidence follow two separate sequences. The states are labeled with A, Q_1, Q_2, arrows indicate the next state A, $Q_1(t+1)$, $Q_2(t+1)$.

TABLE 4-13
Characteristic Table with an Inverter in the Clock Pulse Line

A	Q_1	Q_2	$\bar{A}Q_2 = X$	$X\bar{Q}_1 = Q_1(t+1)$	$A + \bar{Q}_1(t+1) = Y$	$Y\bar{Q}_2 = Q_2(t+1)$
0	0	0	0	0	1	1
0	0	1	1	1	0	0
0	1	0	0	0	1	1
0	1	1	1	0	1	0
1	0	0	0	0	1	1
1	0	1	0	0	1	0
1	1	0	0	0	1	1
1	1	1	0	0	1	0

When $A = 1$, we encounter a different loop, in which the states yield an alternating value of Q_2. The continuing states, 100 and 101, represent a toggle response by Q_2 but a steady value of $Q_1 = 0$. The states 111 and 110 find their way into the loop.

We now see how the circuit behaves when the *JK* flip-flops are clocked differently. The presence of the inverter in the clock pulse line separates the responses of the two *JK* flip-flops. Flip-flop 1 changes state upon the completion of the clock pulse and provides $Q_1(t + 1)$ in accordance with its initial state, while flip-flop 2 does not change. At the start of the next clock pulse (when the transition at flip-flop 2 is high to low) the second flip-flop changes to the state determined by the recently established value of Q_1. The resulting value of $Q_2(t + 1)$ is the state variable of the first flip-flop when it responds next. Table 4-13 is the characteristic table for this time-dependent sequence.

Figure 4-14, the state diagrams for the time-dependent transitions, shows a difference from the case of coincident transitions in Figure 4-13. In this case, the cycle of three is lost, and both final states alternately have the value of Q_2. It is clear that the timing of transitions in separate parts of sequential circuits can affect directly the behavior of the circuits.

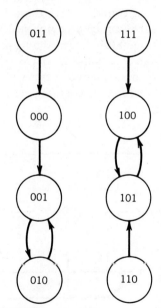

FIGURE 4-14

A State Diagram for a Two-Step Transition

The sequential states of the circuit in Figure 4-12 with the flip-flops clocked in sequence. States are labeled A, Q_1, Q_2; arrows point to the next state A, $Q_1(t + 1)$, $Q_2(t + 1)$.

RIPPLE COUNTERS

Counters or scalers are relatively simple sequential circuits made by interconnecting bistable elements. We will examine a few examples of counters, starting with the simplest arrangement that counts in a binary scale. Presently we will see that this simple arrangement may be altered to count in a BCD (binary-code decimal) scale. We will conclude our study of counters by showing the procedure for designing a circuit to follow a specific sequence.

Ripple counters are made by interconnecting JK flip-flops (or toggle flip-flops) and using the clock pulse line on the first (least significant bit or LSB) flip-flop as the input from the source being counted. Ripple counters are asynchronous and use the output of one flip-flop to trigger the next one.

Figure 4-15 shows a four-stage ripple counter. The outputs Q_3, Q_2, Q_1, and Q_0 provide the binary signal from 0000 to 1111_2, which correspond to the number of pulses entering the LSB flip-flop. After 16-input pulses, the ripple counter returns to its initial state, 0000, and the sequence is repeated. The sequence $(Q_3 Q_2 Q_1 Q_0)$ corresponds to the hexadecimal number system. The utility of the clear line on each JK FF for resetting the counter to 0000 is obvious. All clear lines are tied together so that one action establishes the initial state 0000.

We must be able to stop and start the counter without destroying the accumulated count, so that the counter may be read and the count continued. Count enable can be accomplished in several ways. An AND-gate, which ANDs the enable signal and the incoming pulses, can be used on the incoming pulse line to start or stop counting without clearing the counter. Ripple counters made with JK flip-flops can use a direct method of placing a low on the J and K of the LSB flip-flop to interrupt the counting. The left-hand state diagrams in Figure 4-11 and

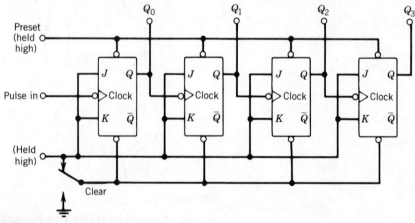

FIGURE 4-15

A Four-Stage Ripple Counter

Asynchronous pulses are scaled; each output Q_n represents 2^n events. The counter is cleared (that is, all outputs are made zero) by grounding the common clear line.

Table 4-9 show that the LSB flip-flop maintains its state with $J = K = 0$. Grounding the J and K inputs of the LSB flip-flop stops the rippling without altering the contents of the counter.

Clock pulses for JK FFs, obtained as MSI (medium-size integrated) circuits in the form of DIPs (dual in-line packages), trigger the flip-flop to $Q(t + 1)$ on the negative-going portion of the clock pulse, that is, when the clock pulse returns to zero, as we pointed out earlier. This property is indicated by a small circle adjacent to the clock pulse input and must be kept in mind when we think out the sequence of events in counters using JK FFs.

The binary length of counters—that is, the power of two by which the incoming pulse is scaled—can be made as large as desired by adding flip-flops. The ability of a counter to count both up and down is frequently used in measuring and control applications. The count can be made to go up (0, 1, 2, etc.), as in Figure 4-15, or down (15, 14, 13, etc.) by making the connections between flip-flops from $\bar{Q}$ to the clock input. This is equivalent to causing all but the LSB flip-flops of the JK flip-flops to change on the positive-going edge of the clock pulse. The count down is made on the same lines (Q_3, Q_2, Q_1, and Q_0) as the count up, as will be illustrated later in this chapter.

Another way of counting down uses the complementary relation of Q and $\bar{Q}$. As $Q_3 Q_2 Q_1 Q_0$ counts up (0000, 0001, 0010, etc), the values given by $\bar{Q}_3 \bar{Q}_2 \bar{Q}_1 \bar{Q}_0$ are 1111, 1110, 1101, which represent counting down. In this case the values are read from $\bar{Q}_3 \bar{Q}_2 \bar{Q}_1$ and $\bar{Q}_0$.

DIVIDE-BY-*N* COUNTER

The four-stage ripple counter of Figure 4-15 is a binary counter with a repeat sequence each $N = 2^4 = 16$ pulses. The number of stages n determines the maximum repeat length in binary counters. The maximum is given by 2^n. Other repeat lengths are desired, however; one of the most frequently needed is a repeat length of 10, or a decade counter.

A four-stage ripple counter may be modified to repeat after each 10 input pulses by interconnecting the flip-flops with logic gates. The decade counter provides outputs from 0000 to $1001_2 = 9_{10}$ and then repeats. The values on the output lines Q_3, Q_2, Q_1, Q_0 limited to this range are called BCD (binary-coded decimal) numbers. A decoder is required to display decimal digits from BCD numbers on seven-segment or other types of displays.

Figure 4-16 shows a BCD ripple counter. Four flip-flops are required even though they have unused states, $1010_2 = 10_{10}$ through $1111_2 = 15_{10}$. BCD counters are less compact than binary counters, and the arithmetic performed with BCD numbers requires more integrated circuits than does binary arithmetic, but the utility of BCD numbers is unquestioned. BCD ripple counters return to 0000 after they reach 1001_2; Figure 4-16 shows how the binary count is reset to zero after counting to $1001_2 = 9$. The connections to the preset and clear terminals in Figure 4-16 are not shown, although they must be kept high during counting.

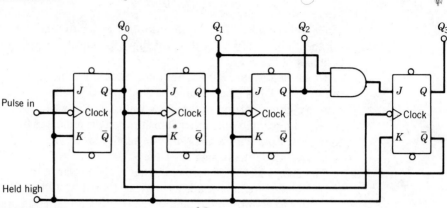

FIGURE 4-16

A BCD Ripple Counter

The asynchronous binary coded decimal (BCD) ripple counter counts from 0 to 9 (0000_2 to 1001_2). Reset and clear lines must be tied high (1) for operation.

Many other arrangements of flip-flops and basic gates may be used to provide BCD counting. Other divide-by-N counters may be made with logic interconnections among flip-flops. The minimum number of binary stages required in a divide-by-N counter is n where $2^n \geq N$. Presently we will design a $N = 5$ counter to show the procedure for obtaining any sequence of states.

Difficulties exist in asynchronous counters and other circuits made with flip-flops and basic logic gates. Some of these arise from unequal propagation times through the bistable elements and gates; others come from voltage transients that arise in state changes or from other sources. These difficulties may make the circuits unreliable, but commercial MSI (medium-size integrated) circuits, such as BCD counters, anticipate and avoid many of the difficulties.

DESIGN OF SEQUENTIAL CIRCUITS

The technique of designing a sequential circuit is illustrated by an $N = 5$ counter. The procedure is general and may be applied to a variety of sequential circuits. The analyses with truth tables and Karnaugh maps, which were demonstrated earlier for combinational circuits, can be extended to time-dependent systems. Here they are used to interpret the state tables or state diagrams that show the sequences of the present and the next states so that they may be implemented by sequential circuits.

Let us describe the sequential behavior we desire. Our example is a divide-by-5 counter, for which the state diagram is shown in Figure 4-17. The five states are to occur in the sequence shown, but another order could be called for. The pulses being counted are to be applied to the clock pulse line, though we could have chosen another means of introducing the pulse. We have also chosen an input x to

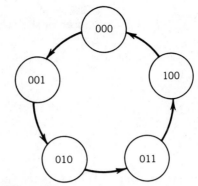

FIGURE 4-17

A State Diagram for a Divide-by-5 Counter

This sequence of five states is to be achieved in the design example of the divided-by-5 counter.

illustrate the manner of including external inputs. We will see that x is a count start-stop or a count enable signal.

We know that $N = 5$ requires $n = 3$ flip-flops to cover the range. The choice of flip-flop types is open; in this example we have chosen JK flip-flops because they are commonly used. Another configuration would result if we chose another type of flip-flop, such as a toggle flip-flop.

The state diagram, Figure 4-17, is used to list the sequence of present states, the next states, and the input requirements for each of the flip-flops. The sequence of five states, 000 to 100, correspond to $Q_2Q_1Q_0$ on the outputs of the $N = 5$ counter. The binary numbers in the circles in Figure 4-17 are the sequences of $Q_2Q_1Q_0$. At the first pulse, for example, Q_0 changes from 0 to 1 while Q_2 and Q_3 remain unchanged from 0. Each transition occurs in the desired order when the proper input states are provided to the flip-flop before the clock pulse. The characteristic equation or the characteristic table of the flip-flop provides the means of fulfilling input requirements. We use the JK flip-flop, whose characteristic equation is

$$Q(t + 1) = J\bar{Q} + \bar{K}Q$$

Table 4-14 is obtained by putting values into this equation. The values designated by Z are "don't care" values. "Don't care" values are those that do not influence $Q(t + 1)$ because $Q(t + 1)$ is established by other terms. A reexamination of Table 4-9 will confirm that it gives "don't care" values. For example, the first two lines in Table 4-9 show that for $Q = 0$ and $J = 0$, $Q(t + 1) = 0$ for both values of K. Table 4-14 is a condensed version of Table 4-9.

We will now form Table 4-15 with the present and next values and use Table 4-14 to specify the inputs J_0, K_0, J_1, and so on. The subscript refers to the flip-flop; J_0 is the J-input of the LSB flip-flop corresponding to the 2^0 binary digit.

TABLE 4-14
Characteristic Table for a JK Flip-Flop

Q	J	K	$J\bar{Q} + \bar{K}Q = Q(t+1)$
0	0	Z	$0 + 0 = 0$
0	1	Z	$1 + 0 = 1$
1	Z	1	$0 + 0 = 0$
1	Z	0	$0 + 1 = 1$

In the first line of Table 4-15 the values of J_0 and K_0 come from line 2 of Table 4-14 for the transition $Q_0 = 0$ to $Q_0(t+1) = 1$. Because Q_1 and Q_2 do not change from zero in this step, their inputs are assigned 0 and Z from the first line in Table 4-14.

The next step is to find the logic functions or logic equations for J_0, K_0, and so on by one of several methods. We use a four-variable Karnaugh map with $Q_0, Q_1, Q_2,$ and x for each input variable, and we mark the J_0, K_0 through J_2, K_2 value in the appropriate square. Figure 4-18 shows the six maps required to find the equations for the input states. For example, the first line in Table 4-15 shows that for $Q_2 = 0, Q_1 = 0, Q_0 = 0, x = 1$, the value to yield $Q_2(t+1) = 0, Q_1(t+1) = 0, Q_0(t+1) = 1$, and $x = 1$ is $J_0 = 1$. 1 is marked for J_0 in the rightmost square in the second row, and each remaining line of Table 4-15 is used to fill in the values of J_0 in the upper left-hand map of Figure 4-18. This procedure is followed for each input variable on its own map.

We use "don't care" values to simplify the diagrams, but we drop them in writing the equations. The equations are given below each map and in Table 4-16.

These equations are satisfied by AND-gates to the flip-flop inputs in Figure 4-19, the circuit for the divide-by-5 counter. Table 4-16 shows that x must be high (1) for the counter to count. With x low, all J and K inputs are low, and the present states are sustained independently of the incoming pulse. This condition confirms that x is a count enable signal. The divide-by-5 counter can be cleared to 000 by making the clear line low.

TABLE 4-15
State Table for a Divide-by-5 Counter

Present State				Next State $(t+1)$				Input Requirements					
Q_2	Q_1	Q_0	x	Q_2	Q_1	Q_0	x	J_0	K_0	J_1	K_1	J_2	K_2
0	0	0	1	0	0	1	1	1	Z	0	Z	0	Z
0	0	1	1	0	1	0	1	Z	1	1	Z	0	Z
0	1	0	1	0	1	1	1	1	Z	Z	0	0	Z
0	1	1	1	1	0	0	1	Z	1	Z	1	1	Z
1	0	0	1	0	0	0	1	0	Z	0	Z	Z	1

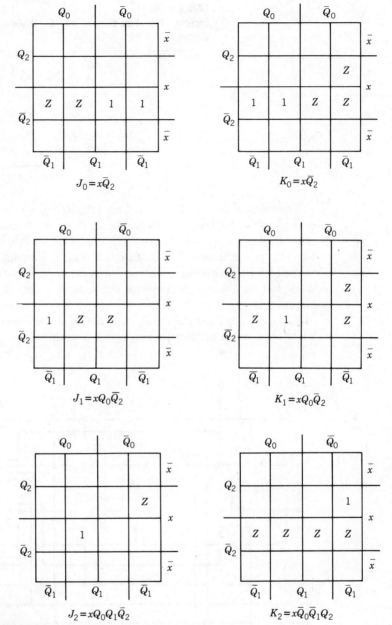

FIGURE 4-18

Karnaugh Maps for Divide-by-5 Counter

Each map solves the logic function for one input state J_0, K_0, J_1, through K_2 of the three JK flip flops connected to yield the sequence of five states shown in Figure 4-17. The marked square shows the value of the input, J_0, K_0, etc. to achieve the next state. Each map is generated from all the lines in Table 4-14.

TABLE 4-16
Inputs for Flip-Flops in a Divide-by-5 Counter

$$J_0 = x\bar{Q}_2$$
$$K_0 = x\bar{Q}_2$$
$$J_1 = xQ_0\bar{Q}_2$$
$$K_1 = xQ_0\bar{Q}_2$$
$$J_2 = xQ_0Q_1\bar{Q}_2$$
$$K_2 = x\bar{Q}_0\bar{Q}_1Q_2$$

The counter of Figure 4-19 counts reliably from 0 to 4_{10} (all five states) and returns to 0. Our task is not completed, however, because we must consider the possibility that the counter may reach one of the unused states (101, 110, and 111) when the power is turned on or if a glitch occurs. We must ensure that it will return to the counting sequence, whatever unused state it may be in. A manual clear will work, but for a person to monitor the counter visibly is a tedious and unworthy human task.

By examining the values of J_0, K_0, and so on for the unused states we can learn whether or not a problem exists, and if a problem does exist, what the solution

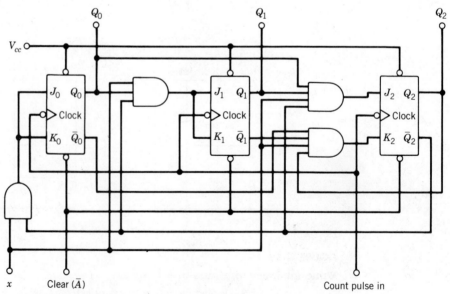

FIGURE 4-19

A Circuit for a Divide-by-5 Counter

The circuit is the solution of the design example for the divide-by-5 counter.

TABLE 4-17
Flip-Flop Inputs with Unused States in a Divide-by-5 Counter

Unused States				Inputs					
Q_2	Q_1	Q_0	x	J_0	K_0	J_1	K_1	J_2	K_2
1	0	1	1	0	0	0	0	0	0
1	1	0	1	0	0	0	0	0	0
1	1	1	1	0	0	0	0	0	0

might be. Table 4-17 shows the values for the inputs J_0, K_0, J_1 to K_2, which we obtained by placing the unused states $Q_0, Q_1,$ and Q_2 in the equations in Table 4-16.

Because all Js and Ks are zero if an unused state is reached, it is clear that the counter will not change from one of those states when a pulse to be counted occurs. An additional circuit must be provided for self-healing. We chose to provide an automatic clear if any of the unused states arise.

From Table 4-17 we can write the logic function (let us call it A) of the states to be avoided. We learn from the first line of Table 4-17 that we want to avoid $Q_2 Q_0$, from the second line $Q_2 Q_1$, and from the third line $Q_2 Q_1 Q_0$; that is,

$$A = Q_2 Q_0 + Q_2 Q_1 + Q_2 Q_1 Q_0$$

which summarizes the unused states. A Karnaugh map or a little Boolean algebra shows that the term $Q_2 Q_1 Q_0$ is redundant. When we check, we find that the logic function A is not satisfied by any of the allowed states. Figure 4-20 shows the self-healing circuit that provides a low (0) on the clear line to each flip-flop and resets the counter to 000 in the allowed sequence if an unused state occurs.

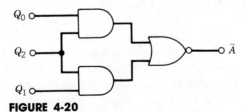

FIGURE 4-20

A Self-Healing Circuit for a Divide-by-5 Counter

The output of this circuit is attached to the clear line in Figure 4-19. If a state not in the sequence of five states arises, the counter is cleared to 000 independently of human intervention so that normal counting can proceed.

We must investigate the possibility of an isolated state or an isolated sequence of states in the design of any sequential circuit. An unwanted branching can be a problem, even if the desired sequence is rejoined. Careful design requires that all possibilities be examined and accommodated.

Divide-by-5 counters are available as MSI circuits, usually as decade counters that allow divide-by-5 or divide-by-2, depending upon the external pin connections. Other MSI circuits provide modulus N (repeat lengths of N) to count lengths that are frequently required, such as 12 for clocks. For nearly all purposes, the MSI circuits are the most reliable and the least expensive means of meeting counting requirements.

SYNCHRONOUS SEQUENTIAL CIRCUITS

Except for the divide-by-five circuit, the sequential circuits discussed above are asynchronous, but now we turn to sequential circuits, in which a common clock pulse causes each stage change (if there is a change) to coincide in time with each of the other state changes. Synchronous circuits are not usually required for counting random events, but they are essential in computers to ensure the time flow of data.

Synchronous sequential circuits have a variety of uses, some of which are revealed by their names. These include data latches, shift registers, serial adders, bidirectional registers, word-time generators, and synchronous counters. Memory units are also clocked synchronously to ensure that the reading from and the writing into them are done at the proper time. (Memory circuits are discussed in Chapter 5.) Synchronism in circuits is of paramount value when many devices are inerconnected. The advantages include the elimination of instabilities arising from transients, differences in propagation time, and other sources of trouble.

SYNCHRONOUS BINARY COUNTER

Figure 4-21 shows a four-stage synchronous binary counter. This counter differs from the counter shown in Figure 4-15, in which the count pulse is propagated through the JK flip-flops rather than being applied simultaneously to each clock pulse line. In the synchronous counter of Figure 4-21 the binary order is established by the AND-gates, which generate the signals on J and K to advance the count in the proper order.

The process demonstrated in the design of the divide-by-5 counter can be applied to the design of synchronous sequential circuits. The reader will note that the divide-by-5 counter in Figure 4-19 is in fact a synchronous counter because the pulses are applied simultaneously to all clock pulse inputs. The AND-gates between the flip-flops in both Figure 4-19 and Figure 4-21 generate the logic states for the proper sequence.

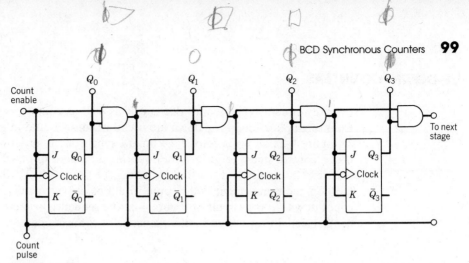

FIGURE 4-21
A Four-Stage Synchronous Binary Counter
The counter uses a common clock pulse line so that the transition of each enabled flip-flip is synchronous. The count enable initiates or stops counting, and the integrate connections establish the order of counting. Clear and present connections are not shown, but must be high (1).

BCD SYNCHRONOUS COUNTERS

BCD synchronous counters may be formed with logic gates between binary flip-flops to govern the sequence 0000 to 1001 and repeat. Synchronous BCD counters can be cascaded to give decimal numbers of the desired length if the carry pulse is provided to each successive set of flip-flops.

In the BCD ripple counter of Figure 4-16, the Q_3 pulse can be connected directly to the pulse-in line of the counter and the length of the decimal number extended. This method works readily because the flip-flops respond on the negative transition of the clock pulse and because Q_3 changes from high to low only once for each 10 pulses. The change in state from 1001 to 0000 triggers the next BCD counter. In the synchronous BCD counter, however, an AND-gate must form $Q_0 Q_3$ from one decade as the input to the next decade. Each time the BCD counter reaches 1001 the next clock pulse will increment the next decade counter synchronously. Another 10 pulses must occur before the next BCD counter registers its next count. The pulse that enables the next counter to record is sometimes called an overflow count.

Modulus N counters are available as MSI circuits; thus the need to design and construct them with basic gates should be infrequent. Both synchronous and asynchronous counters may be purchased as DIPs (dual in-line packages) and in other forms.

UP-DOWN COUNTERS

In many circumstances count decrement as well as count increment is required. Counting up and counting down are useful in controls and in instrumentation.

Figure 4-22 shows a four-stage binary synchronous up-down counter, which uses a control signal to govern the counting direction. In our discussion of ripple counters, it was pointed out that by taking the output as $\bar{Q}$ rather than Q at each flip-flop, the complement of the count would be provided and the sequence would count down. In the circuit of Figure 4-22 the sequence is altered by the up-down control signal rather than by the exchange of connections from Q to $\bar{Q}$.

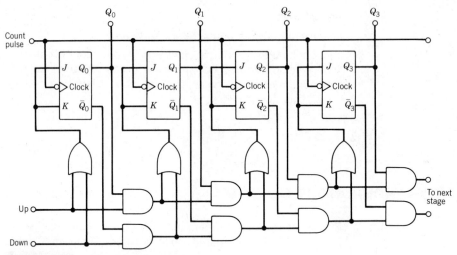

FIGURE 4-22
An Up-Down Binary Counter
The synchronous counter will count up or down depending upon which line, up or down, is high. The counter can be made to count to higher values by adding identical sections.

REVIEW

Thus far in this chapter we have discussed the nature of bistable elements and the means of increasing their usefulness by combining them as master-slave units. We have noted that sets of bistable elements make counters and other useful circuits. Some of the counters in general use (or scalers, as they are sometimes called) have been illustrated here. Two important techniques have been presented. One is the analysis of a circuit formed with bistable elements to establish the pattern of sequential states that it follows. The sequence is illustrated by state diagrams. The second is the process by which a desired sequence of states can be achieved in a sequential circuit composed of common flip-flops. In the first case we illustrated the role of an input variable in selecting one of two possible sequences in the same example that showed the importance of coincident or sequential clocking in establishing the sequence of states. In the second case, we identified states that were isolated from the desired sequential pattern and showed a means of escaping from them. In the next chapter we deal with bistable elements used in a different manner, but now let us examine the other two sequential devices mentioned in the introduction to the chapter: oscillators or clocks and monostable elements.

OSCILLATORS OR CLOCKS

Electronic elements that change state regularly or whose output varies periodically are called *oscillators*. There are two basic kinds of oscillators. One uses an energy storage system made with a capacitance and an inductance that exchange energy when they are powered by a source of electrical energy from an amplifier. The values of the capacitance and inductance establish the frequency of oscillations. The parallel resonance circuit, to be discussed at the end of Chapter 7, is an essential part of this type of oscillator, which typically generates a sine wave output.

The other kind of oscillator also employs energy storage as a means of regulating its exchange of states, but its output is characterized by sudden transitions from a low state to a high state and back again. These oscillators may be made with basic gates and an energy storage arrangement. The oscillator or clock described below is very elemental, but it illustrates the properties of an oscillator made as an unstable, in contrast to a bistable, element.

The oscillator made with open-collector inverters is shown in Figure 4-23. It operates as follows. Assume that the output of inverter A has just become high, V_{cc}. The capacitor, C_1, transfers the high voltage to the input of inverter B, making its output low. This low is transferred by C_2 to the input of inverter A, which confirms that the output of inverter A is high because its input is low.

After the transition the capacitors change their voltages as they are charged through the resistors. The voltage on the input of inverter A rises to approach the value $V_{cc}R_5/(R_5 + R_4 + R_2)$. Similarly, the voltage on the input of inverter B falls to approach

$$\frac{V_{cc}R_6}{R_1 + R_3 + R_6}$$

At some point after the transition the input of one of the inverters reaches a voltage that corresponds to a new input state. That inverter changes its output state, and the voltage distribution described above is exchanged between inverters. The transition is extremely fast because the drop in the output voltage of inverter A reinforces the change in inverter B by driving its input low. The rising output voltage of inverter B reinforces the voltage increase on the input of inverter A, driving its output low. This process is self-reinforcing or regenerative, making the

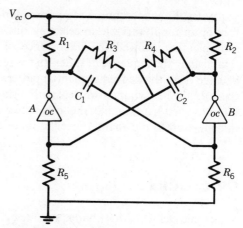

FIGURE 4-23

A Elementary Oscillator

The elementary oscillator changes states abruptly when the voltages on the inputs of the open-collector inverters are relaxed in the RC circuits after a transition.

transition in states abrupt. After a delay determined by the capacitors, resistors, and the inverters, the states are reversed. The system is free-running and generates a voltage-versus-time signal with sharp transitions and a stable frequency.

The waveform taken at an inverter output of the oscillator of Figure 4-23 is not a simple square wave. If it drives another inverter, however, the output of that inverter is a regular high-low chain of pulse. Sometimes it is difficult to find the resistor and capacitor values to make the circuit operate because of the unknown internal currents and resistances of the open-collector inverters.

Oscillators or clocks that operate in this manner are sometimes called *relaxation oscillators* because the abrupt voltage excursions are allowed to decay through the RC circuits until the state is changed. In fact, the energy transferred into and out of the capacitors governs the behavior of the clock.

Integrated circuits called clocks, timers, and voltage-controlled oscillators provide waveforms suitable for synchronizing sequential circuits. For precise frequency control, clocks use the resonance properties of quartz crystals to govern their cycles.

MONOSTABLE OSCILLATORS

A *monostable oscillator* responds to an input pulse by abruptly changing its state. After a predetermined interval established by an RC circuit, it returns spontaneously to its initial state and remains in that state until a new input pulse occurs. As the name implies, the monostable oscillator has a single stable state. The circuit discussed in Chapter 13 is sometimes labelled a "one-shot multivibrator."

The duration of the pulse on the output of an integrated-circuit monostable oscillator is adjustable by the time constant RC of an external capacitor-resistor circuit. The output pulse can be longer than the input pulse that initiates the transition, so that a monostable oscillator may be used to sustain or stretch a pulse. A second input pulse during the extended pulse may retrigger the monostable element; in that case the output pulse will be even longer. The integrated circuit monostable oscillators can be configured, if desired, to ignore a retriggering pulse, so that the pulse length is unchanged. The retriggering pulse is missed in the latter case and the monostable oscillator is said to have a "dead time."

Some commercially available timers and voltage-controlled oscillators may be made into monostable oscillators with external connections.

PROBLEMS

4-1 Make a Karnaugh map for the JK flip-flop and verify the characteristic equation $Q(t+1) = J\bar{Q} + \bar{K}Q$.

4-2 Configure a divide-by-N counter for $N = 7$.

4-3 Show that a count-down counter follows from Figure 4-15 by inverting the clock pulses.

4-4 Design a synchronous BCD counter.

4-5 Design a divide-by-4 counter that counts in the sequence 000, 010, 001, 011.

4-6 Draw a data flip-flop made with NOR-gates.

4-7 Draw a clocked RS FF made with NAND-gates and show how it has the same characteristic table as Table 4-5.

4-8 Verify Table 4-7, the characteristic table of the JK FF, by using the eight combinations of initial states.

4-9 Explain the action in Figure 4-9 when the clear line is made low. What happens if reset and clear are made low simultaneously?

4-10 Show that the term $J\bar{K}$ found by the Karnaugh map (Figure 4-10) and dropped from the characteristic equation for the JK FF adds no value to the relation $Q(t+1) = J\bar{Q} + \bar{K}Q$.

4-11 The inverter, circled in Figure 4-12, might be an XOR-gate with the clock pulse on one input and a control signal on the other. Imagine that an XOR-gate is a controllable inverter and then show how the circuit might be made to be clocked in either manner discussed in the text without physically connecting or disconnecting an element.

4-12 The ripple counter of Figure 4-15 will count down if the connection between flip-flops is made from $\bar{Q}$ to clock. Show that this phenomenon is equivalent to the flip-flop output responding on the positive-going part of the pulse into the LSB flip-flop and that it causes the output $Q_3Q_2Q_1Q_0$ to count down.

4-13 Show how an XOR-gate can be added between flip-flops in a ripple counter to allow a simple control signal to establish the direction of counting.

4-14 Show the connections from Figure 4-16 to the next BCD counter that may be used to continue the count beyond 9_{10}.

4-15 Design a synchronous BCD counter. Show how to generate an overflow count and a way in which it may activate the next counter.

4-16 In Table 4-15 the column for K_2 shows that it must be 1 or Z (either 1 or 0). If K_2 is made 1 by tying it to $+5$ V, will the divide-by-5 counter work? Is the self-healing arrangement required with $K_2 = 1$ always?

CHAPTER 5

DIGITAL MEMORY

Digital data in the form of binary bits have widely differing purposes. A binary word may be a number, an alphabetic letter, a condition code, a program instruction, or many other things.

Digital memory elements must retain a great quantity of binary words. Even the smallest computers require storage capacity for tens of thousands of bits. The user's purpose determines how long binary data will be stored. Digital memory elements are used for temporary, intermediate, and long-term storage of binary data, and the form of the memory element is frequently dictated by the storage time. Some stored data, such as operational codes, govern the computer operation, while other bits are used in a computation and discarded after only a few steps in the program. Still other stored data are long-term records, which may be updated as the occasion requires.

We classify digital memory into four categories. One kind of memory stores permanent digital data that must be preserved and accessed quickly, such as the program that "brings up" a computer to run. This requirement is satisfied by *read only memory* (ROM), because once the binary bits have been stored they cannot be altered or lost, and thus they remain in place. The start-up instructions that direct a computer to find the user's program must be in place when the computer is turned on. ROMs contain memory elements (cells) that can be placed in one or the other of two states by programming them physically or electrically.

Another form of memory is used for temporary storage during intermediate operations, such as binary arithmetic. This is called a *register* or *scratch-pad memory*, and it stores a limited number of bits with immediate access.

A third category includes the massive memory elements that store programs and data that may be accessed quickly. This is called *random access memory* (RAM) because any memory location can be reached at any time and in any order. A more accurate name for this memory is *read and write memory*, because the contents can either be read without being altered or changed by writing new binary words into the randomly addressed memory location. Registers and RAMs are either bistable digital devices made with flip-flops or combinational circuits called latches. Another form of registers and RAMs, dynamic shift registers and dynamic RAMs, are presented in Chapter 10.

The fourth category of memory uses external devices to record binary data. These devices include punched cards, magnetic tapes and discs, laser-read cards and discs, bar codes, and printed records. Because the data in these devices must be addressed serially, considerable time is required to read from them. These long-term storage devices back up RAMs to extend memory capacity and to ensure the availability of programs and data that may be lost in volatile electronic memory devices. A "volatile" memory element is one that loses its contents if the electrical power is lost, even for a fraction of a second. RAMs are volatile; ROMs are not.

A recently developed technology provides another nonvolatile memory device, called a magnetic bubble memory, which will compete with magnetic discs (Intel, 1982). Magnetic bubbles are microscopic regions in a thin film called *domains*, which have the opposite magnetic polarity from the rest of the magnetic material in the thin film. A bubble is a one, and the absence of a bubble where one could exist is a zero. The layer containing bubbles and voids is overlaid by magnetic material arranged in a pattern, which guides the bubbles around a loop. The magnetic bubbles are made to move by a rotating magnetic field, which is produced by a current in two coils placed at right angles to each other. These coils surround the film containing the bubbles and the overlay. The bubbles, the overlaid pattern, and the coils lie between the poles of a permanent magnet.

One loop of bubbles and voids may represent four kilobits. One magnetic bubble chip containing hundreds of loops may store a megabit. For reading or writing the bits are sensed and replicated. For the stored data to be read or changed, the bits in each loop must traverse the entire loop; the data access is serial, as in a magnetic disc. The important advantage of bubble storage over magnetic disc storage is that there are no moving mechanical parts. A power failure stops the circulation, but the permanent magnet preserves the bubble and void patterns.

REGISTERS

Most registers are formed by interconnected bistable elements (flip-flops) that store related bits of data. For example, the related bits may be a binary code for an alphabetic letter to be transferred to an output device, such as a printer. Registers usually have a limited capacity to store bits. Many have a four-bit capacity, but others extend to 8, 16, 32, and 64 bits or more when they are associated with microprocessors or other digital elements that require a particular capacity. Small registers may be tied together to create a larger register. We will examine registers with a few elements, but these have the features of much larger registers.

Registers have a variety of uses, and for that reason they are made in a variety of forms. We note that registers can accept data in a serial mode as bits in time sequence and in a parallel mode as bits accepted in coincidence. Furthermore, registers can unload or transfer the stored bits in both sequential and parallel modes, although not all registers can operate in all modes. Let us first consider a four-bit register with serial-in and parallel-out capability.

SHIFT REGISTERS

Shift registers are sequential circuits similar to those studied in the previous chapter. We are interested here in the applications of shift registers and how they are combined with basic logic gates to perform the assigned tasks.

Figure 5-1 shows a four-bit shift register. It is formed by RS master-slave flip-flops with Q to S and $\bar{Q}$ to R connections and a common clock pulse. That is, the shift register is synchronous, and it is a set of master-slave flip-flops to ensure the proper time sequence in the transfer of data. The input circuit makes the first flip-flop a data flip-flop, so that the bit on its input is stored at Q_3 after one complete clock pulse. The next clock pulse will cause the bit to move to the right to the next flip-flop; a new bit on the serial input line will be registered in the first flip-flop. The sequence of input bits will propagate along the shift register as long as the clock pulses reach the flip-flops. (We assume a continuous regular sequence of clock pulses.) The last (rightmost) flip-flop will discard the bit as it receives another.

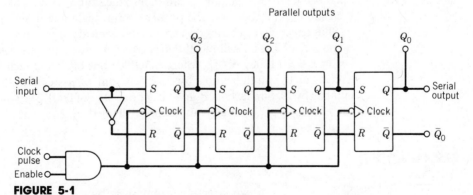

FIGURE 5-1

A Four-Bit Serial-in/Parallel-out Shift Register

RS FFs form a shift register that when enabled shifts bits toward the LSB with each clock pulse. Preset and clear terminals are not shown.

Loading four bits into the shift register requires that the enable signal be on (1) while the four bits are presented sequentially at the serial input. After the enable signal changes to low (0) and inhibits the shift register from accepting new data, the bits are stored and remain available on the parallel output lines.

Another enable signal (1) will allow new binary bits to displace the bits stored earlier. Each bit in the rightmost flip-flop will be transferred out. Thus, Q_0 on the last flip-flop provides a serial output. We can state that a shift register has a serial output regardless of the manner of loading it and whether or not it has parallel outputs. In many IC shift registers the complement of the output of the last flip-flop, $\bar{Q}_0$, is provided on a DIP pin (Figure 5-1).

As an example of the need for a serial-in/parallel-out shift register, consider a typewriter keyboard communicating with a computer. The eight-bit binary code corresponding to the depressed key is transmitted on a single wire to the computer containing a serial-input shift register. In the computer the code is stored in a memory that has parallel inputs. The code can be transferred fully into the shift register when the enable signal is high (1) for only eight clock pulses, and in another step it is transferred by the parallel outputs to the memory.

In addition to the serial-in/parallel-out shift registers there are serial-in/serial-out shift registers, parallel-in/serial-out shift registers, and universal shift registers. Universal shift registers allow parallel-in/parallel-out, shift-right and shift-left, and serial inputs for both directions of shift. One eight-bit universal shift register is available in a 20-pin DIP, and may be operated with a 50-MHz shift frequency. Shift registers may be cascaded to accommodate words of any length.

The parallel-in/serial-out shift register will perform the inverse of the task described above, that of communication between a typewriter and a computer on a single data line. A serial-in/parallel-in/parallel-out shift register, however, will serve to communicate in both directions. However, a MSI circuit known as a UART (Universal Asynchronous Receiver-Transmitter) would be preferred. Printing a letter, for example, can be accomplished by program steps of moving the letter (its binary code) from a memory location into the shift register with parallel inputs and then shifting the code into the printer over a single line in serial fashion.

Shift registers may be used as arithmetic elements for multiplication and division by powers of 2. You will recall that a binary number, such as, $0001\ 0110_2 = 22_{10}$, becomes $0010\ 1100_2 = 44_{10}$ when shifted left one bit. Thus a left shift of one step is equivalent to multiplication by 2. A left shift of n steps multiplies the original number by 2^n. Similarly, a right shift of one step of $0001\ 0110_2 = 22_{10}$ will give us $0000\ 1011_2 = 11_{10}$, which is the number divided by 2.

SERIAL ADDER

Serial addition of two large binary numbers by means of shift registers and a full adder is slower than parallel addition, but it requires much less hardware, such as integrated circuits. Serial addition is used in many applications, perhaps in your handheld calculator. Figure 5-2 shows an arrangement in which two shift registers

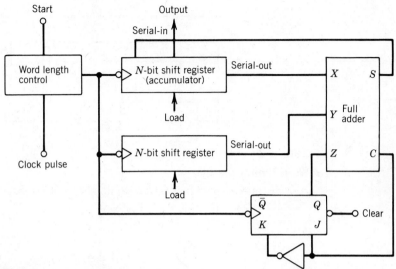

FIGURE 5-2
A Serial Adder
Shift registers and a full adder form the sum of the values stored in the registers. The sum is accumulated by returning the sum to the serial input of one shift register. The word length control terminates the process after all bits have been used. The carry-in Z is provided with a one-bit delay by the *JK* FF.

and one full adder sum the numbers loaded into the shift registers. As the two rightmost bits, one from each shift register, reach the full adder, they and the carry-in Z from the previous bit sum are added to furnish the new sum and the carry-out. The next clock pulse repeats this process until all bits have been transferred out of the shift registers. The manner of loading the numbers, whether parallel or serial, is not important to the manner of addition. The choice of the MSI circuit will allow the desired option for loading the numbers and reading the sum.

In the serial adder of Figure 5-2 the number of registers is minimized by storing (accumulating) the sum in one of the registers by returning the sum bits to the serial input. A similar practice of displacing one number is common even in microprocessors with parallel adders, in which the sum of the two numbers replaces one of the numbers in its register. Hence the term *accumulators* is used for the principal registers in microprocessors.

Figure 5-3 shows a word time generator, which allows the shift registers to be clocked for an exact number of pulses to perform the addition. In the circuit a positive start-pulse initiates the timing interval; the full count of eight clock pulses terminates the interval by resetting the flip-flop. The *JK* flip-flop output is ANDed with the clock pulse to make the word length control shown in Figure 5-2. The word length and the size of the shift registers correspond to the number of bits in the numbers to be added.

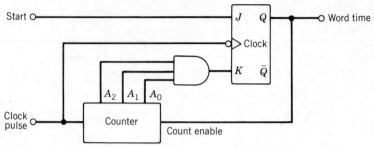

FIGURE 5-3

A Word Time Generator

The octal counter generates a stop signal after eight steps have allowed two bytes to be added in the serial adder of Figure 5-2.

UNIVERSAL SHIFT REGISTERS

The utility of the shift register is increased by providing additional controls to allow different modes of input and output and the direction of shifting. Controls are added by uniting combinational and sequential circuits. Multiplexers and flip-flops together create a universal shift register. Multiplexers, discussed in Chapter 3, allow one of several inputs to be selected by control signals. Four stages of a multibit universal shift register are shown in Figure 5-4. Note that the control signals, S_1 and S_0, and their complements, $\bar{S}_1$ and $\bar{S}_0$, are involved. The multiplexed input to each flip-flop gives versatility with parallel load, hold, and shift in either direction with serial input. The multiplexer circuit for each flip-flop is identical except for the tristate buffer with the one in Figure 3-13.

Table 5-1 lists the function selection for the universal shift register shown in Figure 5-4.

An examination of the multiplexer portion of Figure 5-4 will confirm that the output, Q, of each flip-flop is connected through the appropriate AND-gate to the adjacent flip-flop to provide serial input in either direction in accordance with the function selected by the values in Table 5-1. Logic gates (not shown in Figure 5-4) are installed on the output/input control line to ensure that the parallel input/output can be enabled only when shifting is inhibited. The tristate buffers on the Q-lines isolate the outputs from external circuits except when the parallel output is enabled.

One method in fabricating shift registers uses MOS (metal-oxide-semiconductor) transistors and their intrinsic capacity to store and transfer bit patterns. These are called *dynamic shift registers* because they must constantly transfer the bits to an adjacent register in no more than 1 msec per transfer. The process of transfer recharges the intrinsic capacity so that the content (0 or 1) of the register is preserved between clock pulses. Cells for the dynamic shift register can be made in

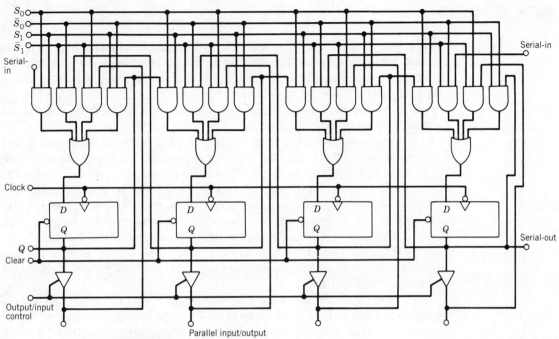

FIGURE 5-4
A Universal Shift Register
A shift register and a multiplexer are combined to allow both right and left shifts with serial input. Parallel input and output are inhibited during shifting. The tristate output buffers isolate the data from a data bus except when enabled by a control signal.

large numbers on a silicon chip. For example, 1024 cells in a long dynamic shift register operating with clock pulses up to 1 MHz circulate a 1024-bit word. A typical use of a long dynamic shift register is in a monitor display or for a serial memory in a hand-held calculator. The bits are read serially as into a magnetic bubble memory, but the dynamic shift registers are volatile. The details of a cell and its refresh cycle are given in Chapter 10.

TABLE 5-1
Function Selection for the Universal Shift Register

Control Signals		
S_1	S_0	Action
0	0	Hold
0	1	Shift right
1	0	Shift left
1	1	Parallel load

DATA REGISTERS

There are many uses for registers that store and transfer binary data. Some registers are used for temporary and immediately accessible bit storage. For example, the net count of a frequency meter is stored at the end of a counting interval, so that it may be displayed while the scalar is reset and counts through another fixed-length period to update the measurement. The data are latched into the display and changed only once each repeat time.

In microprocessors the data in use at the time of an interrupt request are stored temporarily in registers before the interrupt is serviced. They are held there until the microprocessor returns to its task. These data registers are called the *stack*. The stack stores the current binary information from the program counter, the accumulators, the program instruction code, the condition code, and other relevant data so that the interrupt will not invalidate the computation when it can be resumed. The process of preparing for the interrupt is referred to as "pushing" data onto the stack, from which it is "pulled back" when the program resumes. Pushing and pulling are parts of the fixed program to accommodate interrupts in the ROM of the microprocessor itself.

Data registers are parallel-in/parallel-out registers with a control signal input that enables them to hold their contents or to accept new data as desired. Figure 5-5 shows a four-bit data register with the control signal input labeled as a gate. An asynchronous control or gate signal is analogous to a single clock-pulse input. The data registers are referred to as a data latch, as suggested in Chapter 4.

Latches are made in several different configurations of basic logic gates. Two different single-bit latches are shown in Figures 5-6 and 5-7. These gate arrangements are combinational circuits that retain their state independently of the values

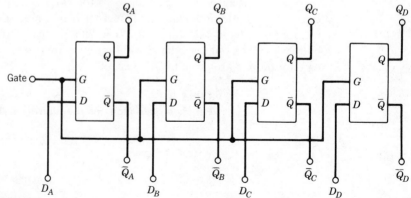

FIGURE 5-5

A Four-Bit Latch

Four data flip-flops accept and store input bits when a gate signal is present. The bits remain in place after the gate signal is removed.

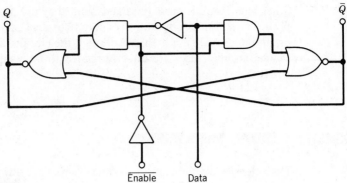

FIGURE 5-6

A Function Diagram of a Latch

The data bit is stored and retained at Q. A low on the enable line is required to place a new bit in the latch. The configuration is that of a data flip-flop.

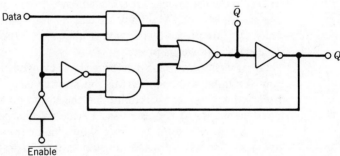

FIGURE 5-7

A Function Diagram of a Latch

A bit on the data line is stored with the enable line low. Both this latch and the one in Figure 5-6 will follow the data bit if it changes while the enable is low (0).

on the inputs until they accept a new datum in response to a low (0) on the enable circuit. The enable pulse of a microsecond or less will latch the datum present at the input. Figure 5-6 will be recognized as a data flip-flop.

The enable or gate pulse allows the bits on the data lines to be read into an array of latches. In one sense the gate pulse corresponds to a clock pulse, but Figures 5-6 and 5-7 show that the circuits forming latches are not master-slave units, which can properly handle sequential data. The data latches do not require master-slave configurations to serve their functions. Problems can arise, however, if the input bits change during the gate pulse, because the stored bits will follow the change. The point is that data registers and other digital memory elements are simple bistable elements for storing bits for a long or short time. Shift registers, on the other hand, handle more data and, thus, they use more complex master-slave flip-flops. The simple configuration of data registers allows many registers to be formed on an integrated circuit chip. Although the circuits of Figures 5-6 and 5-7 are not sequential, their uses are directly related to data input and output of sequential circuits.

HIGH-DENSITY DIGITAL MEMORIES

High-density digital memories come in several forms and have different properties. High-density memories store many kilobytes on a single integrated circuit, and they are evolving into even denser arrays of individual cells, each of which will store a single bit (0 or 1) of binary data. They are arranged so that binary words of desired lengths can be addressed and read or stored.

To achieve the greatest density of bit storage, the binary cell is as simple as it can be made while remaining a reliable storage element. Consequently, the flip-flop is only one kind of storage element; in fact, it is a relatively complex element that makes up the least dense arrays. Memories formed with flip-flops are *static random access memories* (RAM). They are called "static" because they stay in the state in which they are placed by the data storage step without further attention, as long as the electrical power is present.

Other technologies for bit storage have evolved elements that operate differently from flip-flops. *Dynamic* RAMS, bit storage elements whose mechanism will be discussed in Chapter 10, require that the stored bit be refreshed frequently. The bit in a dynamic RAM is present as an electrical charge stored in the cell. The refresh cycle to restore the bit occurs a thousand or more times per second. The term "dynamic" comes from the constant attention to the refresh process. Dynamic memory elements can be placed in denser arrays than static memory elements, but additional circuits must exist to refresh the memory. Some dynamic RAMs have refresh circuits in the same chip as the memory cells.

Static and dynamic memory elements have a characteristic in common: both lose their memory (content of 0 or 1) when the electrical power is removed for even a fraction of a second. That is, they are volatile. In the case of dynamic memory elements, the charge leaks away, and the information is lost. The static memory elements will find themselves in one state or the other (1 or 0) when the power is restored, but the new state is not necessarily the same as it was before the power was interrupted.

A third kind of high-density storage element is actually a combinational circuit that has only one state, 0 or 1, when it is electrically powered. After an interruption of electrical power the element can return only to the same state it was built to have. Because it does not lose its memory, it is a reliable preprogrammed memory. The arrays of these preprogrammed memory elements are *read only memories* (ROM). ROMs are used to provide the set of instructions for computers to perform specific tasks; they remain permanently available and unchanged.

Both ROMs and RAMs have an important property: any set of memory cells can be addressed nonsequentially. The set of memory cells may contain a byte or word of binary data that is retrieved from that memory address. In the case of RAMs the binary data may be changed (i.e., written) as well as read.

STATIC RAM CELL

A static memory cell consists of a flip-flop whose state represents the stored binary bit. Connections to logic gates allow the bit to be written into the cell and to be read when it is addressed. Figure 5-8 is a schematic diagram of a static RAM cell.

The lines labeled $R/\overline{W}$ and $\overline{Enable}$ carry the control signals to activate the read or write process. The organization of memory chips differs from one system to another, but in most instances there is a control line for reading or writing and an enable line that allows the output or input lines to be activated. A few pages ago we briefly mentioned tristate devices, which may be disconnected electrically from the input and output lines that lead to them. In this way many elements may be attached physically to data transmission lines, data buses, while only one element at a time places data on or receives data from the bus.

The binary signal (high or low) that permits reading and writing is indicated by symbolic shorthand. In TTL devices an unconnected pin will float high (1), and the gate will respond as if the unconnected pin were actually connected to $+V_{cc}$. To prevent the write command from changing the state of the cell, a low (0) must be

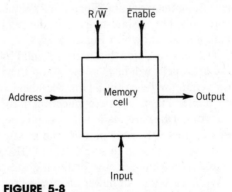

FIGURE 5-8

A Static Random Access Memory Cell

A single flip-flop provides a memory element, which requires the lines shown to operate. As a protection against inadvertent changes, both the $R/\overline{W}$ (read-write) line and the enable line must be low to change the contents (write into) the cell. The power connections are not shown.

provided intentionally before a new datum can be written into the cell. The symbol R/$\overline{\text{W}}$ indicates that the read-write line is in the write mode if the line is low (0). To read from the memory cell, the read-write line must be high (1), as indicated by the R/$\overline{\text{W}}$. To write a binary bit into the memory cell, both $\overline{\text{W}}$ and $\overline{\text{Enable}}$ must be low simultaneously. During the interval in which the two lows overlap, the bit on the input line will be stored in the memory cell. The requirement that both signals be low reflects the caution that is exercised to prevent the contents of memory from being changed (written into) inadvertently. Because the read operation does not erase the stored bit, the protection of the data during the read process is not so critical.

OPERATIVE ELEMENTAL STATIC RAM CELL

Figure 5-9 shows a cell of a static RAM array with the gates that allow the particular cell to be addressed and the data to be written into it or to be read from it. The gates and the inverter form a clocked data flip-flop as the storage element. The write line and the R/$\overline{\text{W}}$ line are connected to all cells, as indicated by the arrows on those lines. The cell is responsive only when it is addressed because the OR-gate with the inverter on the address line and the R/$\overline{\text{W}}$ line are inputs; thus the common connections are permissible.

The data flip-flop can be read only if it is addressed while the R/$\overline{\text{W}}$ line is high. The R/$\overline{\text{W}}$ line is held high at all times until it is desired to write a bit into the cell. To change the contents, the cell is addressed and the R/$\overline{\text{W}}$ line is made low with the new bit on the write line. Then the R/$\overline{\text{W}}$ line is returned to high to protect the newly stored bit. The tristate buffers on the read and write lines and the enable lines to them are not shown in Figure 5-9.

The only read line that carries a valid bit (1 or 0) is the one in the cell addressed. The other cells will have a low (0) on their output lines, which permits the lines to use a common bus; thus all lines can be wired-ORed to that bus. It was pointed out in Chapter 3 that the interconnection of the outputs of gates present a problem and that some gates are made with open collectors to allow this interconnection. The use of wired OR-gates permits the multiple connection of outputs; typically, the read line will be read through a wired OR-gate. Tristate buffers on the cells are used widely as another means of having a common output line with only one cell providing a signal. In newer memories tristate buffers are displacing the wired-OR arrangement.

In Figure 5-9 a 10-bit address (A_0 to A_9) is divided between the row decoder and the column decoder to select any one of the 1024 cells in the array. This arrangement is described as a 1 bit × 1 K RAM. The 1 K = 2^{10} = 1024 cells are selected by the 32 row lines × the 32 column lines. Only six of the 32 lines from each decoder are shown in Figure 5-9.

An eight-bit byte can be stored at any of the 1024 addresses by placing eight 1 × 1 K RAMs in parallel and interconnecting their address lines. Eight write lines

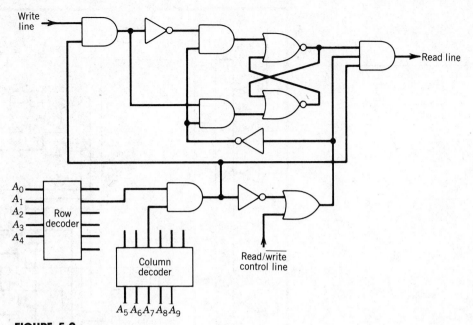

FIGURE 5-9
Operative Elemental Static RAM Cell

The circuit components form a data flip-flop, which uses AND-gates to isolate it from other cells and permits it to be randomly addressed. The dual protection provided by a separate enable signal against writing into the cell, as shown in Figure 5-8, is not present in this case. The 10-bit address, A_0 to A_9, indicates that this cell is one of 1024 arranged in 32 × 32 rows and columns. Only six of the 32 lines from each decoder are shown.

and eight read lines provide parallel inputs and outputs when the address is selected. The R/W̄ lines of the eight RAMs are common (connected together) for the simultaneous enable.

Larger static memory arrays with 4 K (4096) cells, 32 K (32768) cells, and larger bit storage capacity are available. RAMs are available in configurations in which four or eight bits may be addressed at one time. For example, a 4 × 256 RAM as 1 K (1024) cells in which four bits are read or written on four parallel input/output lines when accessed by an eight-bit address and an enable signal. Two of these connected in parallel provide an eight-bit byte at each of the 256 addresses. A schematic of a 4 × 256 RAM is given in Figure 5-10, which shows the logic components for using the memory array. These are internal to the DIP (dual in-line package), whose pin-outs are V_{cc}, *Gnd*, the eight address lines, the four input-output lines, a write enable line, and a chip enable line; a total of 16 pins. The chip enable line selects the memory array from all the other memory arrays in the total memory bank. Sometimes the portions of the total memory selected in this way are called *pages*.

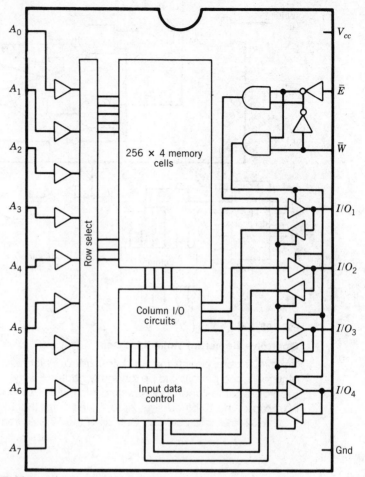

FIGURE 5-10

Schematic of a 256 × 4 Random-Access Memory

The components in a single DIP support the 1024 memory cells arranged in 256 rows of 4 cells each. The tristate input/output buffers isolate the unit from the data bus.

As mentioned above, dynamic RAMs make up larger-capacity and higher-density memory arrays than are possible static RAMs. Static RAMs have shorter read and write times than dynamic RAMS, and they do not require the refresh process. They are preferred for these reasons, but they require more area on the memory board for the same number of bits, and they are more expensive. Technical advances may be expected to increase the density of cells and produce larger memories, even larger than the 256- and 512-kilobit memory chips presently available as dynamic RAMs. The details of a typical dynamic RAM cell are presented in Chapter 10.

READ ONLY MEMORY (ROM)

Read only memories (ROMs) are arrays of unalterable memory cells that store bits permanently. In contrast to RAMs, which are volatile, ROMs are nonvolatile: although ROMs require electrical power to be read, the loss of power does not cause the contents to be altered. ROMs are addressed, and their output signals are provided in the same manner as are RAMs.

ROMs have many uses. Perhaps the most obvious, as mentioned earlier, is the storage of the program of a computer so that it will begin its operation properly when the power is turned on.

ROMs hold the bit patterns that cause a letter, a number, or even a logo to be generated by dots on a monitor screen. The calculation of sines, logarithms, and other mathematical functions frequently uses numerical tables stored in ROMs. In order to calculate a trigonometric function, a binary code corresponding to the angle will address the ROM to find the values for its calculation. An interpolation routine stored in ROM will quickly calculate the value for any angle. Algorithms

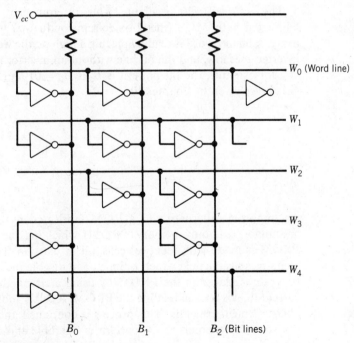

FIGURE 5-11

Read Only Memory (ROM) Array

The ROM is fabricated with the bit pattern in place. The manufacturer omits cells in accordance with the bit pattern supplied by the user. The ROM is unalterable and nonvolatile. Each address (word line) may be randomly selected by the address decoder.

for floating-point arithmetic and other frequently used programs are stored in optimized form in ROMs. ROMs containing the compiling codes enable a user's program written in a higher language, such as Fortran, to be converted into an efficient program code for the computer.

A microcomputer may use a program in ROM to control a process. The ROM supplied by the controller's manufacturer relieves the user from the task of programming the computer-based control system. The computer voice heard in an automobile, which alerts the driver that he or she has failed to extinguish lights or issues other warnings comes from a ROM that contains the bit pattern for voice synthesis.

Finally, it is important to note that ROMs can be used in place of decoders. For example, a seven-segment display can be activated by a bit pattern from a ROM addressed by the binary-coded decimal number to be displayed. The ROM is an alternative to many combinational logic-gate circuits.

Figure 5-11 shows a few cells of a read only memory. The bit pattern is made by the presence or absence of cells in the array formed by the photographic masks of the manufacturer. The ROM can be changed only by starting over with new masks. ROMs are economical when they can be produced in large quantities.

The occupied cells in Figure 5-11 are equivalent to open-collector inverters fabricated with MOS (metal-oxide-semiconductor) technology in a very dense array. A binary word is read by placing a high on the word line. The bit line with an inverter goes low, and the bit line without an inverter remains high. For word line W_3, for example, the bit pattern is $B_2 B_1 B_0 = 010$. (The transistor details of this circuit are given in Chapter 10.)

PROM

An important alternative to a ROM fabricated by the manufacturer is a programmable read only memory (PROM), which may be programmed by the user. A PROM is made with identical cells, all in place in the array. Each memory cell contains a fusible link, which will melt and disable the cell at a bit storage point. A program is placed in the PROM by using a circuit that provides a much higher current than is used in reading the PROM. A cell is addressed, and the high current "blasts" (melts) the link. This process is continued until the desired bit pattern in the array corresponds to the pattern of operable and disabled cells. If an error is made during the programming, the PROM must be discarded and another try must be made.

PROMs are used when the number of ROMs of a single kind is not large enough to justify the cost of making masks for a production run. Figure 5-12 shows an array of PROM cells that has been programmed by melting the fusible links.

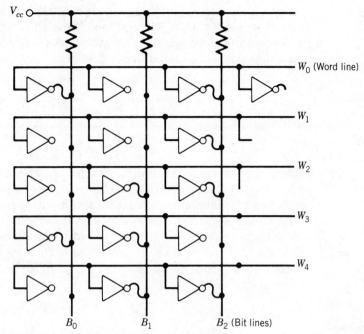

FIGURE 5-12

Programmable Read Only Memory (PROM) Array

The manufacturer provides the PROM with operable cell in each position. The bit pattern in the PROM is established by melting the fuse in the cells. The fuses are melted by addressing the cell and applying an overcurrent. Errors made in forming the bit pattern cannot be corrected.

EPROM

Another level of convenience in nonvolatile randomly accessible bit storage is the erasable programmable read only memory (EPROM). EPROMs store a permanent bit pattern programmed by the user in the memory array. To correct an error, however, or to allow the EPROM to be used for a new purpose, the bit pattern can be erased and another pattern stored. The contents are erased with an intense ultraviolet light through a window above the array, and the entire array is reprogrammed. (The bit storage mechanism is discussed in Chapter 10.)

Electrically alterable read only memories (EAROM) or electrically erasable programmable read only memories (EEPROM) are permanent bit storage devices that can be erased selectively by electrical signals to correct or modify the bit pattern initially stored by the user. As one develops an operating system, these provide further convenience in permanent bit storage because portions of the bit pattern can be altered without modifying other portions. Further technical developments can be expected to grant additional freedom in meeting the needs of users of permanent digital memory devices.

SUMMARY

Several types of memory elements made by different techniques are used for binary bit storage. Some registers (latches) can hold their bit patterns for intervals of time while others (shift registers) can also transfer bits in accordance with control and clock signals. Some high density digital memory elements (ROMs) have permanent bit patterns to be read but not changed while others (RAMs) may have their contents changed as well as read. The ability to fetch and store the data nonsequentially, that is, in a random access manner, is the key that opened the way to programmable computers.

REFERENCES

1982   Intel, *A Primer on Magnetic Bubble Memory*, Intel Corporation, Santa Clara, Calif.

CHAPTER 6

MICROPROCESSOR BASICS

Microprocessors evolved from integrated circuits in a predictable (in hindsight) sequence of events. The idea that the connections between transistors could be put on the same substrate as the transistors led to small-scale integrated circuits (SSIC) that eliminated external wiring between transistors, reduced costs, and increased reliability. These circuits became more sophisticated and diverse as a greater number of specific functions satisfied by transistors and their internal connections were provided for users. Jack Kilby of the Texas Instruments Company is credited with making the first integrated circuit in 1958 (see Preston 1983).

The microprocessor concept was initiated in 1971 when a four-bit computer on a chip was introduced by Intel Corporation (Gupta and Toong, 1983). It was decided to fabricate a general-purpose integrated circuit that could be used for a variety of purposes (including the one they were implementing) when directed by programs stored in memory. The early microprocessors, a name first used in 1972, with a four-bit data bus were quickly recognized as a major advance in the technology, which itself was generating more and more specialized, single-purpose integrated circuits.

The power and versatility of microprocessors has been increased by the improvements in silicon and germanium purification and fabrication and by the refinements in photo masking and transistor fabrication on the semiconductor materials. At this writing, 16-bit microprocessors are common, and 32-bit microprocessors are available. The bit size of a microprocessor is determined by the number of bits accepted by the microprocessor in one step on the input/output data terminals through the data bus. Important internal registers in microprocessors called *accumulators* can usually handle simultaneously twice as many bits as the data bus. Some confusion exists in the "size" description of a microprocessor because some equipment supppliers do not adhere to the definition based on the capacity of the microprocessor to accept or provide bits in a single load or store command, that is, the bit width of the data bus.

Microprocessors are extensive arrays of digital logic gates formed by tens of thousands of transistors in an integrated circuit. Because the logic gates are grouped to perform specific tasks, a microprocessor may be described as an interconnected arrangement of functional blocks. The internal connections are also called *buses*, and as a sequence of digital operations occurs, the digital signals are transferred along a bus from one functional block to another. The order of digital operations is determined by a program or instruction set prepared for the task. The microprocessor must receive each instruction in sequence and perform the digital operation that it directs.

Our task is to understand what a microprocessor is and how it operates. For this purpose we look at a relatively simple computation and see how the computation is done by transferring data, by numerical operations, and by the digital logic that governs each step. In addition, we will find out how the program code actuates the logic in the microprocessor. By following the steps in the computation we can determine the functional blocks that must be present in the microprocessor and what they do. In a sense we will be designing a primitive, imaginary microprocessor, but we will find that it is remarkably like actual microprocessors. We won't be surprised to find that the only digital elements we need are AND-gates, OR-gates, inverters, combinational circuits like the full adder, and sequential circuits like the shift register, which themselves are made with the elementary gates. The only difference is that a capable microprocessor will contain many elementary gates.

A microprocessor is one component of a computer. In order to function, the microprocessor must have the input from several other electronic components. These include input/output devices, such as a keyboard and video terminal, and read and write memory (RAM). In our examination of a microprocessor we assume that all the necessary components and buses exterior to the microprocessor are present. Because of its relation to the other components the microprocessor is frequently called the *central processing unit* (CPU). Figure 6-1 shows the relation of the CPU to external elements of a microprocessor-based computer.

As a means of developing the architecture of a microprocessor, we will calculate the square root of a number using a well-known algorithm. In the calculation we will identify the functional groups that must be present in the CPU.

An algorithm is a sequence of steps that accomplishes an arithmetic process. The algorithm for the square root requires multiplication and division, each of which involves an algorithm itself. Thus we find that a number of steps exist, some of which are "nested" within others. Each nested algorithm is a subroutine to which

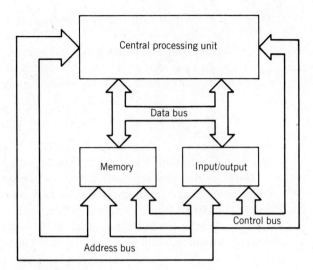

FIGURE 6-1
A Block Diagram of a Computor
The central processing unit (the microprocessor) with the external memory and input/output devices connected with data, address, and control buses form a computor. An electrical power source (not shown) is required.

the programmer can transfer at need rather than reprogramming it each time the operation (multiplication, for example) is required. The square root routine is itself a subroutine that, once programmed, is called into the program sequence whenever it is needed.

In order to follow a program, elements in the system must take certain actions and make certain choices. Before starting an addition, for example, the digital adder must be cleared; that is, any previous contents of the adding circuits must be removed. Clearing may or may not require a separate program step, but it must be done. A means of concluding the process must also be provided. If the process is well planned, the termination will take place efficiently, without meaningless steps. Several of these features are illustrated in the material that follows, and the necessary gate logic is shown.

SQUARE ROOT ALGORITHM

The square root algorithm is a simple iterative computation, which illustrates the nature of the microprocessor. The square root algorithm is often the first example used in demonstrating how to program a computer in its own (machine) language.

That is, it is the first example of software that a novice writes in the operational code of the microprocessor.

The square root algorithm is a good iterative procedure because it converges quickly to give the root even when it is started with a poor first guess of the square root. Let N be any positive number whose square root is desired. N can be written as

$$N = (x + a)^2 = x^2 + 2ax + a^2 \qquad (6\text{-}1)$$

It can be seen that if a is very small then $N^{1/2} \approx x$. The algorithm seeks to reduce a to a very small number so that x is very close to $N^{1/2}$.

The algorithm proceeds as follows. The first step is to recognize that if a is small, then a^2 is quite small and can be discarded from Equation 6-1. If the first choice for x is quite different from $N^{1/2}$, a will not be small for the first few iterations, but if the algorithm works, a will become small as we come close to finding the root. In any case, a^2 is dropped, and Equation 6-1 becomes

$$N \approx x^2 + 2ax \qquad (6\text{-}2)$$

which may be rearranged to give

$$a = \frac{N - x^2}{2x} \qquad (6\text{-}3)$$

To have a general starting point, we always choose x to be 1.0,

$$x = x_1 = 1.0$$

x_1 is put into Equation 6-3 to give the first value of $a = a_1$. Now we improve our guess by letting the second value of x be

$$x_2 = x_1 + a_1$$

and this can be used in Equation 6-3 to find a_2,

$$a_2 = \frac{N - x_2^2}{2x_2}$$

Then $x_3 = x_2 + a_2$, and the sequence is repeated to yield $x_i = x_{i-1} + a_{i-1}$, where $i = 2, 3, 4, \ldots$.

We have not yet considered one aspect of the routine, however. How do we stop the iteration, which otherwise will continue indefinitely? Equation 6-3 suggests a means of terminating the iteration. We merely need to decide when the magnitude $|N - x_i^2|$ is small enough so that x_i can be used as $N^{1/2}$. This step is taken by comparing $|N - x_i^2|$ with a predetermined small positive number ε and ending the iteration when $|N - x_i^2| < \varepsilon$. This comparison test is made in each iteration until it is valid, and the iteration is then halted. When the comparison test is satisfied, a digital signal called a *flag* is generated to indicate that some other action may be taken, such as displaying the root.

A numerical example of the iterative procedure is given in Table 6-1 for $N = 7.000$ and $\varepsilon = 10^{-4}$.

TABLE 6-1
Square Root of 7.000 by the Iterative Procedure

Iteration	$x_i = x_{i-1} + a_{i-1}$	a_i	x_i^2	$N - x_i^2$	$|N - x_i^2| < \varepsilon$
1	1.0000	3.0000	1.0000	6.0000	No
2	4.0000	−1.1250	16.0000	−9.0000	No
3	2.8750	−0.2201	8.2656	−1.2656	No
4	2.6549	−0.0091	7.0485	−0.0485	No
5	2.6458	−0.00002	7.0001	0.0001	No
6	2.64575	10^{-6}	7.0000	0.0000	Yes

PROGRAM FOR A MICROPROCESSOR

The exact sequence of steps must be listed for the microprocessor to follow. In contrast to a program written in a high computer language, such as Fortran, these steps are written in the operational code or op-code of the microprocessor. In higher languages the program statement to find a square root, for example, is a single line. The single-line statement calls for the same sequence of steps as the operational code. Later in this chapter we will write out these steps in detail as a sequence of hexadecimal numbers in the lowest level language, the *machine language*.

Assembly language is the next higher language used to write programs for microprocessors. Assembly language uses mnemonics, similar to those in Table 6-2, to give the steps in the program. The program is converted into machine language, a sequence of operational code numbers, by a sophisticated program called an *assembler*. The assembler not only generates the operational code numbers but it also assigns the program step numbers to them, and it can link different program parts written separately into a longer program.

A program written in a still higher language such as Fortran, must be compiled to give the exact sequence of steps for each part of the total program. This process is carried out by a computer program called a *compiler*, which is written for the microprocessor that is being used.

A compiler does much the same thing that an assembler does, but it starts with a set of statements that are independent of the microprocessor that is to be used. One advantage of high level languages is that the person writing a Fortran program, for example, does not need to know any details of the microprocessor or computer on which it will be run. However, the compiler must contain the details of the system being used. The compiler takes each statement in the higher language and converts it into the operational code of the microprocessor. Then each statement is nested with the others to form a coherent program.

The compiler stores the program in machine language (i.e., binary bits of the op-code) in the random access memory (RAM) that is peripheral to the microprocessor. For the square root subroutine the compiler also stores ε, the precision parameter, in the RAM and keeps track of its location in memory. The value of N, however, must be entered from a keyboard or be generated in the larger calculation. Here we assume that N also has a value and that it is stored in a known location in the RAM as result of a read-and-store sequence.

OPERATIONAL CODE

Program Decoder

We will see that each element of the operational code itself is a statement for a specific action by the microprocessor. The microprocessor has a functional block called the *program decoder*, which interprets the operational code and directs the action. The program decoder is a read only memory or a programmable logic array (PLA). (PLAs are discussed in Chapter 13.) Many different operational codes exist for each microprocessor, and each one calls for a unique action, but some of the actions may be grouped into subsets. For instance, the basic add instruction may have five different operational codes to specify the location of the number to be added, that is, an add instruction may use five different addresses. The operational code corresponding to ADD, for example, may mean "add the next value in the program listing to what arithmetically preceded it," while the operational code corresponding to ADDA may mean "add the number that is stored at the memory location whose address is next in the program listing."

A microprocessor can perform 50 or more basic operations. More than half of these have different kinds of addressing modes, so that a microprocessor has more than 150 operational codes. For the convenience of the person writing the program in the operational code, the code is written in hexadecimal rather than in binary numbers. Two hexadecimal numbers can specify 256 different operational codes, but not all combinations are required. To help distinguish the various operations, a mnemonic is associated with each operational code, but unfortunately the mnemonics are not standardized.

Table 6-2 gives only the terms of the operation code for our hypothetical microprocessor used in the calculation of the square root of N. The table lists the mnemonics, the operation code in hexadecimal numbers, and the comments that explain each step. A complete operational code, of course, is much larger than the one given here.

Several different kinds of program steps are listed in the table. They differ in their relation to subsequent steps. The simplest ones do not refer to another step; for example, the program code WAT (25_{hex}) merely stops the sequence of steps without regard to any other item in the program listing. This *inherent* command requires the smallest number of clock pulses to be implemented. A more involved program step is LDA (61_{hex}). This step loads the accumulator A with the value, called an operand, which is expressed as a hexadecimal number in the next program listing. LDA is an *immediate* program step. At the next level of complexity is a direct program step such as LDAA ($A3_{hex}$), which requires the value at the

TABLE 6-2
Operational Codes for Hypothetical Microprocessor

Mnemonics	Op-Code$_{hex}$	Comments
For accumulators		
ABVA	18	Absolute value of the contents of ACCA.
ADDA	04	Adds the contents of ACCA.
ADDB	05	Adds the contents of ACCB.
ADAB	1F	Adds the contents of ACCB to the contents of ACCA.
CLA	81	Clears ACCA.
CLB	91	Clears ACCB.
LDA	61	Loads ACCA with the value in the next program step.
LDB	63	Loads ACCB with the value in the next program step.
LDAA	A3	Loads ACCA with the value from the memory address in the next program step.
LDBA	A5	Loads ACCB with the value from the memory address in the next program step.
PUSA	1A	Places the contents of ACCA on stack.
PULA	6A	Retrieves the value from stack into ACCA.
SBA	68	Subtracts the contents of ACCB from ACCA.
STAA	A8	Stores the contents of ACCA at the memory address in the next program step.
STBA	AF	Stores the contents of ACCB at the memory address in the next program step.
STO	A6	Stores the contents of the ALU at the memory address in the next program step.
SLA	AA	Shifts the contents of ACCA left one bit.
SRA	AD	Shifts the contents of ACCA right one bit.
SLB	AB	Shifts the contents of ACCB left one bit.
SRB	34	Shifts the contents of ACCB right one bit.
TSTA	12	Tests the magnitude of the contents of ACCA.
TSTB	14	Tests the magnitude of the contents of ACCB.
For arithmetic logic unit		
ACL	EE	Clears ALU.
For logic and counting		
CMPX	44	Compares the index register immediate.
DECA	64	Decrements ACCA.
INCA	13	Increments ACCA.
Action codes		
BLZ	38	Branches if less than zero.
BR	87	Branches always.
BRCC	37	Branches if carry clear.
BRZ	66	Branches if zero.
BSR	7A	Branches to subroutine.
JMP	57	Jumps immediate.
JSR	77	Jumps to subroutine.
NOP	41	No operation (program step is skipped).
RTN	35	Returns to the program sequence.
WAT	25	Waits (stop).

memory location given in the next program step to be fetched and loaded into the accumulator A. Still more complex operational codes require the microprocessor to save the contents of its registers before taking the action. Jump-to-subroutine JSR (77_{hex}) is an example of this kind of code and requires a number of clock pulses to be implemented.

When we write programs, the list of steps we take must comply with the nature of the operational code. If a command like LDAA ($A3_{hex}$) is used, for example, a memory location must be given in the next line. A printout of the operational code in hexadecimal numbers quickly becomes confusing because the program steps, the values to be used immediately, the memory address, and other details are mixed together. The program must be annotated or documented so that it can be understood.

Address Bus

Addresses are binary numbers, also written as hexadecimal numbers for convenience. You will recall that a byte (8 bits or 2 hexadecimal numbers) has 256 different values, a word (16 bits or 2 bytes or 4 hex) has 65,536 (64 K) values, and a long word (24 bits or 3 bytes or 6 hex) has 16,777,216 values, so that the number of memory locations that can be addressed is determined by the size of the address code. The external address bus will have as many lines as there are bits in the address code provided by the microprocessor, so that any memory location can be addressed in one step. Some microprocessors, however, have fewer external address bus lines and use a process called *paging* to address one portion or another of a large memory bank. The microprocessor may use an address it places on the address bus to communicate with an input/output device, such as temperature sensor. This is a clever way to communicate with an external device because the microprocessor "pretends" that the bit transmission is the same as if the bit were being transmitted to or from a memory location. If this technique is used, the external devices are said to be *memory-mapped*.

Data Buses

The data bus, which carries the number stored in the external memory location, has the number of lines by which the microprocessor is characterized. An eight-bit microprocessor has an eight-line data bus; a 16-bit microprocessor has a 16-line data bus. The capacity in bits of each external memory location corresponds to the number of lines in the data bus. The internal data buses between data processing components within the microprocessor have the same width as the accumulators, typically twice the width of the data buses to the external memory.

Thus far we have noted three components of the microprocessor: the program decoder, the data bus, and the address bus. The pinouts of the microprocessor package connect the internal buses of the microprocessor through a buffer to the external buses of the memory boards. Before we list the specific operational code (the microprocessor program) for the iterative calculation, we must describe several other components of the microprocessor, so that the purpose of the steps will be clear.

Accumulators

At least two registers are designated as *accumulators*. These are the registers to which and from which the data are transferred to be available for arithmetic and other operations. The accumulators may be similar to the shift register shown in Figure 5-4. They have parallel inputs and outputs, and shift in both directions. One will be designated as accumulator *A* (ACCA) and the other as accumulator *B* (*ACCB*).

Arithmetic Logic Unit

Another necessary functional block is the *arithmetic logic unit* (ALU), in which the adding, subtracting, multiplying, and dividing are performed. Actually, adding is the only arithmetic operation carried out by the ALU. Subtraction is the addition of the complement of the number to be subtracted; multiplication is successive adding with shifting; division is successive subtraction with shifting (A different and faster method of finding a quotient is discussed in Appendix A.) The arithmetic logic unit is an array of full adders with digital gates to provide the control logic directed by the operational code. The arithmetic logic unit is examined later in detail.

Program Counter

The orderly flow of computational steps requires a register to keep track of the progress through the program. The *program counter* does this; it can be set ahead or behind to execute a jump from one sequence of steps to another when a branch or jump operational code is encountered. The program must specify the number of steps to back up or to move ahead or must indicate where the next program step is found, as in the example of jump to a subroutine. The program counter must hold the program step number in the main program sequence so that the program may be rejoined upon the return from the subroutine.

MULTIPLICATION ALGORITHM

An examination of Equation 6-3 shows that it requires one subtraction, two multiplications, and one division to yield *a*. In our microprocessor no operational code exists for either multiplication or division. The only operational codes are such commands as add, complement, compare, store, shift, jump, and return; thus we must generate a subroutine using these and other digital operations. Once the subroutine for the multiplication algorithm is formed, it can be used whenever two

TABLE 6-3
Decimal and Binary Multiplication

	a $\times b$	23_{10} 11_{10}		a $\times b$	$0001\ 0111_2$ $0000\ 1011_2$
Step					
1		00	Clear product register.		0000 0000
2		23	Add $1 \times a$.		0001 0111
		23			0001 0111
3		23	Shift a to left.		0 0010 111
4			Add $1 \times a$ shifted.		
		253			0 0100 0101
5			Shift a to left.		00 0101 11
6			Add $0 \times a$ shifted.		00 0000 00
7			Shift a to left.		00 0100 0101
8			Add $1 \times a$ shifted.		000 1011 1
	$ab = 253_{10}$				$ab = 000\ 1111\ 1101_2$

numbers are to be multiplied by jumping from the program to the subroutine and then returning to the program when the product is found.. Sixteen-bit (and their compatible eight-bit) microprocessors have operational codes for multiplication and division, which use internally programmed sequences.

The multiplication algorithm is similar for decimal and for binary numbers. To illustrate the algorithm, let us multiply two numbers. If $a = 23_{10}$ and $b = 11_{10}$, then the product $ab = 253_{10}$. The a is called the *multiplicand*, and the b is called the *multiplier*.

Table 6-3 illustrates several points about multiplication, which will be useful in writing the microprocessor program for the multiplication algorithm. First, we examined the rightmost number in the multiplier, multiplied the multiplicand with it, and added it to the cleared register. Then we shifted the multiplicand one step to the left. At the same time we examined the second number from the right in the multiplier. Since the number in each case was one, we multiplied the multiplicand with it and added the product from its shifted position to the register contents. In the case of the decimal multiplication we stopped because there were no nonzero digits to the left in the multiplier.

In the binary multiplication, however, we did not reach the condition of no nonzero bits to the left. We continued by shifting the multiplicand once again. When we examined the third bit from the right in the multiplier, we found that it was zero. We multiplied the shifted multiplicand by zero and added it (zero) to the partial sum. After we made another shift and examined the fourth bit from the right, we added the multiplicand at its shifted position to the partial sum. Then we stopped, as we did for the decimal multiplication, because there were no nonzero bits to the left.

SUBROUTINE FOR
MULTIPLICATION ALGORITHM

The program can be developed more easily if we prepare a flowchart, which outlines the steps and shows the program branches. Figure 6-2 is the flowchart for the multiplication of binary numbers. Note that the first step is to initialize (clear) the component to be used, in this case the ALU. Then we follow a sequence of testing for nonzeros on the left of the multiplier, shifting the right-hand bit to the carry register, testing the carry bit, adding the multiplicand (carry bit = 1) or

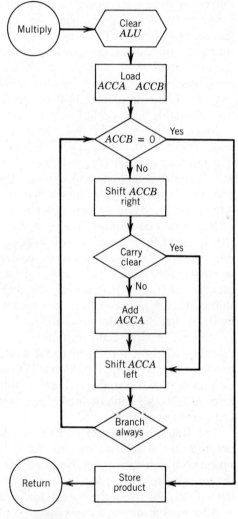

FIGURE 6-2
A Flow Chart for a Multiply Algorithm

skipping addition (carry bit = 0), shifting the multiplicand left, and returning to repeat the steps until the full product is found. The last step is to return to the main program.

Now we will implement the binary multiplication algorithm in a program, using the operational codes of the microprocessor. The implementation must be general, so that any multiplicand and any multiplier can be used, and it should be efficient, meaning that it should avoid any unnecessary steps.

An accumulator has an additional one-bit register, called the *carry register* or *overflow register*. The carry register can be examined to see whether it contains a one or a zero. If, for instance, it contains a one, a *carry flag* is said to be set to signal that condition. Several operational codes exist to shift the contents of an accumulator one bit to the right and place the least significant bit, the one just shifted out, into the carry register. The program code we use here does this and places a zero in the most significant bit position (the leftmost register) each time a bit is shifted into the carry register, so that the accumulator fills up with zeros. Another type of operational code may rotate the contents out of the accumulator and back in again so that it does not become empty (all zeros), but this kind of shift command would not be suitable for the multiplication subroutine.

In our program, Table 6-4, the multiplicand is loaded into ACCA and the multiplier into ACCB as a first step in multiplying the two numbers. For the moment we assume that each number is an eight-bit number, as we did in Table 6-3. The arithmetic logic unit is a 16-bit register with its own carry register. The product of two eight-bit numbers may be 16-bits long and have a carry; thus the ALU must be able to accommodate a 16-bit product and a carry.

Now it may be clear how the algorithm should be programmed. Shift ACCB one step to the right to put its least significant bit into the carry register. If that bit is one, the carry flag indicates that the contents of ACCA should be added, and then its inputs to the ALU should be shifted one step to the left. If the bit shifted into the carry register is zero, then nothing should be added, but the multiplicand should be shifted. This sequence should be repeated until ACCB is filled with zeros, which indicates that the product is complete.

Suppose that the eight-bit multiplier itself is zero, so that the result of eight repetitions of the sequence will be the product zero. It will be much more efficient to test the contents of ACCB each time before going through the sequence and ending as soon as ACCB contains only zeros. If the test shows that ACCB contains only zeros (0000 0000) initially or after any sequence, the program should jump to the end of the subroutine.

The multiplication subroutine is located in a region of memory; each of its program steps sequentially fills the memory. A subroutine is called by the operational code "jump to subroutine," or JSR (77_{hex}). The address of the first step in the subroutine is given as the next hexadecimal number in the program immediately following the operational code for JSR.

After the subroutine is completed, the program must continue from the point it had reached when the subroutine was called. To allow reentry at the correct point

TABLE 6-4
Program for Multiplication Algorithm

Program Step Number[a]	Op-Code	Mnemonic	Comments
E0$_{hex}$	EE$_{hex}$	ACL	Sets (clears) arithmetic logic unit to zero.
E1	A3	LDAA	Loads ACCA with the contents (byte *a*) in the
E2	65		memory location 65$_{hex}$.
E3	A5	LDBA	Loads ACCB with the contents (byte *b*) in
E4	67		memory location 67$_{hex}$.
E5	14	TSTB	Tests the contents of ACCB and sets the flag if zero.
E6	66	BRZ	Branches if zero; otherwise goes to next operation code, step E8.
E7	08		Number of program numbers (8) to jump ahead on branch.
E8	34	SRB	Shifts value in ACCB to right-LSB to carry.
E9	37	BRCC	Branches if carry is clear (i.e., zero).
EA	02		Number of program numbers (2) to jump ahead.
EB	04	ADDA	Adds the contents of ACCA.
EC	AA	SLA	Shifts the contents of ACCA left.
ED	87	BR	Branches always to repeat sequence.[b]
EE	F5		Number of program steps (−11) to backup.[c]
EF	A6	STO	Stores product (*ab*) in ALU at memory location
F0	69		69$_{hex}$.
F1	35	RTN	Returns from subroutine.

[a] The program step number is the address of the memory location holding the operational code. The program steps that do not contain a mnemonic are memory addresses or jump intervals.
[b] Of course, the "branch always" will not occur when this program step is skipped at the end of the calculation.
[c] This is considered a negative number (−11 = F5), as explained in Appendix A. Count from program step EF, which will be in the program register when the branch code is acted upon to back up to the E5.

in the program, the program step number and other data, such as the contents of the accumulators, must be remembered. For this purpose these data are stored in a "scratch pad" memory area of the microprocessor. As mentioned in Chapter 5, this memory is called the *stack*; an operational code calling for an interrupt or jump to subroutine will first store ("push") the data into the stack before taking the program action. Upon return from interrupt or subroutine the data are retrieved ("pulled") from the stack to continue the program.

A different kind of jump may be used to repeat a sequence in a program by jumping to an earlier program step. In this case the stack is not used; only the program step number is modified and a portion of the program is repeated. A similar action occurs when a branch is called by a test for equality between two values or for a zero or other magnitude relation. A branch or jump may go forward in the program as well as backward.

PROGRAM FOR SQUARE ROOT ALGORITHM

The square root program may be considered a higher-level subroutine than the multiplication subroutine because it repeatedly calls the more elementary multiplication subroutine in seeking the square root. In a full calculation there are several levels or hierarchies of subroutines and program blocks. Programs written to take advantage of this feature are called *structured programs*. In structured programs all the elements can be written and verified as good subroutines and program blocks before they are put together as the complete program. Structured programs are also valuable with higher languages, such as Fortran. The sequence of steps and branches is shown in Figure 6-3.

In the square root subroutine it is assumed that the subroutine for the division algorithm is available to be called, just as the multiplication subroutine is called. (See Appendix A for a division algorithm and a different square root algorithm.) Table 6-5 is the implementation of the square root algorithm in the operational code of our hypothetical microprocessor.

The program step numbers (memory locations) for the square root subroutine start at 01 CO_{hex}. This notation indicates that two bytes are used to address a single memory location; thus there are more than $2^8 = 256_{10}$ external memory registers. The additional byte, $hex_3\,hex_2 = 01$, suggests that as few as $2^9 = 512$ or as many as $2^{16} = 256 \times 256 = 65{,}536 = 64\,K$ memory addresses may exist. It can be seen that elementary programs use up memory locations at a fast rate, so that any reasonable computer requires a great deal of external memory.

The beginning program steps in Table 6-5, 01 C0 to 01 C9, initialize the subroutine and accomplish the first iteration, using $x_1 = 1$. Division by 2 is performed efficiently by shifting one bit to the right rather than by using the division subroutine. Because the first steps in the program involve simple numbers, that is, $x_1 = 1$, it is more time efficient to start the routine and then iterate. A few more memory locations are used in this fashion, but in present-day microcomputers memory space is not usually a limitation.

Additional iterations use the program steps 01 CA through 01 EB until the precision criterion has been satisfied. The memory locations 65, 67, and 69_{hex} are used repeatedly for temporary storage; the multiplication and division subroutines will seek values from these locations. Temporary storage could have been programmed for the microprocessor to use internal memory, scratch pad memory, or the stack. In any case the result of the calculation is stored ultimately in an accumulator, which is the pivot point for most actions.

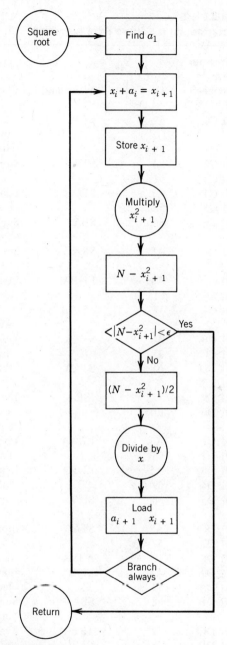

FIGURE 6-3
A Flow Chart for a Square Root Algorithm

TABLE 6-5
Program for a Square Root Algorithm

Program Step Number	Op-Code	Mnemonic	Comments
01 C0 $_{hex}$	63 $_{hex}$	LDB	Loads ACCB with value in the next program line (1).
01 C1	01		Initial value $x_1 = 1.0$
01 C2	AF	STBA	Stores the contents (1) of ACCB in memory location 51_{hex}.
01 C3	51		
01 C4	A3	LDAA	Loads ACCA with the contents (N) in memory location 55.[a]
01 C5	55		
01 C6	68	SBA	Subtracts the contents (1) of ACCB from the contents (N) of ACCA = ($N - 1$).
01 C7	AD	SRA	Shifts the contents of ACCA right = ($N - 1$)/2 = a_1.
01 C8	A5	LDBA	Loads ACCB with the contents (1) in memory location 51.
01 C9	51		
01 CA	1F	ADAB	Adds ACCB to ACCA = ($x_i + a_i$) = x_{i+1} in ACCA.
01 CB	A8	STAA	Stores x_i (i redefined in each iteration) in memory location 65.[c]
01 CC	65		
01 CD	A8	STAA	Stores x_i in memory location 67.[c]
01 CE	67		
01 CF	77	JSR	Jumps to (multiplication) subroutine.
01 D0	E0		First step in multiplication subroutine, which stores product in 69.
01 D1	A3	LDAA	Loads ACCA with contents (N) in memory location 55.
01 D2	55		
01 D3	A5	LDBA	Loads ACCB with contents (x_i^2) in memory location 69.
01 D4	69		
01 D5	68	SBA	Subtracts ACCB from ACCA = ($N - x_i^2$) in ACCA.
01 D6	A8	STAA	Stores ($N - x_i^2$) in memory location 57.
01 D7	57		
01 D8	18	ABVA	Absolute value of ($N - x_i^2$) in ACCA.
01 D9	A5	LDBA	Loads ACCB with contents (ε) in memory location 53.[b]
01 DA	53		
01 DB	68	SBA	Subtracts ACCB from ACCA = $|N - x_i^2| - \varepsilon$ in ACCA.
01 DC	12	TSTA	Tests magnitude of $|N - x_i^2| - \varepsilon$.
01 DD	38	BLZ	Branches if less than zero.
01 DE	0E		Number of program steps (14) to skip.
01 DF	A3	LDAA	Loads ACCA with the contents ($N - x_i^2$) in memory location 57.
01 E0	57		
01 E1	AD	SRA	Shifts the contents of ACCA to the right = ($N - x_i^2$)/2.

01 E2	A8	STAA	Stores the contents of ACCA $(N - x_i^2)/2$ in memory location 65.[c]
01 E3	65		
01 E4	77	JSR	Jumps to (division) subroutine, which starts in memory location CO_{hex} and places the quotient (a_i) in 69.
01 E5	C0		
01 E6	A3	LDAA	Loads ACCA with the contents (a_i) in memory location 69.
01 E7	69		
01 E8	A5	LDBA	Loads ACCB with the contents (x_i) in memory location 67.
01 E9	67		
01 EA	87	BR	Branches always.
01 EB	DD		Number of program steps (34) to backup.
01 EC	35	RTN	Returns from subroutine.

[a] N is assumed to be placed in memory location 55_{hex} in preparation for finding its square root.

[b] ε, the precision parameter, is assumed to be stored in memory location 53_{hex}.

[c] Both the multiplication subroutine (Table 6-3) and the division subroutine (not shown) use the same temporary storage registers. The multiplicand is placed in 65_{hex} and the multiplier in 67_{hex} before the multiplication subroutine is called. Similarly, the dividend is placed in 65_{hex} and the divisor in 67_{hex} before the division subroutine is called. The product or the quotient is stored by the subroutines in 69_{hex}. It was not necessary at program step 01 E3 to store x_i, the divisor, because it was still in 67_{hex} as a result of step 01 CD.

MICROPROCESSOR ARCHITECTURE

The square root program is used to illustrate the process of calculation with a microprocessor that can be directed to perform elementary steps. In devising the program we made some assumptions about the elements that are part of the microprocessor and about the binary number manipulations that they perform electronically. In other words, we have created a configuration of elements and have made assumptions about how they function. Earlier we discussed the program decoder, memory bus, data bus, accumulators, and arithmetic logic unit; now we can explain the other essential parts of our hypothetical microprocessor. The configuration of the elements is the architecture of the microprocessor, and the way the elements function is a matter of combinational and sequential digital gate logic. First we will describe further the architecture of our hypothetical microprocessor; then we will examine the gate logic required in the program steps that govern the arithmetic logic unit.

Bus Interface

The components of the microprocessor are interconnected by an internal bus structure. Figure 6-4 shows that the bus allows the transfer of bits from any element to any other element. Two buses, however, are unidirectional: the bus from the timing and control elements exterior to the microprocessor and the address bus to the exterior random access memory. The interface elements on the data and

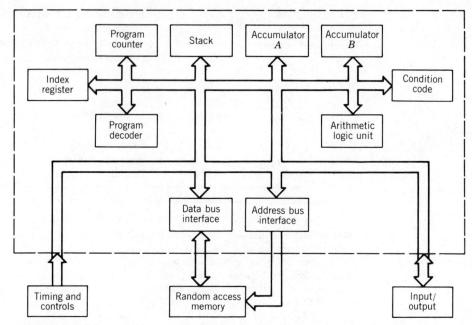

FIGURE 6-4

The Architecture of a Hypothetical Microprocessor

Each separately designated element within the dashed lines fulfills an important function in the microprocessor and establishes the architecture of the microprocessor.

address buses are essentially buffer amplifiers to ensure that the electrical connections to the exterior elements do not overload the internal circuits of the microprocessor.

Timing and Control

The timing and control block contains a pulse generator, called the *clock*, which provides the synchronizing signals for the microprocessor and the external manual control signals such as power on-off, reset, start, and stop. The input/output block may be a keyboard with a key decoder and a video display. The input–output devices will be asynchronous; that is, they will not operate at the clock frequency of the microprocessor; thus a universal asynchronous receiver-transmitter (UART) will be involved in communicating with the microprocessor. The many control functions can be performed through the input–output block and through the timing and control block.

Condition Code Register

Several of the blocks in Figure 6-4 have been explained, but some need additional comments to show their relation to the internal functioning of the microprocessor.

The condition code register is sometimes considered a part of the arithmetic logic unit because the conditions of overflow, zero, negative, and others are generated usually in the ALU. Because these conditions may initiate branches in the program sequence, they are retained by being pushed onto the stack in the case of an interrupt or a jump to subroutine, so that the program can be rejoined where it was left.

Index Register

Our square root routine did not require an index register, but this element facilitates many calculations. The index register stores the value that the program indicates is to be used to find the next operand (the number to be acted upon). For example, the sum of a number of values stored sequentially in the memory can be found by a program that calls the first address. Subsequent memory addresses are generated by indexing, that is, by adding a specific increment to the memory location, until the entire list has been reached. The process is ended when the steps are counted and the count matches the number in the list or by a branch at the last memory location. When an index register is available, we can save ourselves the trouble of writing a great many program steps in adding the values in 100 sequential memory locations, as we would, for example, in obtaining the grade point average of 100 students. Without an index register we would have to list each of the 100 memory addresses alternately with another 100 program steps; each program step would call for the value in the next memory address to be added to the sum.

Stack

The role of the clock pulses in the operation of all elements of the microprocessor can be illustrated in the process of pushing onto the stack. The stack is a column of parallel registers, as shown in Figure 6-5. The individual registers are master-slave units; thus the contents of the register are determined by the input only after a full clock cycle. (The process of bit transfer in master-slave flip-flops is discussed in Chapter 4.)

To push onto the stack, the quantities, like the contents of the accumulators, the condition code register, or the program counter, are placed consecutively on the bus lines $A_7 \cdots A_0$. An enable signal and an up signal are established by the program decoder. The clock pulse transfers the first byte into the lowest registers. The second byte is placed on the bus at the conclusion of the same clock pulse that stored the first byte. The first byte travels on the input lines to the next-from-the-bottom registers in the stack; that is, each column of registers is a shift register. At the completion of the second clock pulse the first byte is in the next-from-the-bottom register, and the second byte is in the bottom register. This process occurs with each clock pulse, and the bytes climb up the stack until the necessary data are preserved in preparation for an interrupt.

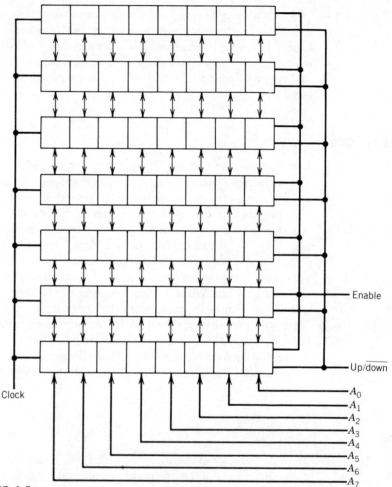

FIGURE 6-5

Registers of the stack

The stack of shift registers accepts accumulator contents, program counts, condition codes, and other data in sequence, raising them in the stack until essential data are preserved for an interrupt.

We have established that the accumulators, the stack, the index register, and the condition code register are all shift registers, and that the program counter is a scaler that is incremented as each program step is executed and is set to a new value upon a branch or a jump. The program decoder is a read only memory (ROM) or programmable logic array (PLA), which is addressed by the operational code byte. Now let us examine the arithmetic logic unit (ALU) to see how the binary word from the program decoder actuates the arithmetic process.

ARITHMETIC LOGIC UNIT

The two subroutines, Tables 6-4 and 6-5, enabled us to identify the elements of the hypothetical microprocessor. One of the most important elements is the arithmetic logic unit (ALU). In examining the ALU, we are not interested in the way it adds, but in the role of the control signals generated by the program decoder in directing the ALU. The control signals are provided by the program decoder upon the receipt of the operational code from the program step.

The arithmetic logic unit is a good element through which to study the control process, although similar features occur elsewhere in the microprocessor. For example, we know from our earlier chapters that to read from or write into a memory requires both that the address be supplied and that the control signals be placed on the enable lines to cause the action. Similarly, we noted that the appropriate control signals had to be provided to the multiplexer for data to be placed on the bus lines or made available to a register. Shift registers likewise respond to control signals, which cause them to accept new data, to shift right or left, and to allow serial input.

Figure 6-6 shows an arithmetic logic unit formed with logic gates and full adders and the control signal lines. The figure does not show the multiplexers and other elements that place data on the inputs A_i and B_i and transfer the sum along the bus to the accumulators or memory locations. It is understood that these functions are provided and that they are synchronized by the clock pulses. The arithmetic logic unit, however, is a combinational rather than a sequential circuit, but the data paths to it and from it are controlled sequentially.

An examination of Figure 6-6 shows control signal lines labeled L_2, L_1, and L_0, but a fourth line exists as well: the carry-in Z to the least significant bit (LSB) full adder. Ordinarily, addition will not involve a carry-in to the LSB adder and a half adder can be used, but the input Z is important in several processes that occur in the arithmetic logic unit. These will be explained below.

The ALU has a number ($2n$) of parallel adders, where n is the number of the bits that the accumulators can hold. In addition and subtraction only n full adders are required, but multiplication of two n-bit numbers can produce a $2n$-bit product.

Table 6-6 lists the actions of the arithmetic logic unit in responding to the control signal states. The numbers in ACCA and ACCB are the input bits $A_i \ldots A_2 A_1 A_0$ and $B_i \ldots B_2 B_1 B_0$. Each bit may be zero or one. The bits will be added in the ALU only if they are passed to $a_i \ldots a_2 a_1 a_0$ and $b_i \ldots b_2 b_1 b_0$ by the control signals.

In preparation for examining Table 6-6 we emphasize two specific actions. Let both L_2 and L_1 be high (1) so that $b_i \ldots b_2 b_1 b_0$ are all ones, because $B + \bar{B} = 1$. First we add the number $A = A_i \ldots A_2 A_1 A_0$. We use a specific value to illustrate the result, letting $A = 0001\ 0101$, for example.

$$
\begin{array}{rcc}
A & 0001\ 0101_2 & 21_{10} \\
B + \bar{B} & 1111\ 1111 & \\
\hline
\text{Sum} & 1\ 0001\ 0100_2 & 20_{10}
\end{array}
$$

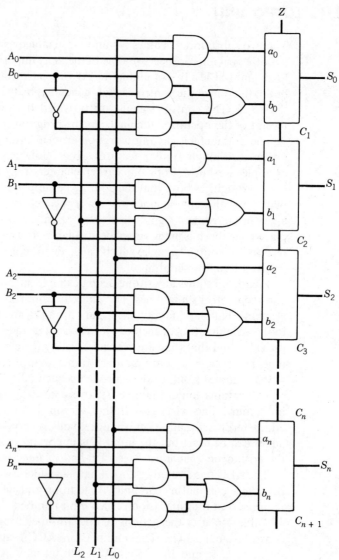

FIGURE 6-6

An Arithmetic Logic Unit

The arithmetic logic unit is an array of full adders with logic gates to implement the control signals, which cause a variety of arithmetic actions.

TABLE 6-6
Action of Control Signals on the Arithmetic Logic Unit

Line Number	ALU Control Signals L_2 L_1 L_0 Z	Inputs AB	Full Adder Values a_i b_i C_{i+1} S_i	Comments
0	0 0 0 0	AB	0 0 0 $\;\;0$	Clears ALU (sets to zero).
1	0 0 0 1	AB	0 0 0 $\;\;1$	Sum = 1.
2	0 0 1 0	AB	A 0 0 $\;\;A$	Transfers A to S.
3	0 0 1 1	AB	A 0 x $\;A+1$	Increments A.
4	0 1 0 0	AB	0 B 0 $\;\;B$	Transfers B to S.
5	0 1 0 1	AB	0 B x $\;B+1$	Increments B.
6	0 1 1 0	AB	A B x $\;A+B$	Adds B to A.
7	0 1 1 1	AB	A B x $\;A+B+1$	Increments $A+B$.
8	1 0 0 0	AB	0 $\bar{B}$ 0 $\;\bar{B}$	Complement of B to S.
9	1 0 0 1	AB	0 $\bar{B}$ x $\;\bar{B}+1$	Increments $\bar{B}$.
10	1 0 1 0	AB	A $\bar{B}$ x $\;A+\bar{B}$	A plus 1's complement of B.
11	1 0 1 1	AB	A $\bar{B}$ x $\;A-B$	Subtracts B from A.
12	1 1 0 0	AB	0 1 0 $\;\;1$	Sum = 1...111.
13	1 1 0 1	AB	0 1 1 $\;\;0$	Sets carry.
14	1 1 1 0	AB	A 1 x $\;A-1$	Decrements A.
15	1 1 1 1	AB	A 1 1 $\;\;A$	Transfers A to S.

x is used to indicate that the value of C_{i+1} cannot be predicted because it depends upon the specific values of A and B.

When we discard the carry-out, the result is $A - 1$; that is, A has been decremented one bit. Thus we find that independent of the value of A, the simultaneous highs (1) on L_2, L_1, and L_0, and $Z = 0$ cause A to be reduced by one bit. This finding corresponds to line 14 in Table 6-6. In this action Z, the carry-in to the LSB adder, is zero, but the reader will recognize that if $Z = 1$, then A will be transferred to the sum lines $S_i \ldots S_2 S_1 S_0$, corresponding to line 15 in Table 6-6.

The second specific action is that of subtracting B from A. The difference is found by adding the two's complement of B to A. The two's complement is obtained by changing each bit in B to its complement (zeros to ones and vice versa), to obtain the one's complement and adding one to the one's complement to obtain the two's complement. We use specific values for A and B to illustrate subtraction. Let $A = 21_{10}$ and $B = 17_{10}$. Then $A - B$ is found as follows.

$$
\begin{array}{ll}
A & 21_{10} = 0001\ 0101_2 \\
-B & 17\ \ \ = 0001\ 0001
\end{array}
\qquad\qquad = 0001\ 0101_2
$$

$$\bar{B}\quad 1110\ 1110 \quad \text{(1's complement)}$$
$$\phantom{\bar{B}\quad}1 \quad\quad\quad\quad\quad\quad\quad Z$$

$$\bar{B} + Z\quad \overline{1110\ 1111} \quad \text{(2's complement)} = 1110\ 1111$$

$$A - B = \overline{4}_{10} \qquad\qquad\qquad\qquad A + \bar{B} + Z = \overline{0000\ 0100}_2$$

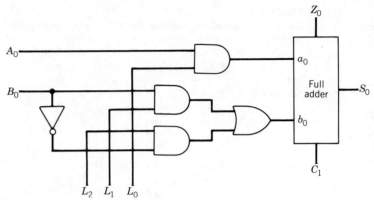

FIGURE 6-7
The First Element of an Arithmetic Logic Unit
A full adder with control lines illustrates the means of comparing the relative magnitude of two bits.

The control signals to achieve the subtraction of B from A are $L_2 L_1 L_0 Z = 1011$. This is the action specified in line 11 of Table 6-6. The steps for subtracting B from A when A is smaller than B are given in Appendix A.

Another feature we require of the arithmetic logic unit is that it can recognize when the magnitude of A is less than the magnitude of B. We used this condition in the square root routine to stop the iterations. Line 01 DD_{hex} in Table 6-5 branches on the condition that $|N - x^2| < \varepsilon$, using the program operational code BLZ (38_{hex}), "branch if less than zero."

This feature can be illustrated by using only one full adder; A and B are assumed to be one bit each. The arrangement is shown in Figure 6-7. The reader can verify that this is no less general than if more bits per value were used. Taking A_0 and B_0, we examine the difference $A_0 - B_0$ and use line 11 ($L_2 L_1 L_0 Z = 1011$) to find the difference. Four possibilities are found by adding the two's complement of B_0 to A_0.

We note in Table 6-7 that the branch condition $A < B$ results in $C_1 = 0$, $S_0 = 1$, while all the other magnitude relations have $C_1 = 1$. After the subtraction the carry-out is examined; if $C_{out} = 0$, the branch is executed. With more bits each S_i

TABLE 6-7
Magnitude Comparison

	A_0	B_0	C_1	S_0
$A > B$	1	0	1	1
$A = B$	1	1	1	0
$A = B$	0	0	1	0
$A < B$	0	1	0	1

will not necessarily be 1, so it is not used in the test. The reader will recall that subtraction with the two's complement requires that $C = 1$ be discarded after the test to get the difference when $A > B$.

We have not shown all the possible arrangements of logic gates and control lines in the ALU. With additional logic gates, for example, the ALU can perform Boolean algebra operations on the inputs A and B, such as A AND B (AB), A OR B ($A + B$), and A XOR B ($A \oplus B$). The design of the ALU and of any other element of the microprocessor is similar to the design of any array of logic gates made to perform a specific task.

SUMMARY

In concluding this examination of microprocessor basics, we note that the control signals provided by the program decoder initiate a variety of actions in the microprocessor. These include the transmission of data among internal and external registers and the processing of the data in the arithmetic logic unit. The programmed action may be relatively simple, like "stop" or "wait," or it may be quite involved, like a branch or interrupt in which the stack is involved. The operation of the microprocessor is directed step by step by a program written to achieve a particular result. Developments in the art of programming will allow microprocessors to appear more intelligent, and they will become easier to use.

The trend in microprocessors is to expand them to include the functions or components that were on external chips for earlier microprocessors. At this writing, 16-bit microprocessors are being upgraded by increasing the number of transistors on the chip from approximately 30,000 to 130,000, and 32-bit microprocessors with 450,000 transistors on a chip have been produced. The newer, larger chips replace about 20 chips that surrounded and supported the earlier microprocessors. In one model, the new microprocessor and two latch chips for input/output constitute a microcomputer that needs only external memory banks and a terminal to function.

Another trend is the use of coprocessors. Coprocessors are large integrated circuits that directly complement the microprocessor. Arithmetic operations become far more precise (18 decimal digits), and arithmetic is much faster when a *numerical data processor* is used as an extension of the microprocessor. Other coprocessors, including one called an *operating systems processor*, handle interrupts and management tasks that permit multiterminal access to the computer facility, supported by the microprocessor and the exterior memory banks.

The evolution of microprocessors is moving in two directions. The first is the gathering of the peripheral digital elements within the microprocessor, the expansion of the bit length, and the use of coprocessors to approach the performance of mainframe computers, as mentioned above. The second is the development of very capable dedicated microprocessors with extensive on-board processing (fast Fourier transforms, for example) of data from external sensors to provide sophisticated analysis and process control. A major advance is being made as analog devices are included as on-board elements of the microprocessor.

Our purpose has been to show that there is nothing magical or mysterious about the complicated and extensive device known as a microprocessor. We hope that this discussion has stripped some of the mystery from microprocessors, so that they may be understood and used well.

PROBLEMS

6-1 Write a program for the hypothetical microprocessor to calculate $N!$ (N factorial).

6-2 Write a program to divide D/d by subtract and shift steps.

6-3 Calculate $20^{1/2}$ by iteration. Find the number of iterations for $\varepsilon = 10^{-3}$.

6-4 Calculate $0.9^{1/2}$ by iteration. What is the accuracy after four iterations?

6-5 Write a microprocessor program for $\exp(x) = 1 + x + x^2/2! + x^3/3! + \cdots$.

6-6 Divide $D/d = 910/55$ by Wallace's algorithm (see Appendix A), which uses multiplication to find a quotient.

6-7 For $Q = D/d$ show that in Wallace's method that a_i converges to 1.0 and b_i converges to $1/d$ as δ^n, $n = 2^i$. Take $p_0 = (1/d) + \delta$, where δ is the initial error in estimating $1/d$ and i is the iteration number.

6-8 Write a program to divide by Wallace's algorithm.

6-9 Calculate $20^{1/2}$ by Wallace's square root algorithm in Appendix A.

REFERENCES

1983 Gupta, A., and Toong, H. D., "Microprocessors—The First Twelve Years," *Proceedings of the IEEE*, **71** 1236 (1983).

1983 Preston, Glenn W., "The Very Large Scale Integrated Circuits," *American Scientist*, **71** 466 (1983).

CHAPTER 7

BASIC ELECTRICAL CIRCUIT THEORY

Until now in our study of digital electronics, we have avoided electrical circuit theory and have devoted our attention to the flow of digital signals. Now, as preparation to study analog circuits and other aspects of electronics, such as power sources, we consider the fundamentals of electrical circuit analysis. As a dividend we will learn some important things about the internal electronics of logic gates and will be able to unite analog and digital elements.

Two nearly self-evident principles are the foundations for electrical circuit analysis. These two principles are called Kirchhoff's laws. One law states that the electrical current coming to any point on an electrical conductor must equal the current leaving that point. In other words, electrical charge cannot accumulate at any point on a conductor in a circuit.

If two or more conductors come together at a junction in a circuit, the same principle applies at that junction. The motion of the charges constitutes an electrical current. All the charge that enters the junction must also leave the junction. (The junction of electrical conductors is also called a *node*.) We will write this Kirchhoff law as a node equation.

The second Kirchhoff law concerns the voltage around a circuit. Voltage or electromotive force (EMF) is the cause of the current. Charged particles (electrons or ions) move under the effect of an electromotive force.

The second Kirchhoff law states that the sum of all the EMFs encountered around a closed circuit must be zero. That is, when the circuit is traversed and the point on the circuit from which the traversal began is once again reached, the voltage at that point is identical to its initial value. This law requires that the sum of all the increments of voltage around the loop or path through the circuit must equal zero. The second Kirchhoff law is written as a loop equation.

In many electrical circuits, several different paths may be taken to traverse a circuit completely from a starting point. Kirchhoff's second law states that the sum of the voltage increments must be zero regardless of the path.

For ease in using Kirchhoff's laws to analyze electrical circuits, the laws are applied systematically as we write the loop equations and the node equations. We will adopt conventions that ensure a systematic approach in finding the equations. In order to state the conventions and illustrate the use of Kirchhoff's laws, we first consider the current through and the voltage across a resistor in a simple electrical circuit.

OHM'S LAW

Ohm's law states that the voltage across a resistor and the current through it are directly proportional. The resistance R is the proportionality constant. This linear relationship may be written as the equation

$$V = RI \tag{7-1}$$

where V is the voltage in volts, I is the current in amperes, and R is the resistance in ohms.

The simple circuit in Figure 7-1 consists of a voltage source (a battery) and a resistance connected to it by conductors (wires). In drawing Figure 7-1 we have

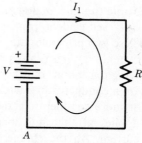

FIGURE 7-1

A Simple Resistor Circuit

The directions of the current and of the traversal of the circuit are chosen in applying Kirchhoff's laws.

assumed that all the resistance is concentrated at the resistor R and that there is no resistance in the battery or in the conductors. By making this assumption we have elected to use *discrete element* electrical circuit theory. (Distributed element electrical circuit theory, which we will not use in this work, must be employed when our assumption is not realistic.)

Discrete element theory has the advantage that any point on a wire *between* elements can be considered electrically equivalent to all other points in that sector of the wire. Later we will examine the assumption of no resistance in the battery, but for now we will simply work with it.

CONVENTIONS FOR USING KIRCHHOFF'S LAWS

The first step in analyzing an electrical circuit is to *assume* the direction of the current in each part of the circuit. If a current is in the direction that is guessed, the value calculated with Kirchhoff's laws will be positive, and we will know that we guessed correctly. If the direction we assumed initially for a current is the opposite to its actual direction, the calculated value will be negative. In complicated circuits there is no sure way of guessing the actual direction of the current.

The second step is to choose a direction around the circuit in which a loop equation is to be written. Each circuit element must be included at least once in the loop equations. There is *no single right way* of doing this, but experience may help to avoid extra algebra.

Figure 7-1 shows our choice for the direction of the current I_1. Because there is only one path, there are no other currents; thus we do not write a node equation. We elect to traverse the circuit in a clockwise direction.

Proceeding from A around the circuit of Figure 7-1, we traverse the battery in the direction of the voltage. The convention we adopt is:

I. The voltage is taken to be positive if the circuit is traversed through a voltage source in the direction of the voltage polarity, that is, from negative to positive.

The positive end of the resistor corresponds to the positive end of the battery, so that according to our choice of current direction the current in the resistor goes from its positive end to its negative end. Because we are traversing the circuit in the same direction, we see that the voltage changes from positive to negative as we pass the resistor. The voltage drops through the resistor in the direction of the current. We use this observation to establish a second convention in writing Kirchhoff's loop equations. The convention is:

II. The voltage increment across a resistor is negative if the resistor is traversed in the assumed direction of the current.

We now may write the loop equation for the simple circuit of Figure 7-1. Starting at point A, the voltages that must total zero in accord with Kirchhoff's law are added together as they are encountered in the clockwise traversal.

$$V_{\text{battery}} + V_{\text{resistor}} = 0 \tag{7-2}$$

This may be restated as

$$+V - I_1 R = 0$$

where convention I indicates that $V_{\text{battery}} = +V$ and convention II states that $V_{\text{resistor}} = -I_1 R$.

Solving Equation 7-2 for the current I_1, we obtain

$$I_1 = +V/R$$

The positive answer for the current I_1 confirms that we guessed correctly the direction of the current.

To illustrate and state the three additional conventions for using Kirchhoff's laws, we must consider slightly more complicated resistor circuits. The five conventions will be listed together presently when we restate Kirchhoff's laws. First, however, let us look at a circuit with two resistors in series.

SERIES RESISTORS

Figure 7-2 shows a single loop circuit with two resistors in series, one designated R_L and the other R. Traversing the circuit from A, we find

$$V + V_{R_L} + V_O = 0$$
$$V - I_1 R_L - I_1 R = 0. \tag{7-3}$$

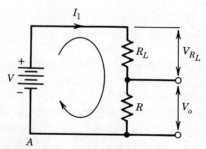

FIGURE 7-2

A Series Resistor Circuit

The series resistors divide the applied voltage in proportion to their relative resistances.

We have used convention II to equate $V_{R_L} = -I_1 R_L$ and $V_O = -I_1 R$.
The current is obtained from Equation 7-3; it is

$$I_1 = \frac{V}{R + R_L}$$

The current found with Equation 7-3 and Ohm's law can give the voltage, V_O, across the resistor R.

$$V_O = I_1 R = \left(\frac{V}{R + R_L}\right) R = V \frac{R}{R + R_L} \qquad (7\text{-}4)$$

Equation 7-4 is identical to Equation 1-1 because the circuit of Figure 7-2 and the circuit of Figure 1-1 are the same, except that in Figure 1-1 the battery is not shown and its voltage is denoted by V_{cc}.

The circuit of Figure 7-2 is useful as a means of providing a portion of the voltage across the two resistors. It is called a *voltage divider*; the fraction of the voltage is given by $\frac{R}{R + R_L}$.

PARALLEL RESISTORS

The circuit of Figure 7-3 has two resistors in parallel so that the current I through the battery can divide and return by two different paths. At node B the current entering the junction must equal the currents leaving the junction or node, and

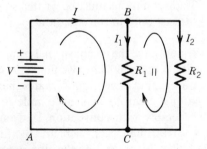

FIGURE 7-3

A Parallel Resistor Circuit

Each element in the circuit must be included in writing the equations for Kirchhoff's law. The selections of the circuits traversals and current directions are arbitrary.

similarly at node *C*. To ensure the systematic application of Kirchhoff's laws we adopt the convention for currents at a node. It is:

III. The current is taken to be positive if its assumed direction is that of entering the junction or node, and it is taken to be negative if its assumed direction indicates that it is leaving the node.

Using convention III we can write the node equation at *B*:

$$I - I_1 - I_2 = 0 \tag{7-5}$$

In accordance with Kirchhoff's law that charge cannot accumulate at a node or elsewhere in a conductor in a circuit, a similar equation can be written for node *C*:

$$I_1 + I_2 - I = 0$$

A comparison of these two node equations reveals that one can be obtained from the other by multiplying by -1. These equations certainly do not contain different information about the currents; one is redundant and therefore not needed. If there were more nodes we would require more node equations, but if an equation were written for each node, one of the equations would not contain any information about the currents that is not already contained in the others. In recognition of this fact we state convention IV:

IV A node equation must be written for each junction in the circuit except one.

The fifth convention in applying Kirchhoff's laws recognizes that all voltages and currents in a circuit are interdependent and that all elements, such as the battery and the resistor, influence them. The contribution of each element must be considered in analyzing the circuit. The fifth convention is:

V Loop equations must be written so that each element in the circuit is included at least once in the equations.

We now have the means of analyzing the parallel resistor circuit of Figure 7-3. In applying Kirchhoff's laws we first choose the directions for the currents. There are two nodes, but convention IV states that only one node equation must be written. Equation 7-5 can be applied to node *B*.

To comply with convention V, we must choose the paths and a direction of traversal through each element. In the circuit in Figure 7-3 there are several possibilities for traversing the circuit. Our choices are shown in the figure.

The summation of the voltages around the circuit yield

$$V - I_1 R_1 = 0 \tag{7-6}$$

and

$$-I_2 R_2 + I_1 R_1 = 0 \tag{7-7}$$

In this example we have written each equation in a single step rather than the two steps used in Equation 7-3. Note that the signs of the terms in Equation 7-7 agree with convention II.

If V, R_1, and R_2 are known, there are three unknowns, I, I_1, and I_2; thus we require the three independent equations, Equations 7-5, 7-6, and 7-7, to find the currents.

The elimination of variables among the equations gives

$$I = \frac{V}{R_1} + \frac{V}{R_2} \tag{7-8}$$

for the current provided by the battery, and the currents in the two paths are

$$I_1 = \frac{V}{R_1} \quad \text{and} \quad I_2 = \frac{V}{R_2}$$

Now let us state Kirchhoff's laws and the rules or conventions for applying them systematically to analyze electrical circuits. Although they are shown for direct current circuits, we will see later that they may be used to analyze alternating current circuits and the electrical behavior of electronic devices, such as amplifiers and logic gates.

KIRCHHOFF'S LAWS AND CONVENTIONS

Kirchhoff's node law — The sum of currents at a junction or node must equal zero.

$$\sum I_j = 0$$

Kirchhoff's loop law — The sum of all voltages encountered in a complete traversal of a circuit is zero.

$$\sum V_i = 0$$

Convention I — The voltage is taken to be positive if the circuit is traversed through a voltage source in the direction of the voltage polarity, that is, from negative to positive.

Convention II — The voltage increment across a circuit element (resistor) is negative if the element is traversed in the assumed direction of the current.

Convention III The current is taken to be positive if its assumed direction is that of entering the junction or node; it is taken to be negative if its assumed direction indicates that it is leaving the node.

Convention IV A node equation must be written for each junction in the circuit except one.

Convention V Loop equations must be written so that each element in the circuit is included at least once in the equations.

SIMPLIFICATION OF RESISTOR CIRCUITS

In analyzing the circuits in Figures 7-2 and 7-3, we first assumed a direction for the currents; then we chose ways to traverse the circuit in such a manner as to include each element. Thus we applied Kirchhoff's laws in accordance with the conventions. For the circuit of Figure 7-2 we found (Equation 7-3)

$$I_1 = \frac{V}{R + R_L}$$

We now recognize that this can be written in a simpler manner if we set

$$R + R_L = R_{equivalent} \qquad (7\text{-}9)$$

The two series resistors are replaced by an equivalent resistor, and Figure 7-2 reduces to the simpler circuit of Figure 7-1, that is,

$$I_1 = \frac{V}{R_{eq}}$$

Because a number of resistances in series can be combined in pairs and the pairs combined, we conclude that any number of resistors in series can be combined to form an equivalent resistor that is equal to the sum of the individual resistances. We may write

$$R_{eq} = \sum_{i=1}^{n} R_i \qquad (7\text{-}10)$$

where the resistors R_i are in series, and we may state this general rule:

Resistors in series may be added together to give the value of an equivalent resistor.

The current I in the parallel resistor circuit of Figure 7-3 (Equation 7-8) is

$$I = V\left(\frac{1}{R_1} + \frac{1}{R_2}\right)$$

In this case, too, the current through the battery will be the unchanged if we require that

$$I = V\left(\frac{1}{R_{equivalent}}\right)$$

that is,

$$\frac{1}{R_{eq}} = \frac{1}{R_1} + \frac{1}{R_2} \tag{7-11}$$

or

$$R_{eq} = \frac{R_1 R_2}{R_1 + R_2}$$

We see that Figure 7-3 can be reduced to Figure 7-1 by using an equivalent resistor to replace the two parallel resistors in Figure 7-3 if its value is found from Equation 7-11. This process may be continued when there are several resistors in parallel. Thus we may write

$$\frac{1}{R_{eq}} = \sum_{i=1}^{n} \frac{1}{R_i} \tag{7-12}$$

when resistors R_i are in parallel, and we may state this general rule:

For resistors in parallel the reciprocal of the equivalent resistor is the sum of the reciprocals of the individual resistors.

For many circuits made with voltage sources and resistors, the most direct way to find the currents is to combine the resistors into equivalent resistors as the circuit is simplified. We illustrate this technique with the circuit of Figure 7-4. (Later we consider a circuit on which this technique cannot be used.) Kirchhoff's laws, however, are sufficient to solve all kinds of circuits.

In each element of the circuit in Figure 7-4, we wish to find the currents by combining the resistors. We find the equivalent resistor for the parallel combination of R_2 and R_3 in Figure 7-4a,

$$\frac{1}{R_{23}} = \frac{1}{R_2} + \frac{1}{R_3}$$

which yields

$$R_{23} = \frac{R_2 R_3}{R_2 + R_3}$$

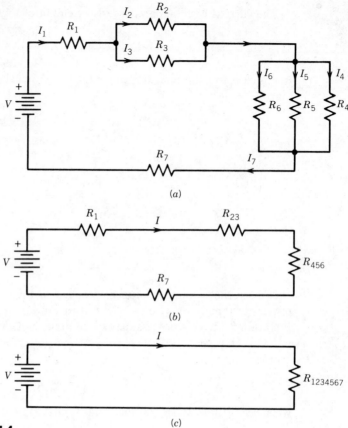

FIGURE 7-4

A Resistor Circuit

Parallel and series resistances may be replaced by their equivalent resistances, and the circuit may be simplified to find the current and voltages.

Similarly, the equivalent resistor to the parallel combination of R_4, R_5, and R_6 is

$$R_{456} = \frac{R_4 R_5 R_6}{R_5 R_6 + R_4 R_6 + R_4 R_5}$$

Figure 7-4b shows the simplified circuit. The resistors in series may be simplified further by finding the equivalent resistor to the series resistors,

$$R_{1234567} = R_1 + R_{23} + R_{456} + R_7$$

Figure 7-4c shows the result of simplifying the original circuit.

The current $I = I_1 = I_7$ is found by Ohm's law to be

$$I = \frac{V}{R_{1234567}}$$

The values of the currents $I_2, I_3, I_4, I_5,$ and I_6 are obtained first by finding the voltages across the equivalent resistors, R_{23} and R_{456}, respectively, and then by applying Ohm's law.

$$V_{23} = IR_{23} = \frac{VR_{23}}{R_{1234567}}$$

and

$$V_{456} = IR_{456} = \frac{VR_{456}}{R_{1234567}}$$

Now the currents are $I_2 = V_{23}/R_2$ and $I_3 = V_{23}/R_3$, and $I_4 = V_{456}/R_4$, $I_5 = V_{456}/R_5$, and $I_6 = V_{456}/R_6$.

We can now make an important observation: currents in resistors in parallel divide in *inverse* proportion to the respective values of resistance. For example, in Figure 7-4

$$V_{23} = I_2 R_2 = I_3 R_3$$

so that

$$\frac{I_2}{I_3} = \frac{R_3}{R_2}$$

If the resistance value of R_3 is larger than the resistance value of R_2, then the current through R_3 is smaller than the current through R_2.

Resistances in series (Figure 7-2) make a voltage divider with a portion of the voltage occurring across each resistor (see Equation 7-4). Here we find that parallel resistors make a *current divider*, that is, they divide the current to them inversely proportional to their resistances.

A numerical example is useful to illustrate the method of combining resistances to find the currents in the circuit of Figure 7-4. Let $V = 10$ V, and let the resistors be

$$R_1 = 5 \, \Omega \qquad R_4 = 300 \, \Omega \qquad R_7 = 33 \, \Omega$$
$$R_2 = 30 \, \Omega \qquad R_5 = 100 \, \Omega$$
$$R_3 = 20 \, \Omega \qquad R_6 = 150 \, \Omega$$

We find $R_{23} = 12 \, \Omega$, $R_{456} = 50 \, \Omega$, and $R_{1234567} = 5 + 12 + 50 + 33 = 100 \, \Omega$.

The current $I = V/R_{1234567} = 10/100 = 0.1$ A $= I_1 = I_7$. To find the other currents, we find the voltages across the equivalent resistors. $V_{23} = IR_{23} = 0.1 \times 12 = 1.2$ V and $V_{456} = 0.1 \times 50 = 5.0$ V. The current $I_1 = V_{23}/R_2 = 1.2/30 = 0.04$ A and $I_3 = 0.06$ A. We check that $I_2 + I_3 = I = 0.1$ A. Similarly, $I_4 = 0.0167$ A, $I_5 = 0.050$ A, and $I_6 = 0.0333$ A.

We can check our values by calculating the voltages around the circuit and summing them. We should find that

$$V - (V_1 + V_{23} + V_{456} + V_7) = 0$$

We have

$$V - (IR_1 + IR_{23} + IR_{456} + IR_7) = 10 - (0.5 + 1.2 + 5.0 + 3.3) = 0$$

In solving the circuit of Figure 7-4, we did not directly use Kirchhoff's laws, but the means of finding equivalent resistances and the way the currents divide at junctions or nodes are the fruits of these laws.

We conclude this section on simplifying resistor circuits by observing a very practical point in measuring voltages in electronic circuits because of the behavior of currents with parallel paths. To measure accurately the voltage across a resistor, the voltmeter should not modify the current through the resistor by bypassing a portion of it through the voltmeter. The resistor and the voltmeter in parallel, however, form an equivalent resistor whose value will be less than either the resistor or the resistance of the voltmeter.

The effect of the voltmeter will be small if its resistance is many times larger than the resistor across which the voltage is being measured. Instruments with extremely high input resistances are preferred for measurements in electronic circuits so that the effect of the voltmeter on the measurements will be negligible. In Chapter 13 we will show how an inexpensive (relatively low-resistance) voltmeter may be made to have an extremely high input resistance so that it does not significantly modify a circuit in a voltage measurement.

BRIDGE CIRCUIT

When we examine some circuits to find the parallel and series resistance combinations, we observe that it is impossible to place one or more resistors in either parallel or series arrangements. Figure 7-5 shows a simple and useful circuit that cannot be reduced by combining resistors, as was done above. This is the bridge

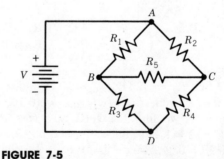

FIGURE 7-5
A Bridge Circuit

The bridge circuit requires Kirchhoff's laws for its analysis because the resistors cannot be placed into parallel or series equivalents.

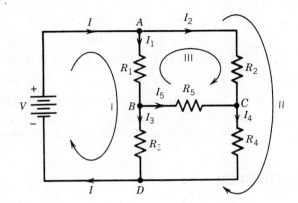

FIGURE 7-6
The Bridge Circuit
The bridge circuit is redrawn, and the choices of current and loop directions are made to apply Kirchhoff's laws.

circuit, which is used to compare an unknown resistance (R_3, for instance) with known ones (R_1, R_2, and R_4) to measure the unknown R_3 resistance value. By adjusting one known resistor (R_4, for instance) until there is no voltage difference between the ends of R_5, we find that the resistors are related as follows:

$$\frac{R_1}{R_2} = \frac{R_3}{R_4}$$

and R_3 is found.

In general, there is a current in R_5 so we use Kirchhoff's laws to analyze this circuit, which cannot be reduced by combining resistors. To do this we use the loops and nodes shown in Figure 7-6, which is merely Figure 7-5 redrawn. The Kirchhoff equations are:

Node equations
$$I - I_1 - I_2 = 0 \quad \text{at node } A$$
$$I_1 - I_3 - I_5 = 0 \quad \text{at node } B$$
$$I_2 + I_5 - I_4 = 0 \quad \text{at node } C$$

Loop equations
$$V - I_1 R_1 - I_3 R_3 = 0 \quad \text{for loop I}$$
$$V - I_2 R_2 - I_4 R_4 = 0 \quad \text{for loop II}$$
$$I_1 R_1 - I_2 R_2 + I_5 R_5 = 0 \quad \text{for loop III}$$

The six unknown currents are found by the simultaneous solution of these six equations. For even this simple system the algebra is extensive. The equation for I_5, found by successive elimination, is

$$V\left(\frac{R_3}{R_1 + R_3} - \frac{R_4}{R_2 + R_4}\right) - \left(R_5 + \frac{R_2 R_4}{R_2 + R_4} + \frac{R_1 R_3}{R_1 + R_3}\right)I_5 = 0$$

which reduces to

$$I_5 = \frac{V(R_2 R_3 - R_1 R_4)}{R_1 R_2 R_5 + R_2 R_3 R_5 + R_1 R_4 R_5 + R_3 R_4 R_5 + R_1 R_2 R_3 + R_1 R_3 R_4 + R_1 R_2 R_4 + R_2 R_3 R_4}$$

The equations for the other currents are similarly complicated.

We note that the current I_5 is zero when the resistors R_1, R_2, R_3, and R_4 have the ratio stated above for finding an unknown resistance.

THEVENIN'S THEOREM

Another means of finding a current in a single branch of a circuit, including the bridge circuit, is to apply Thevenin's theorem. Thevenin's theorem states:

> *A two-terminal linear circuit can be represented as a Thevenin voltage source and an internal resistance (impedance).*

Figure 7-7 shows the simple configuration postulated by Thevenin. We will now show how to find the two Thevenin components, the internal resistance and the Thevenin voltage source.

First we find the internal resistance of the source by "shorting" all the internal voltages sources and calculating or otherwise determining the resistance between the terminals B and C. By "shorting" we mean that conceptually we remove the internal voltage sources and replace the voltage sources with conductors (wires). Second we find the equivalent or Thevenin voltage by calculating the voltage difference at the terminals with no current between the terminals; that is, with no connection between B and C. This also is called "open-circuit" voltage.

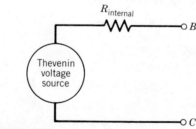

FIGURE 7-7

The Equivalent of a Two-Terminal Circuit According to Thevenin's Theorem

All electrical circuits with two terminals are equivalent to a Thevenin voltage source and an internal impedance (resistor).

A two-terminal circuit is like the one shown in Figure 7-7, in which an electrical circuit element, such as a resistor, can be placed between B and C. The term "linear" indicates that if the circuit element between the terminals B and C changes (for example, if the resistance of the element inserted there increases), the circuit will respond proportionally.

We will use the bridge circuit (Figure 7-5) as our example in applying Thevenin's theorem. To find the current through R_5 in the bridge circuit, we first open the branch containing R_5, as shown in Figure 7-8a. As shown in Figure 7-8b, we short the voltage source, and for clarity the circuit is redrawn (Figure 7-8c). The reader can confirm the rearrangement by tracing the several paths in Figure 7-8b from terminal B to terminal C, recalling that all points on the conductors between elements are equivalent. The equivalent internal resistance between terminal B and C is

$$R_{\text{internal}} = \frac{R_1 R_3}{R_1 + R_3} + \frac{R_2 R_4}{R_2 + R_4} = R_{\text{Thevenin}}$$

Next we find the open-circuit voltage across the terminals. We note on Figure 7-8a that R_1 and R_3 form a voltage divider for the battery voltage, V, so that

$$V_B = V \frac{R_3}{R_1 + R_3}$$

Similarly,

$$V_C = V \frac{R_4}{R_2 + R_4}$$

so that

$$V_{\text{Thevenin}} = V_{\text{open circuit}} = V_B - V_C = V\left(\frac{R_3}{R_1 + R_3} - \frac{R_4}{R_2 + R_4}\right)$$

Now, with R_5 in the circuit we see from Figure 7-8d that the Kirchhoff's loop equation of the circuit with R_{int} and R_5 in series is

$$V_{\text{Thevenin}} - (R_5 + R_{\text{int}})I_5 = 0$$

This may be written

$$V\left(\frac{R_3}{R_1 + R_3} - \frac{R_4}{R_2 + R_4}\right) - \left(R_5 + \frac{R_2 R_4}{R_2 + R_4} + \frac{R_1 R_3}{R_1 + R_3}\right)I_5 = 0$$

which is identical with the equation found by successive elimination with the Kirchhoff equations. Thus, the current in R_5 is given by an equivalent arrangement, in which R_5 is placed across the two terminals of a Thevenin voltage source with an internal resistance.

The significance of this section is not that we have another method of finding the current in a branch of a circuit, although this can be helpful in avoiding tedious calculations. The important point is that, in general, *any two-terminal linear circuit*

164 Basic Electrical Circuit Theory

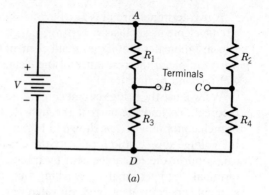
(a)

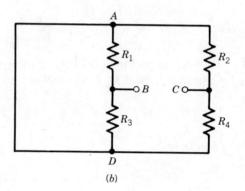
(b)

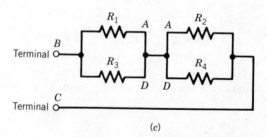
(c)

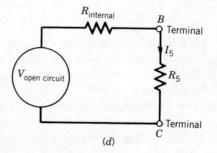
(d)

can be represented by an internal (Thevenin) voltage source and a series internal resistance (impedance). In fact, if the open circuit voltage and the series internal resistance are known by measurement or by calculation, the details of the circuit may not be useful.

The details of a device, such as a photovoltaic source, may be incompletely known, or the mechanism by which it produces a voltage may not even be understood, but it can be represented as a "black box" containing an open-circuit voltage source and a series internal resistance.

A couple of simple measurements can yield the open-circuit (Thevenin) voltage V_{oc} and the internal resistance R_{int} of a circuit or an electrical device. The open-circuit voltage is measured across the open terminals by a voltmeter, which draws negligible current.

The internal resistance can be found by placing a known load resistor, R_L, across the terminals and measuring the voltage V_O across it. The internal resistance may be calculated from R_L, V_{oc}, and V_O.

Without developing the details, we can also state a corollary to Thevenin's theorem, known as Norton's theorem. Norton's theorem states:

A linear two-terminal circuit can be represented as a current source and a shunt (bypass) resistor.

Figure 7-9 shows the two-terminal network that is characterized in accordance with Norton's theorem.

We conclude this section on Thevenin's theorem by noting that a battery or other voltage source, being a two-terminal circuit, is both a voltage source and an internal resistance. The voltage produced by chemical energy in a battery is diminished when current is drawn from the battery by the voltage drop IR_{int} in the internal resistance. The net voltage across the terminals $V_o = V_{oc} - IR_{int}$ depends upon the amount of current and the "quality" of the battery. In "shorting" the voltage sources to find the internal resistance of the Thevenin source, we must

FIGURE 7-8

Application of Thevenin's Theorem to Bridge Circuit

(*a*) Bridge circuit with an open branch. (*b*) Bridge circuit with a shorted voltage source. (*c*) Bridge circuit with a shorted voltage source redrawn. (*d*) Two-terminal circuit with R_5.
The steps in applying Thevenin's theorem are:

1. Open the branch to create the two terminals and find the voltage, $V_{open circuit}$, between them.
2. Remove the voltage source and replace it by a conductor,
3. Find the resistance between the terminals,
4. Use the Thevenin voltage and internal resistance to find the current through the element, R_5, between the terminals.

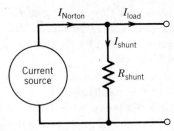

FIGURE 7-9

Equivalent of a Two-Terminal Circuit According to Norton's Theorem

An alternate equivalent to all two-terminal electrical circuits is a current source and an impedance (resistor) that bypasses (shunts) a portion of the current.

include the internal resistance of the sources themselves; the "shorting" conductors are really the respective internal resistances.

A good voltage source has low internal resistance, so that the voltage remains nearly constant when the current drawn lies within the rated values. (The property of maintaining nearly constant voltage at the output is called *regulation*. A regulated power supply provides unchanging voltage from the minimum to the rated amount of current. In Chapter 14 we will consider power sources and the means by which they may be regulated.)

It is important to realize that in integrated circuits, too, the voltage output depends upon the current demand. For this reason the number of integrated circuits that may be driven by the output of a single integrated circuit is limited. Manufacturers specify the current load or the number of integrated circuits that may be attached to the output. If more are used, the output voltage may be diminished so much that the performance of the system is unsatisfactory. Buffers and open-collector gates are used to offset the current demands on integrated circuits.

TIME-DEPENDENT CIRCUITS

In the next phase of electrical circuit analysis we examine the other circuit elements, capacitance and inductance. These elements depend on time because they respond in a transient manner to a voltage or current. First we illustrate the transient responses of capacitance and inductance to show their contributions as circuit elements; later we consider them in alternating current (ac) circuits. The notation for currents and voltages will change to lowercase letters to emphasize the time variation involved. We will also introduce E and e to have another option in designating voltage. E is associated with electromotive force in volts.

In many electrical circuits and devices the capacitive and inductive effects are negligible, and the intrinsic capacity and inductance can be ignored. In some circumstances, however, the capacitance and inductance in a circuit or in an electronic device influence strongly the performance of the electrical system, and in these cases they must be taken into consideration.

CAPACITANCE

We are interested here in a discrete capacitance that is placed in an electrical circuit. Figure 7-10 shows a voltage source that may be connected by a switch to a resistor and a capacitor in series. The fundamental relation for any capacitor is

$$Q = CE_C \tag{7-13}$$

where the charge Q in the capacitor is expressed in coulombs, the voltage across the capacitor E_C in volts, and the capacitance C in farads.

A capacitor is made by two electrical conductors separated by a nonconducting medium such as air, oil, plastic film, or other dielectric material. The amount of capacitance is proportional to the area of the conductors (plates) and the dielectric value of the medium, and inversely proportional to the separation of the plates. Because typical capacitance in electronic circuits and devices is extremely small, as expressed in farads, the values will be stated in microfarads (1 μF = 10^{-6} F) or picofarads (1 pF = 10^{-12} F).

We shall assume that the switch in Figure 7-10 is closed at time $t = 0$ when no charge is stored in the capacitor; thus, according to Equation 7-13, there is no voltage across it. As we did for the other circuits, we write the Kirchhoff loop equation for the voltages. Although it is not shown in Figure 7-10, we traverse the circuit in a clockwise direction.

$$E + E_R + E_C = 0$$

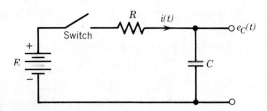

FIGURE 7-10

A Circuit with a Resistor and Capacitor

The lowercase letters $i(t)$ and $e_c(t)$ indicate time dependent quantities, in contradistinction to the constant values represented by capital letters. E and $e(t)$ are used for voltages rather than V.

or

$$E - iR - \frac{q}{C} = 0 \qquad (7\text{-}14)$$

No Kirchhoff node equation is required.

We use the symbol i for the time-varying current rather than I for the steady current, as before. The symbol q represents the time-varying charge. The current is the time rate of change of charge, while charge is the sum over time of the current. They are related by

$$i = \frac{dq}{dt}$$

and

$$q = \int_0^t i\, dt \qquad (7\text{-}15)$$

Current is the time derivative of charge, and charge is the time integral (sum) of current.

The Equation 7-14 is solved to give the charge, the current, and the voltage across the capacitor as functions of time.

$$i(t) = \frac{E}{R} \exp\left(-\frac{t}{RC}\right) \qquad (7\text{-}16)$$

$$e_C(t) = E\left[1 - \exp\left(-\frac{t}{RC}\right)\right] \qquad (7\text{-}17)$$

$$q(t) = Ce_C(t) = EC\left[1 - \exp\left(-\frac{t}{RC}\right)\right] \qquad (7\text{-}18)$$

The term $q(t)$ is the charge accumulated in the capacitor as time passes after the switch is closed. Eventually the charge q reaches the value $q = CE$ and the current $i(t = \infty) = 0$. Figure 7-11 gives the graphs of $i(t)$ and $e_C(t)$ for the series RC circuit. The validity of the solutions can be confirmed by substituting them into Equation 7-14.

The product of the resistance times the capacitance is called the *time constant*.

$$\tau = RC \text{ sec} \qquad (7\text{-}19)$$

The time constant, τ, is a measure of the rate at which the quantities can change. As time passes after the switch is closed, it reaches $\tau = RC$, and in that same interval the current $i(t)$ has dropped to 37 percent of its initial value. [In Equation 7-16 for $t/\tau = 1$ the term $\exp(-1) = 0.368$.] In the same interval the voltage across the capacitor has risen to 63 percent of its final value ($1 - 0.368 = 0.632$), as given by Equation 7-17.

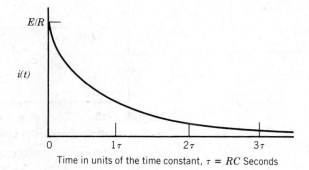

FIGURE 7-11

Graphs of i(t) and $e_c(t)$ for an RC Circuit

The time dependances of $i(t)$ and $e_c(t)$ following the switch closure are shown. Initially the capacitor was uncharged. The rate is governed by the time constant $\tau = RC$.

For $C = 1\ \mu F$ (10^{-6} F) and $R = 1\ M\Omega$ ($10^{+6}\ \Omega$) the time constant $\tau = RC = 1$ sec, and the changes in $i(t)$ and $e_c(t)$ stated above will occur in 1 sec. After 2 sec the current $i(t)$ will be 13.5 percent of its original value of $i(t = 0) = E/R$ because $\exp(-2) = 0.135$, and the voltage $e_c(t)$ will be 86.5 percent of its final value since $(1 - 0.135 = 0.865)$. Other combinations of resistance and capacitance will yield other time constants, which can be much shorter or longer. For digital circuits the time of transition from one state to the other is made as small as possible; time constants of the order of $\tau = 10^{-9}$ sec are involved. Efforts in integrated circuit design are directed toward making the time constants governing the transitions between states even smaller.

It is important to understand that every electrical circuit or device has a response time, and that ultimately the response time will limit the rate at which digital or analog signals can be processed.

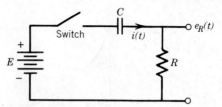

FIGURE 7-12

A Circuit with a Capacitor and Resistor

The arrangement of the capacitor and resistance, which is different from that of Figure 7-10, emphasizes the decay in time of the voltage across the resistor, as shown in Figure 7-13.

Figure 7-12 results from rearranging the circuit of Figure 7-10 so that the output voltage is taken across the resistor. The same equation, Equation 7-14, applies; the solutions of the equation are the same, but now the output voltage is

$$e_R(t) = Ri(t) = E \exp(-t/RC) \qquad (7\text{-}20)$$

The term $e_R(t)$ is not plotted because it has the same form as $i(t)$ in Figure 7-11. The ordinate must be multiplied by R to give the proper scale in accordance with Ohm's law.

It is instructive to examine the voltage across C in Figure 7-10 and the voltage across R in Figure 7-12. The input is the same for both arrangements: the closing of a switch that connects a constant voltage E. The input voltage is a step function in which the voltage rises from zero to E at the instant at which the switch is closed and then remains at E. The step function of voltage and the output voltages $e_R(t)$ and $e_C(t)$ are plotted in Figure 7-13, where it is observed that their sum is E at all times.

The circuit of Figure 7-10 has an output voltage $e_C(t)$, which grows after the switch is closed as the charge accumulates in time on the capacitor. For this reason the simple circuit is called an *integrating circuit*. Later we see how a true integrating circuit can be made. The circuit of Figure 7-12 differs from the circuit of Figure 7-10 in that the output voltage $e_R(t)$ is taken across the resistor. This circuit is called a *differentiating circuit*.

An important property of capacitors is revealed by the analysis. *The voltage across a capacitance cannot change instantaneously.* Because the voltage across a capacitor is proportional to the charge, $v(t) = q(t)/C$, the charge must change instantaneously for the voltage to change instantaneously. This situation would require an infinitely large current for an infinitesimally short time, a physical impossibility.

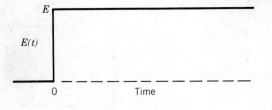

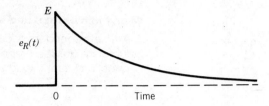

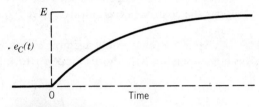

FIGURE 7-13
Voltage Functions of Time

The sum of the voltages $e_R(t)$ and $e_c(t)$ equals E, the step voltage. The time constant, $\tau = RC$, establishes the rate of change of the voltages.

INDUCTANCE

Along with resistance and capacitance, inductance is the third element of every circuit. We examine inductance as a discrete element to display its properties.

The most effective means of producing a discrete inductance is to make a tight coil of wire. Inductance is denoted by L and measured in henrys.

When the current in an inductance changes, it creates a voltage across the inductance to oppose the change. The voltage associated with an inductance is given by

$$e_L(t) = -L\frac{di}{dt} \tag{7-21}$$

where e_L is expressed in volts, L in henrys, i in amperes, and t in seconds; di/dt, the rate of change of current, is expressed in amperes per second. The negative sign

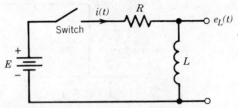

FIGURE 7-14

Circuit with Resistor and Inductance

The time dependence arises because the inductance has the property of opposing a change in current through it.

denotes that the voltage $e_L(t)$ induced across the inductance has the polarity to oppose the external electromotive force (voltage), making the current increase or diminish.

As before, we shall study a simple circuit with a resistance, an inductance, and a switch. Applying Kirchhoff's loop law to the circuit of Figure 7-14, we have

$$E + E_R + E_L = 0$$

$$E - iR - L\frac{di}{dt} = 0 \quad (7\text{-}22)$$

This differential equation is solved to give

$$i(t) = \frac{E}{R}\left[1 - \exp\left(-\frac{R}{L}t\right)\right] \quad (7\text{-}23)$$

and

$$e_L(t) = E \exp\left(-\frac{R}{L}t\right) \quad (7\text{-}24)$$

when the initial condition is that the switch is closed at $t = 0$. The solutions may be verified by using them in Equation 7-22.

Figure 7-15 shows that the current grows with time in accordance with Equation 7-23. Finally the current reaches the value it would have if the circuit had only a resistor. The time constant

$$\tau = \frac{L}{R} \quad (7\text{-}25)$$

measures the rate of change in this circuit with inductance.

The voltage across the inductance decays exponentially with the time constant $\tau = L/R$ seconds. Conversely, if the switch in Figure 7-14 is opened when there is current through the inductance, the current cannot drop instantaneously to zero. In accordance with Equation 7-21 the inductance will generate a voltage to sustain

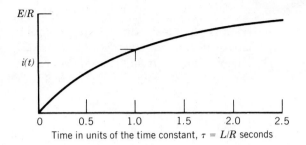

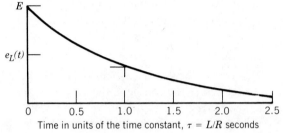

FIGURE 7-15

Current and Voltage of Inductance Versus Time

The rate of change of currents and voltages is governed by the time constant, $\tau = L/R$ seconds.

the current. Typically, a spark will develop across the switch as it opens. This phenomenon makes it difficult to switch large currents in circuits containing inductances.

We can now make an important statement about the nature of inductances, which is analogous to the statement about capacitances made above: *the current through an inductance cannot change instantaneously.* A step change in current would produce an infinite voltage. In practice, large voltage spikes are produced by using this property of an inductance in which the current is interrupted, as in the ignition systems of automobiles. In Chapter 14 we see how this property of inductances is used in making efficient power supplies for electronic systems.

CURRENT-VOLTAGE RELATIONS IN PASSIVE ELEMENTS

Let us now summarize the time relation of current and voltage in the three passive electrical circuit elements: resistance, capacitance, and inductance.

We must distinguish between passive and active electrical circuit elements. Active elements are devices connected to a power source that can make their response to a voltage or current input quite different from that of passive elements.

For example, an electronic amplifier and a logic gate are active elements that can provide an output many times more powerful than the voltage or current input and perhaps of different form because the device uses the small signal input to guide the conversion of energy from the power source. In contrast, passive elements respond in proportion to the voltages and currents applied to them.

Resistance

In a resistor the current and voltage vary simultaneously, as suggested by Ohm's law, $e(t) = Ri(t)$, in which the current and the voltage are proportional to each other. The proportionality constant is the resistance R. Because the time dependence of current and voltage is the same, we say that they are *in phase*.

Capacitance

The voltage across a capacitor *follows* the current because the voltage cannot change unless the charge in the capacitor has changed. That is, the charge, related to the current by Equation 7-15, must first accumulate in the capacitor for its voltage to increase; charge must be removed from the capacitor for its voltage to decrease. "Follow" means that the capacitor voltage *lags* in time behind the current to the capacitor. That is, they are *out of phase*.

Inductance

The voltage across an inductor *leads* the current through it because the inductance voltage arises as the current changes. (Equation 7-21 suggests this point.) The voltage and current are *out of phase* in inductances.

These time relations will be important in the study of alternating current and voltage circuits, which contain the three passive elements. In any circuit the resultant phase—the time relation of the voltage to the current—will depend upon the particular circuit configuration of resistances, capacitances, and inductances, and upon the frequency of the alternating currents and voltages.

ALTERNATING CURRENTS

Circuits in which the voltages and currents vary with time are generally referred to as *ac* (alternating current) *circuits*. When the variations are periodic, rather than singular time events like the switch closing used above, the circuits may be studied by considering the voltages and currents to vary sinusoidally with time. That is, the variable is represented as a fixed magnitude times a sine wave. For example, an ac voltage is given by

$$e(t) = E \sin \omega t = E \sin 2\pi f t \qquad (7\text{-}26)$$

where E is the peak voltage amplitude, ω is the angular frequency in radians per second, f is the frequency in hertz, and t is time in seconds.

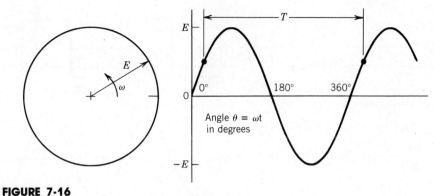

FIGURE 7-16
A Sine Wave Generated by a Rotating Vector
The time projection of a rotating vector generates a sine wave. The period T is the interval between identical points on the sine wave and the interval of one rotation of the vector.

Figure 7-16 shows this voltage sine wave as a function of the angle $\theta = \omega t$. The units of θ may be either radians or degrees. Figure 7-16 also shows that the sine wave is generated by the projection of the end of a radial line, whose length represents the peak voltage amplitude, as it rotates at an angular frequency ω. In the repeat interval or period of T, seconds, the sine wave passes through a complete cycle from one point (amplitude and slope) on the wave to an identical point 2π radians or 360° later. This interval corresponds to one complete turn of the rotating line.

With the relations $\omega = 2\pi f$ and $T = 1/f$ we find, for example, that with a repeat interval $T = 1/100$ sec that the frequency is 100 Hz (cycles per second) and the angular velocity $\omega = 200\pi = 628.3$ radians/sec, or 36,000°/sec. The signals of interest in integrated-circuit applications range in frequency from zero (dc, direct current) to gigahertz (10^9 Hz) and higher.

In some applications the electrical signals are simple sine waves, but in digital circuits, for example, the clock pulses form a square wave of voltage amplitude versus time. Periodic wave can be decomposed by Fourier analysis into a series of sine waves, called the Fourier series, which, when added together, form the original wave. In the breakdown of a complicated periodic wave into its Fourier components it is found that the frequency of each component is a multiple of a fundamental or lowest frequency of the wave being studied. Figure 7-17 illustrates how a complicated wave, in this case a square wave, begins to be formed by only the first two of its sine wave (Fourier) components.

A very large number of components, in theory an infinite number, must be used to reconstruct exactly the square wave form. Both the amplitude and the time relation, or phase, of each sine wave component are specified by the Fourier analysis. The wave at the bottom of Figure 7-17 shows the effect of a shift from the

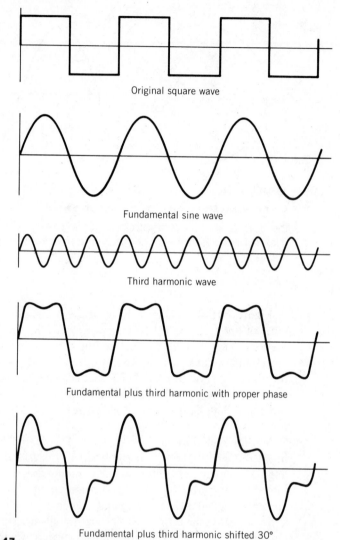

FIGURE 7-17

Square Wave and First Fourier Components

A square wave can be constructed by adding sine wave components with the proper amplitude and phase. The fundamental and the third harmonic begin to approximate the original wave, but a shift in phase destroys the approximation.

phase specified to yield a square wave. A sawtooth wave rather than a square wave is formed when the phase is changed.

Because we do not need Fourier analysis in our work, it will not be discussed here at any length. We refer to it because it allows us to analyze the response of circuits to all periodic signals by examining only the properties of electronic circuits as revealed by their response to sine waves. To judge the response of an electronic device to the complicated input wave, however, we must consider the full range of frequency components in a complicated wave. In other words, the use of sine waves in studying electrical circuits is sufficient to reveal the circuit response to any kind of periodic wave.

PASSIVE ELEMENTS AND SINE WAVE SIGNALS

Let us now examine the current-voltage relationships in the three passive elements when the voltage across them takes the form of the sine wave given by $e(t) = E \sin \omega t$ (Equation 7-26).

Resistance

If, in Figure 7-1, the alternating voltage source $e(t)$ were substituted for the battery V, we would find that in a resistance the current-voltage relation is given by Ohm's law, $e(t) = Ri(t)$, and that the current $i(t)$ has the same time dependence as the voltage source $e(t)$. We find that

$$i(t) = \frac{e(t)}{R} = \frac{E}{R} \sin \omega t = I \sin \omega t \tag{7-27}$$

The current amplitude is $I = E/R$ ampere. In this case the current and voltage are said to be in phase with each other. The current and voltage in Figure 7-18, which illustrates this relation, are seen to vary together.

Capacitance

The loop equation for the circuit of an alternating voltage source and a capacitor, as shown in Figure 7-19, is

$$e(t) - \frac{q(t)}{C} = 0 \tag{7-28}$$

The voltage-charge relation in a capacitance is Equation 7-13.

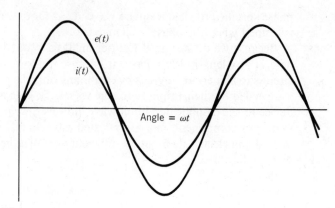

FIGURE 7-18

The Current-Voltage Relation in a Resistor

The alternating current and the voltage in a resistor are related by Ohm's law, $e(t) = Ri(t)$; they are in phase.

We find the current by solving Equation 7-28 for $q(t)$ and use it along with the definition of current,

$$i(t) = \frac{d\,q(t)}{dt}$$

to obtain

$$i(t) = \frac{d\,Ce(t)}{dt} = \frac{d\,CE\sin\omega t}{dt} = \omega CE\cos\omega t$$

$$= \omega CE\sin(\omega t + 90°) \tag{7-29}$$

We have used the trigonometric identity $\cos\omega t = \sin(\omega t + 90°)$ in Equation 7-29

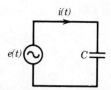

FIGURE 7-19

A Circuit with an Alternating Voltage Source and Capacitor

The circuit allows the phase relation of voltage and current in a capacitor to be evaluated.

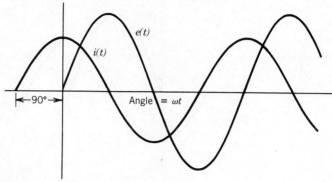

FIGURE 7-20

The Current-Voltage Relation in a Capacitor

The current $i(t)$ leads the voltage $e(t)$ by 90°. The phase relation confirms that the voltage in a capacitor arises from the accumulation of charge.

to reveal that the current *leads* the voltage by 90°. This is the expected result, because the charge must accumulate in the capacitor for the voltage to appear across it. Figure 7-20 shows the current and voltage with the *phase angle* of 90° between them.

Inductance

Figure 7-21 shows a simple circuit with an inductance and an ac driving voltage. The loop equation

$$e(t) - L\frac{di(t)}{dt} = 0 \qquad (7\text{-}30)$$

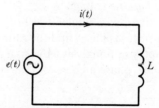

FIGURE 7-21

A Circuit with an Alternating Voltage Source and Inductance

The circuit allows the phase relation of voltage and current in an inductance to be evaluated.

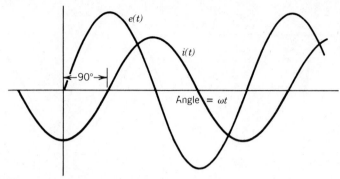

FIGURE 7-22

The Current-Voltage Relation in an Inductance

The current $i(t)$ lags the voltage $e(t)$ by 90°. The phase relation confirms that the rate of change of the current is the source of the voltage in an inductance.

may be solved to find $i(t)$. By rewriting Equation 7-30 and integrating with respect to time, we have

$$i(t) = \frac{1}{L} \int e(t)\, dt$$

$$= \frac{E}{L} \int \sin \omega t\, dt$$

$$= -\frac{E}{\omega L} \cos \omega t$$

$$= +\frac{E}{\omega L} \sin (\omega t - 90°) \qquad (7.31)$$

We have used the trigonometric identity $-\cos \omega t = \sin (\omega t - 90°)$ to find that the current *lags* the voltage by 90°. The relation between $e(t)$ and $i(t)$ is shown in Figure 7-22 with a phase angle of $-90°$.

Our next task is to find the ratios $e(t)/i(t)$ for each of the three passive elements. To simplify our analysis let us first point out a mathematical shorthand to account for the 90° phase angles.

ROTATIONAL OPERATOR j

In the theory of complex variables, an important relation known as the Euler relation states that

$$A \exp(j\omega t) = A \cos \omega t + jA \sin \omega t \qquad (7\text{-}32)$$

where $j = \sqrt{-1}$. We multiply Euler's relation by j to obtain

$$\begin{aligned} jA \exp(j\omega t) &= jA \cos \omega t - A \sin \omega t \\ &= A[j \sin(\omega t + 90°) + \cos(\omega t + 90°)] \\ &= A \exp j(\omega t + 90°) \\ &= A \exp(j\omega t) \exp(j\, 90°) \end{aligned} \quad (7\text{-}33)$$

In Equation 7-33 we first substituted the trigonometric identities, $\cos \omega t = \sin(\omega t + 90°)$ and $\sin \omega t = -\cos(\omega t + 90°)$ and then compared the result with Equation 7-32. Finally we use the relation $\exp(A + B) = \exp(A) \exp(B)$ to separate the arguments ωt and $90°$.

We find that multiplying by the *operator j* is equivalent to a rotation of $+90° = \pi/2$ radians.

In studying electrical circuits, it has become customary to use $j = \sqrt{-1}$ rather than $i = \sqrt{-1}$ because using j reduces the possibility of confusing the rotational operator with the current $i(t)$.

If a sinusoidally varying quantity, such as voltage or current, is multiplied by the operator j, the angle variable is increased by $90°$. Each repeated use of j provides an additional *rotation* of $90°$. For example,

$$\begin{aligned} j \cdot j = j^2 = -1, & \quad \text{a rotation of } 180° \\ j \cdot j \cdot j = j^3 = -j, & \quad \text{a rotation of } 270° \\ j \cdot j \cdot j \cdot j = j^4 = +1, & \quad \text{a complete cycle} \end{aligned}$$

A complete cycle of $360°$ is equivalent to $0°$ in relative phase angle.

REACTANCES

The ratio $e(t)/i(t) = R$ for a resistance. This result, given in Equation 7-27, is Ohm's law. The electrical energy in a resistor is converted into heat energy, and this process is frequently referred to as Ohm heating. If the current and voltage are out of phase by $90°$, however, the electrical energy is not converted into heat energy. Because this is the case for a capacitor and an inductor, we distinguish the ratios $e(t)/i(t)$ for capacitance and inductance from the ratio for resistance by calling them reactances. We now find the reactances for capacitors and inductors by examining their $e(t)/i(t)$ ratios.

CAPACITIVE REACTANCE

We use the rotational operator j to rewrite Equation 7-29 $i(t) = \omega CE \sin(\omega t + 90°)$, which becomes $i(t) = j\omega CE \sin \omega t$. We use the latter to form the ratio $e(t)/i(t)$.

$$\frac{e(t)}{i(t)} = \frac{E \sin \omega t}{j\omega CE \sin \omega t} = \frac{1}{j\omega C} = -j\frac{1}{\omega C} = -j\frac{1}{2\pi f C}$$

$$= -jX_C \quad (7\text{-}34)$$

X_C is called the *capacitive reactance*. The unit of X_C is ohm, which is analogous to resistance, but electrical energy is not converted to heat by the reactance. Capacitive reactance also differs from resistance in another important way: X_C is *frequency-dependent*. In fact, X_C is inversely proportional to frequency. The rotational operator $-j$ is written separately, but it must be used with X_C to convey the 90° phase angle by which the current leads the voltage in a capacitor.

The frequency dependence of $X_C = 1/2\pi f C$ causes reactance to become very large for low frequencies; for zero hertz (dc) the reactance becomes infinite. Because the magnitude of the current is the voltage divided by the reactance at dc (constant voltage), the current is zero and the capacitor is equivalent to an open or disconnected circuit. Capacitors are used to block dc signals.

Conversely, at high frequencies the capacitive reactance X_C becomes small, and for very high frequencies a capacitor may be thought of as a "short circuit." The frequency dependence of X_C suggests that capacitors may be used to pass time-varying signals from stage to stage in circuits while blocking direct currents and direct voltages between stages. Alternatively, capacitors may be used to lessen the ac components in signals by bypassing them to ground. These points will be illustrated in the later chapters.

INDUCTIVE REACTANCE

We use Equations 7-26 and 7-31 to obtain the ratio $e(t)/i(t)$. With the j operator, Equation 7-31 becomes $i(t) = -j(E/\omega L) \sin \omega t$, so that

$$\frac{E(t)}{i(t)} = \frac{E \sin \omega t}{-j(E/\omega L) \sin \omega t} = \frac{1}{-j(1/\omega L)} = j\omega L = j2\pi f L = +jX_L \quad (7\text{-}35)$$

X_L is called the *inductive reactance*, expressed in ohms. This reactance is also analogous to resistance, but it does not convert electrical energy to heat, and *it is frequency-dependent*. The rotational operator j is written separately, and it must be used with X_L to account for the 90° phase angle by which the current lags the voltage across the inductance.

At low frequencies the inductive reactance X_L is small; it vanishes for zero hertz or dc signals. At high frequencies $\omega L = 2\pi f L$ becomes large and impedes the ac signals in the circuit. At very high frequencies it virtually prohibits the flow of electrical energy. This property is used to isolate portions of circuits from high-frequency signals, the inverse of the use of capacitors to isolate portions of circuits from direct currents.

AC RESISTANCE-CAPACITANCE CIRCUIT

Having found X_C and X_L we have a great advantage in analyzing ac circuits. We can now find the relation between voltage and current by applying Kirchhoff's laws

AC Resistance-Capacitance Circuit

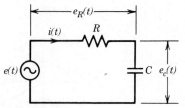

FIGURE 7-23

The Resistor-Capacitor with an Alternating Voltage Source

The relation between $e(t)$ and $i(t)$ found from the circuit shows the contribution of each element in impeding the current.

for sinusoidal voltage sources. The frequency dependence is contained intrinsically in the reactances as we use them in the Kirchhoff loop equations written for the circuit. It is tacitly assumed that the loop and node equations are written at an instant in time, as if we are taking an instantaneous photograph of the currents and voltages to be found. The solutions of the Kirchhoff's law equations contain the amplitudes and the phase relations of the sinusoidal currents and voltages.

To illustrate how resistance and capacitive reactance combine, we seek the voltage-current relation in the simple circuit of Figure 7-23, in which the voltage source is time-varying or an alternating-current voltage $e(t) = E \sin \omega t$. We write the Kirchhoff loop equation by using the resistance and the capacitive reactance, as the circuit is traversed clockwise.

$$e(t) + e_R(t) + e_C(t) = 0$$
$$e(t) - i(t)R - i(t)(-jX_C) = 0$$
$$e(t) - i(t)(R - jX_C) = 0$$
$$e(t) - i(t)Z\underline{/\theta} = 0 \qquad (7\text{-}36)$$

In this case there is no need to write a node equation for the current, but it is necessary to assume a direction for the current. The conventions apply for ac as well as dc circuits, but the current directions are not confirmed by positive values; rather, they are given by the phase angles in the solution.

We acknowledge that the resistance and capacitive reactance correspond to a real and an imaginary value, respectively, by writing them as

$$Z\underline{/\theta} = R - jX_C \qquad (7\text{-}37)$$

The vectorlike sum of resistance and reactance is called the *impedance*. Because it is analogous to a vector, $Z\underline{/\theta}$ has both a magnitude and an angle, so that it must be written that way or in its component form. Figure 7-24 shows R on the real axis, X_C on the imaginary axis, and the impedance $Z\underline{/\theta}$ at the angle $\underline{/\theta}$.

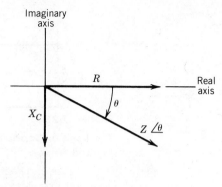

FIGURE 7-24

The Impedance $Z\underline{/\theta} = R - jX_c$ in a Resistor-Capacitor Cicuit

The vectorlike addition of resistance and capacitive reactance yields the circuit impedance $Z\underline{/\theta}$; j is the rotation operator.

The magnitude of $Z\underline{/\theta}$ is the length of the hypotenuse of the right triangle formed by the resistance and the reactance. Using complex multiplication,

$$Z^2 = (R - jX_C)(R - jX_C)^* = (R - jX_C)(R + jX_C)$$
$$= (R^2 + X_C^2)$$

we find that the magnitude of Z is

$$Z = (R^2 + X_C^2)^{1/2} \ \Omega \tag{7-38}$$

In finding Z^2 we used the fact that the square of the magnitude of a complex quantity is obtained by multiplying the quantity, in this case $(R - jX_C)$, by its complex conjugate $(R - jX_C)^* = (R + jX_C)$. The complex conjugate is obtained by changing the sign of the imaginary term. The magnitude Z is obtained by taking the square root of the real, positive Z^2. The angle is given by

$$\theta = \arctan \frac{-X_C}{R} = \arctan -\frac{1}{2\pi f CR} \text{ radians or degrees} \tag{7-39}$$

As an example of the current-voltage relation in a specific case, let $f = 3$ kHz., $C = 0.1 \ \mu\text{F}$ and $R = 1000 \ \Omega$. Then

$$Z\underline{/\theta} = \left[10^3 - j\left(\frac{1}{2\pi 3 \times 10^3 \times 10^{-7}}\right) \right]$$
$$= (10^3 - j531)$$
$$= 1132\underline{/-28°} \ \Omega$$

We find $\underline{/\theta} = \tan^{-1}(-531/1000) = 332° = -28°$.

AC Resistance-Capacitance Circuit

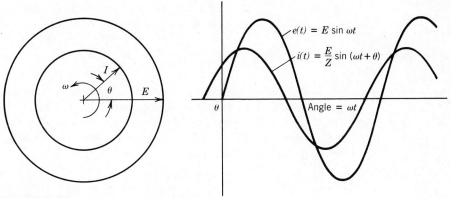

FIGURE 7-25

Voltage and Current in an *RC* Circuit

The current leads the voltage by the angle θ, established by the relative values of resistance and capacitive reactance.

The combined effect of resistance and capacitance is to give a phase angle between 0 and $-90°$ for the impedance. The angle and magnitude of the impedance are frequency-dependent because of the nature of X_C.

The current I is found from Equation 7-36 by assuming that at the instant of calculation the angle of E is $0°$; that is, E is directed instantaneously along the real axis, as shown in Figure 7-25.

$$I\underline{/\theta_i} = \frac{E\underline{/0°}}{Z\underline{/\theta°}}$$

The angle of the current relative to the angle of the voltage, $\underline{/0°}$, is found by "dividing" by the angle $\underline{/\theta}$.

To see how an angle is used as a "divisor", we recall Euler's relation, Equation 7-32. $Ae^{j\theta} = A\cos\theta + jA\sin\theta = A\underline{/\theta}$. We can confirm that this may be written simply as $A\underline{/\theta}$ by plotting $A\cos\theta$ on the real axis and $A\sin\theta$ on the imaginary axis and noting that the quantity A is formed by vector addition of the components at the angle θ.

For $A\underline{/\theta_1}$ and $A\underline{/\theta_2}$ we write $A\exp(j\theta_1)$ and $A\exp(j\theta_2)$ and form the ratio of these complex quantities.

$$\frac{A\underline{/\theta_1}}{A\underline{/\theta_2}} = \frac{A\exp(j\theta_1)}{A\exp(j\theta_2)}$$

$$= \exp(j\theta_1)\exp(-j\theta_2)$$

$$= \exp j(\theta_1 - \theta_2)$$

$$= 1\underline{/\theta_1 - \theta_2} \qquad (7\text{-}40)$$

The "ratio" of two angles is the difference between the angles, as shown in Equation 7-40. The angle of $I\underline{/\theta_i}$ is found by $\underline{/0°} - \theta_z = \underline{/-\theta_z}$; that is,

$$I\underline{/\theta_i} = \frac{E}{Z}\underline{/-\theta_z}$$

The sinusoidal current is $i(t) = I \sin(\omega t - \theta_z)$. As you may recall, θ_z is negative (see the numerical example following Equation 7-39), so that $\underline{/-\theta_z}$ is a positive angle. This is plotted in Figure 7-25. Both $e(t)$ and $i(t)$ rotate with the angular velocity ω, as shown in Figure 7-25 with $i(t)$ leading $e(t)$. We extend our numerical example by letting $E = 10$ V to find that

$$I\underline{/\theta} = \frac{E\underline{/0°}}{Z\underline{/-28°}} = \frac{10}{1132}\underline{/+28°} = 8.8 \times 10^{-3}\underline{/28°}\text{ A}$$

$$= (7.8 + j4.15) \times 10^{-3}\text{ A}$$

In our example $i(t) = 8.8 \times 10^{-3} \sin(\omega t + 28°)$ A, which confirms that the current leads the voltage in a capacitor circuit. The voltage across the capacitor is found by

$$E_C = (I\underline{/\theta})(-jX_C)$$
$$= (8.8 \times 10^{-3}\underline{/28°})\,531\underline{/-90°}$$
$$= 4.69\underline{/-62°}\text{ V}$$

The "product" of two angles is the sum of the angles, for example, $A\underline{/\theta_1} \times B\underline{/\theta_2} = A \times B\underline{/\theta_1 + \theta_2}$. The reader should verify this with Euler's relation.

Similarly,

$$E_R = (I\underline{/\theta})R\underline{/0°}$$
$$= (8.83 \times 10^{-3}\underline{/+28°}) \times 10^3\underline{/0°}$$
$$= 8.83\underline{/28°}\text{ V}$$

Figure 7-26 shows that these components add graphically to give the magnitude of the source voltage E. To add the voltage algebraically, we require their components,

$$E_C = 4.69\underline{/-62°} = 2.20 - j4.14$$
$$E_R = 8.83\underline{/28°} = 7.80 + j4.14$$

Adding the real components and the imaginary components separately, we find

$$E_C + E_R = 10.00 + j0\text{ V}$$

The magnitude of the sum $E_C + E_R$ equals the magnitude of source voltage E, and Equation 7-36 verifies that their sum is zero, as required by Kirchhoff's loop law.

We observe that to multiply two terms, the product of their magnitudes and the sum of the angles give the product and its angle. To add or subtract terms, their components are added or subtracted separately. Accordingly we find that the quotient of two terms is the quotient of their magnitudes and the difference of their angles. In solving ac circuits, conversions using $A\underline{/\theta} = A \cos\theta + jA \sin\theta$ (Equation

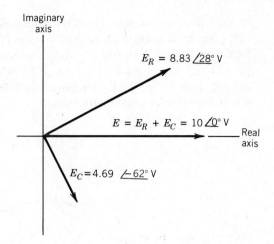

FIGURE 7-26

Voltage Summation in RC Circuit

The rotating voltage vectors in the RC circuit always add to give the magnitude of the rotating source voltage.

7-32) from polar components (magnitude and angle) to rectangular (real and imaginary components) and vice versa, using Equations 7-38 and 7-39, are frequent operations. This fact is recognized by makers of hand-held calculators, which facilitate the transformations by simple key strokes.

FREQUENCY DEPENDENCE OF RC CIRCUIT

The frequency dependence of the resistance-capacitance circuit of Figure 7-23 is revealed by calculating the ratio of the voltage across the capacitor, $e_c(t)$, to the source voltage, $e(t)$. The phase angle between the two voltages is also found. Because $e_c(t)$ and $e(t)$ are both sine waves, the ratio of their magnitudes E_C and E are studied.

$E_C(f)$ may be found by multiplying the current and the capacitive reactance.

$$E_C(f) = I(f)(-jX_c) = \frac{E}{R - jX_c}(-jX_c)$$

The relative amplitude $A(f) = e_c(t)/e(t) = E_c(f)/E$ is

$$A(f) = \frac{-jX_c}{R - jX_c} = \frac{X_c^2 - jRX_c}{R^2 + X_c^2} \qquad (7\text{-}41)$$

and the phase angle is

$$\theta(f) = \tan^{-1}\left(-\frac{RX_c}{X_c^2}\right) = \tan^{-1}(-2\pi fCR) \qquad (7\text{-}42)$$

In Equation 7-41 we multiplied the numerator and the denominator by $R + jX_c$ (the complex conjugate of the denominator) to place the fraction in a more useful form; that is, a fraction with a real denominator. A complex quantity like $A(f)$ can be written $A(f) = \mathcal{R}e\, A(f) + j\,\mathcal{I}m\, A(f)$, where $\mathcal{R}e\, A(f)$ is the real part of the quantity and $\mathcal{I}m\, A(f)$ is the imaginary part. Because they have a real denominator, the two components are easily written as

$$\mathcal{R}e\, A(f) = \frac{X_c^2}{R^2 + X_c^2}$$

and

$$\mathcal{I}m\, A(f) = \frac{-RX_c}{R^2 + X_c^2}$$

the phase angle of $A(f)$ is

$$\theta(f) = \tan^{-1}\left(\frac{\mathcal{I}m\, A(f)}{\mathcal{R}e\, A(f)}\right)$$

as shown in Equation 7-42.

Again we use $R = 1000\ \Omega$ and $C = 0.1\ \mu F$ and calculate $A(f)$ and $\theta(f)$ for various frequencies, as shown in Figure 7-27. At low frequencies the relative

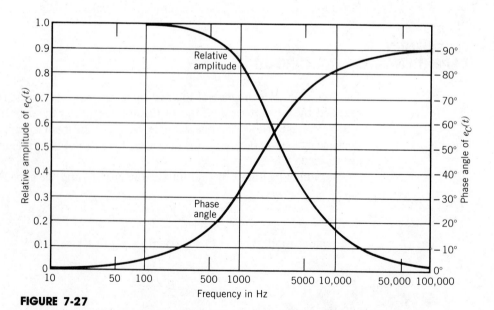

FIGURE 7-27

The Response of a Low-Pass Filter

The relative amplitude of $e_c(t)$ (Figure 7-23) is the ratio of $e_c(t)$ at any frequency to its value at 0 Hz. As the frequency increases, the relative amplitude decreases.

amplitude is near unity and the lagging phase angle is small, but at higher frequencies the relative amplitude falls off and the lagging phase angle increases. This behavior causes the resistor-capacitor circuit of Figure 7-23 to be called a *low-pass filter*, because low-frequency voltages are relatively unattenuated while high-frequency voltages are strongly attenuated.

AC RESISTANCE-INDUCTANCE CIRCUIT

Figure 7-28 shows an inductance and resistor in series driven by an alternating voltage source. The Kirchhoff's loop equation is

$$e(t) - i(t)(R + jX_L) = 0$$

and the impedance in this case is

$$Z\underline{/\theta} = R + jX_L = R + j2\pi fL \qquad (7\text{-}43)$$

The reader can verify that for $R = 10^3 \ \Omega$, $L = 10^{-3}$ H, and $f = 3 \times 10^3$ Hz:

$$Z\underline{/\theta} = 10^3 + j18.85$$
$$= 1000.18\underline{/1.08°} \ \Omega$$

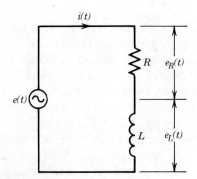

FIGURE 7-28

An Inductance-Resistor Circuit with an Alternating Voltage Source

The resistor and the inductance impede the current $i(t)$ and its phase relative to the driving voltage $e(t)$ in a frequency-dependent manner.

The relative amplitude of the voltage across the inductance is

$$A(f) = \left[\frac{e_L(t)}{e(t)}\right] = \frac{E_L(f)}{E}$$

$$= \frac{X_L^2 + jRX_L}{R^2 + X_L^2} \tag{7-44}$$

The frequency dependent angle between E_L and the supply voltage E is

$$\theta(f) = \tan^{-1}\frac{RX_L}{X_L^2} = \tan^{-1}\frac{R}{2\pi fL} \tag{7-45}$$

The voltage across the inductance $E_L(f)$ is very small at low frequencies. This is indicated by $A(f)$ because $X_L = 2\pi fL$ occurs in both terms in the numerator. For high frequencies the ratio $A(f)$ approaches unity. Because low-frequency voltages are attenuated and high-frequency voltages occur across the inductance, this circuit is called a *high-pass filter*.

The reader should note that at low frequencies E_R, the voltage across the resistor in Figure 7-28, is nearly E, the supply voltage. The circuit is a frequency-dependent voltage divider. For zero hertz, dc, the inductance makes no contribution to the circuit; that is, the inductance passes direct currents unattenuated. For very high frequencies, X_L is very large and the very high-frequency currents are blocked. An inductance in a circuit can isolate the rest of the circuit from very high-frequency currents. This function complements that of a capacitor, which passes high-frequency currents and blocks direct currents.

SIMPLIFICATION OF CIRCUITS WITH IMPEDANCE

Many complicated circuits can be reduced in the manner used earlier for combining parallel and series resistances. Not all circuits can be handled in that manner, however, but all circuits can be solved with Kirchhoff's laws and other aids, such as Thevenin's theorem. The impedance of an ac circuit is found by adding the resistances and reactances as they are related through the circuit configuration while maintaining the complex properties through the use of the rotational operator j. For example, we find the impedance of the circuit of Figure 7-29 by successive reductions of impedances.

In the upper elements of the circuit we expect that the resistance R_1 and the reactance $X_1 = 1/2\pi fC_1$ should combine, as do resistances in parallel. To verify that this is the case, we apply Kirchhoff's laws to the portion across which the voltage $e_1(t)$ exists.

$$i(t) - i_R(t) - i_C(t) = 0$$

$$e_1(t) - i_R(t)R_1 = 0$$

$$e_1(t) - i_C(t)(-jX_1) = 0 \tag{7-46}$$

Simplification of Circuits with Impedance

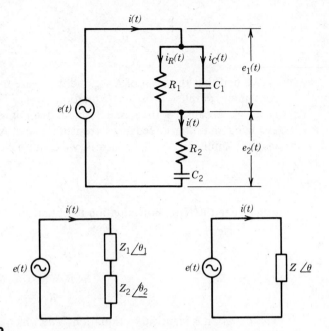

FIGURE 7-29
The Circuit for Simplification by Successive Reduction of Impedance
The technique of reducing parallel and series resistor combinations is applied to impedance reduction. The impedances and phase angles are accounted for by using the rotation operator *j* with the reactances.

These equations may be combined to give

$$i(t) = \frac{e_1(t)}{Z_1 \underline{/\theta}} = \frac{e_1(t)}{R_1} + \frac{e_1(t)}{-jX_1}$$

which yields

$$\frac{1}{Z_1 \underline{/\theta}} = \frac{1}{R_1} + \frac{1}{-jX_1} \tag{7-47}$$

Additional algebra gives

$$Z_1 \underline{/\theta} = \frac{-jX_1 R_1}{R_1 - jX_1}$$

To express this equation in better form, we multiply the top and bottom by $R_1 + jX_1$, the complex conjugate of the denominator, to obtain

$$Z_1 \underline{/\theta} = \frac{R_1 X_1^2 - jX_1 R_1^2}{R_1^2 + X_1^2}$$

$$= R_{1,eq} - jX_{1,eq} \tag{7-48}$$

where

$$R_{1,eq} = \frac{R_1 X_1^2}{R_1^2 + X_1^2} \quad \text{and} \quad X_{1,eq} = \frac{X_1 R_1^2}{R_1^2 + X_1^2}$$

We find that the form of $R_{1,eq}$ and $X_{1,eq}$ are the same as we expect if we were combining parallel resistors (Equation 7-11). Two terms exist for this form, however, because both a real term and an imaginary term are involved. By assuming that the impedances (resistance and reactance) would combine in a manner similar to parallel resistors, we could have written

$$\frac{1}{Z_1 \underline{/\theta}} = \frac{1}{R_1} + \frac{1}{-jX_1}$$

rather than the Kirchhoff equation to obtain

$$Z_1 \underline{/\theta} = \frac{R_1(-jX_1)}{R_1 - jX_1}$$

$$= \frac{R_1 X_1^2 - jX_1 R_1^2}{R_1^2 + X_1^2}$$

This equation is identical to Equation 7-48, which confirms that some parallel and series circuits involving resistance and reactance can be reduced in the same way that some resistor circuits can be simplified.

The resistance R_2 and the reactance X_2 add in series to give

$$Z_2 \underline{/\theta} = R_2 - jX_2 \tag{7-49}$$

The total impedance for the circuit of Figure 7-29 is

$$Z \underline{/\theta} = Z_1 \underline{/\theta} + Z_2 \underline{/\theta} = (R_{1,eq} + R_2) - j(X_{1,eq} + X_2) \tag{7-50}$$

In adding complex quantities, we obtain the sum by adding the real parts and the imaginary parts separately. The magnitude of $Z \underline{/\theta}$ and its angle $\underline{/\theta}$ are found with Equations 7-38 and 7-39.

SERIES RESONANCE CIRCUIT

Now let us examine two arrangements of all three passive elements in a common circuit. The first is formed with resistance, capacitance, and inductance in series, as shown in Figure 7-30. The loop equations are written using the rotational operator j with the reactances. We have

$$e(t) + e_R(t) + e_L(t) + e_C(t) = 0$$

$$e(t) - i(t)(R + jX_L - jX_C) = 0$$

$$e(t) - i(t)Z\underline{/\theta} = 0 \tag{7-51}$$

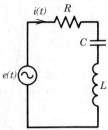

FIGURE 7-30

A Series Resonance Circuit

The series combination of resistance, capacitance, and inductance has a minimum impedance at the resonance frequency, $f_r = (1/2\pi)(LC)^{-1/2}$.

The complex impedance $Z\underline{/\theta}$ is rewritten to show explicitly the frequency dependence.

$$Z\underline{/\theta} = R + j(X_L - X_C)$$
$$= R + j\left(2\pi fL - \frac{1}{2\pi fC}\right) \qquad (7\text{-}52)$$

The resistance may or may not be a discrete element, but some resistance is always present in the wires connecting the elements and in the wire with which the inductance is wound and is included in R.

An examination of Equation 7-52 reveals that the two reactances vary according to frequency in a reciprocal manner. For given values of L and C, the inductive reactance $X_L = 2\pi fL$ will grow as the frequency is increased from a low value while the capacitive reactance $X_C = 1/2\pi fC$ will decrease as the frequency is increased. We may conclude that at a particular frequency the reactances will reach equal magnitudes, and at that frequency the reactance part of the impedance will become zero. The impedance is wholly resistive at this *resonance frequency*, f_r.

From Equation 7-52 the resonance frequency, at which the inductive reactance X_L and the capacitive reactance X_C have equal magnitudes, is

$$f_r = \frac{1}{2\pi}\sqrt{\frac{1}{LC}} \qquad \text{Hz (sec}^{-1}) \qquad (7\text{-}53)$$

At all other frequencies either the capacitive reactance or the inductive reactance is larger, and the impedance is no longer purely resistive. At low frequencies the capacitive reactance dominates so that the current leads the voltage; at frequencies above the resonance frequency, where the inductive reactance is larger than the capacitive reactance, the voltage leads the current.

Figure 7-31 gives the impedance of the series resonance circuit versus frequency for two different values of R, 10 and 100 Ω. The values of L and C are chosen to give

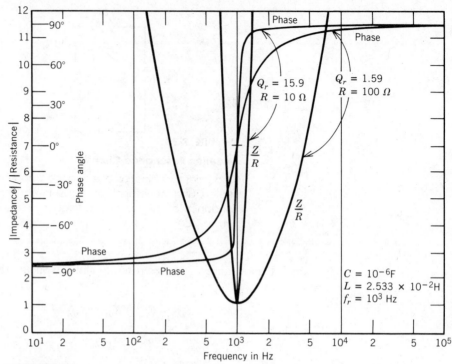

FIGURE 7-31

Impedance and Phase Versus Frequency in a Series Resonance Circuit

The ratio of the magnitude of the impedance to the magnitude of the resistance and the relative phase angle is shown for two values of the resistance in the circuit. The resonance frequency is 1000 Hz, at which the impedance is only resistive and equal to R. The quality, Q, of the circuit depends upon the resistive energy loss. High-Q circuits have sharper resonance responses than those with more energy loss.

a resonance frequency $f_r = 10^3$ Hz. In Figure 7-31 the magnitude of the impedance is divided by the resistance R to normalize the impedances to unity at the resonance frequency. The ratio Z/R is plotted with two values of R to emphasize the change in the resonance response with different resistances.

The sharpness of the resonance curves is an important property of the circuit and is related to the electrical energy that is dissipated (converted to heat in the resistance) in each cycle. A circuit that has low losses is termed a *high Q* circuit. Q is the "quality factor";

$$Q = \frac{2\pi f L}{R} \quad \text{or} \quad Q = \frac{1}{2\pi f C R} \tag{7-54}$$

Because Q is frequency-dependent, it is usual to define the quality factor at the resonance frequency, where Q has a single value.

$$Q_r = \frac{2\pi f_r L}{R} = \frac{1}{2\pi f_r C R} = \frac{1}{R}\sqrt{\frac{L}{C}} \tag{7-55}$$

It is instructive to calculate the voltages across the three passive elements in the series resonance circuit at the resonance frequency. First $e_R(t) = i(t) R = e(t)$; that is, the voltage across the resistance is the source voltage. To find the voltages across the inductance and the capacitance, we multiply the reactances with the current $i(t) = e(t)/R$,

$$e_L(t) = i(t)j(2\pi f_r L) = e(t)j\left(\frac{2\pi f_r L}{R}\right) = jQ_r e(t)$$

and

$$e_C(t) = e(t)\left(-\frac{j}{2\pi f_r CR}\right) = -jQ_r e(t) \qquad (7\text{-}56)$$

Thus the magnitudes of the voltages across the inductance and the capacitance are Q_r times the source voltage; they have 90° phase differences with the source voltage. For a high-quality series resonance circuit with a high value of Q_r, the voltages across the reactances at resonance are many times the source voltage.

PARALLEL RESONANCE CIRCUIT

We now examine the other fundamental resonance circuit in which the inductance and capacitance are in parallel. The branch containing the inductance shows the resistance, which is inherent in the wire forming the inductance or is added as a discrete element. Figure 7-32 shows the parallel resonant circuit. The performance of the circuit is affected by the resistance, regardless of its origin in the wires of the inductance or as an added resistor.

The impedance of the parallel elements is found from

$$\frac{1}{Z\underline{/\theta}} = \frac{1}{Z_1\underline{/\theta_1}} + \frac{1}{Z_2\underline{/\theta_2}}$$

where $Z_1\underline{/\theta_1} = R + j2\pi fL$ and $Z_2\underline{/\theta_2} = -j/2\pi fC$. This yields

$$Z\underline{/\theta} = \frac{\dfrac{R}{(2\pi fC)^2} - j\left[\dfrac{R^2}{2\pi fC} + \dfrac{L}{C}\left(2\pi fL - \dfrac{1}{2\pi fC}\right)\right]}{R^2 + \left(2\pi fL - \dfrac{1}{2\pi fC}\right)^2} \qquad (7\text{-}57)$$

Equation 7-57 reveals the vast difference between the parallel resonance circuit and the series resonance circuit, whose impedance is given by Equation 7-52. In the series resonance circuit the lowest impedance occurs at its resonance frequency (Equation 7-53); in the parallel resonance circuit the highest impedance occurs near the resonance frequency.

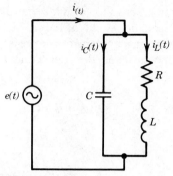

FIGURE 7-32
A Parallel Resonance Circuit
The parallel resonance circuit has an impedance that peaks as the frequency is changed.

In fact, three frequencies pertain to a parallel resonance circuit like the one shown in Figure 7-32. One is f_r (Equation 7-53), the frequency identified with resonance in series resonance circuits with the same R, L, and C values. The second is the one we designate below as f_0, where the impedance is purely resistive $Z\underline{/\theta} = Z\underline{/0°}$. The third, f_{max}, is the frequency at which the parallel resonance circuit has its maximum impedance. The f_0 is always at a lower frequency than f_r; f_{max} lies between them. The frequencies are quite close together in a high-Q circuit.

At f_r the denominator contains only the resistance term R^2, which is the least value of the denominator. The impedance is not purely resistive at f_r, however, but it is given by

$$Z\underline{/\theta} = \frac{1}{2\pi f_r C}\left(\frac{1}{2\pi f C_r R} - j\right) = X_C\left(Q_r - j\right) \tag{7-58}$$

We denote f_0 as the frequency at which the parallel resonance circuit impedance is purely resistive; that is, if the imaginary term in Equation 7-57 is zero, we find the relation between f_0 and f_r to be

$$f_0^2 = f_r^2\left(1 - \frac{R^2 C}{L}\right) = f_r^2\left(1 - \frac{1}{Q_r^2}\right) \tag{7-59}$$

Q_r in Equations 7-58 and 7-59 is defined by Equation 7-55.

In the parallel resonance circuit the currents in the branches at f_0 are approximately Q_r times the current from the voltage source. Figure 7-33 gives the resonance behavior of the parallel resonance circuit for two values of Q_r. The inserts show the current relation at f_0.

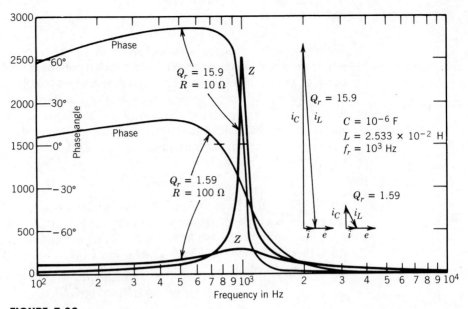

FIGURE 7-33

Impedance and Phase Versus Frequency in Parallel Resonance Circuit

The typical resonance curve occurs as the frequency is changed. The high-Q circuit, which has lower resistive energy loss, has the sharper resonance response. Three distinct frequencies in the parallel resonance circuit are described in the text.

The frequency, f_{max}, at which the impedance Z (Equation 7-57) has its maximum value can be found most easily by numerical calculation. In plotting Figure 7-33 the same two sets of circuit parameters used in the series resonance circuit are used for the parallel resonance circuit. They are $L = 2.533 \times 10^{-2}$ H and $C = 10^{-6}$ F and the frequencies that result are:

R	Q_r	f_r	f_{max}	f_o
10 Ω	15.9	1000 Hz	999.6 Hz	998.0 Hz
100 Ω	1.59	1000	993.8	777.5

The resistance in the parallel resonance circuit broadens the resonance curve and shifts the frequency of its peak toward lower values as it dissipates the electrical energy by converting it into heat.

ELECTRICAL POWER

We conclude our treatment of electrical circuit theory by considering the electrical power in circuits. Because power is the time rate of doing work, joules per second or watts, the nature of the work on electrical charges reveals the relations we seek.

Work measured in joules is the product of the force on the charge times the distance the charge moves in the direction of the force. The force on an electrical charge is the electric field times the magnitude of the charge, $\mathbf{F} = q\mathscr{E}$, where the force $\mathbf{F}$ is measured in newtons, the charge q in coulombs, and the electric field $\mathscr{E}$ in volts per meter. The work is

$$W = |F||D| \cos(F, D) \tag{7-60}$$

where $|F|$ is the force magnitude, $|D|$ is the amount of displacement, and $\cos(F, D)$ is cosine of the angle between the two. If F and D are in the same direction, the angle (F, D) is zero and $\cos(F, D) = 1$ and the greatest work is done. Any other angle between F and D yields less work.

We may rewrite W as

$$\begin{aligned} W &= |q\mathscr{E}||D| \cos(q\mathscr{E}, D) \\ &= |q||\mathscr{E}\,D| \cos(q\mathscr{E}, D) \\ &= |q||E| \cos(q, E) \end{aligned} \tag{7-61}$$

The last line in Equation 7-61 uses the fact that the electromotive force or voltage E is the electric field times the displacement, volts per meter times meters equals volts.

The electrical power, P, becomes

$$P = \frac{dw}{dt} = \left|\frac{dq}{dt}\right||E| \cos\left(\frac{dq}{dt}, E\right)$$

$$= IE \cos(I, E) \tag{7-62}$$

where I is the current or the flow of charges measured in coulombs per second or amperes, E in volts, and P in joules per seconds or watts.

In direct current circuits the phase angle between current and voltage is zero degrees. The dc power is

$$\begin{aligned} P_{dc} &= IE \\ &= I^2 R \end{aligned}$$

or

$$= \frac{E^2}{R} \quad \text{Watts} \tag{7-63}$$

In ac circuits, capacitive and inductive effects cause the phase angle to depart from zero degrees and the power factor, the cosine of the phase angle, is less than unity. The current and voltage in a capacitance or an inductance have a 90° phase angle. This makes the power factor $\cos(I, E) = \cos 90° = 0$, so there is no work done in them. It was for this reason that the ratio between voltage and current for them is called reactance rather than resistance.

For other ac circuits, which include resistance and reactance, the power factor is between zero and unity, and the power is less than the product of the magnitudes of the current and voltage.

Another useful relation can be found by considering a resistive circuit with an ac voltage source. Since the current and voltage are in phase, that is, $\cos(I, E) = 1$,

$$P_{ac}(t) = i(t)e(t)$$
$$= I \sin \omega t \, E \sin \omega t$$
$$= EI \sin^2 \omega t = E_{ac} I_{ac} \sin^2 \omega t \quad \text{Watts} \quad (7\text{-}64)$$

The $\sin^2 \omega t$ is a positive quantity that varies between 0 and 1; average over a complete cycle is 0.5 as shown in Chapter 14. The average ac power in the resistor, the rate of conversion of electrical energy into heat, is

$$P_{ac} = \frac{E_{ac} I_{ac}}{2} \quad \text{Watts} \quad (7\text{-}65)$$

Here we label the magnitude of the ac voltage and current whereas before they were unlabeled.

The same resistor in a dc circuit will convert electrical energy into heat at the rate

$$P_{dc} = E_{dc} I_{dc} \quad \text{Watts} \quad (7\text{-}66)$$

In order for the rate of energy conversion to be the same, P_{ac} must equal P_{dc} or

$$\frac{E_{ac} I_{ac}}{2} = E_{dc} I_{dc}$$

This will occur if $E_{ac} = E_{dc}\sqrt{2}$ and similarly $I_{ac} = I_{dc}\sqrt{2}$. Where E_{ac} and I_{ac} are the peak values of the ac sine waves. When the ac peak voltages and currents have this relation, a light bulb (resistor) will be equally bright when connected to either an ac or a dc power source.

$E_{ac}/\sqrt{2}$ and $I_{ac}/\sqrt{2}$ are called root-mean-square values and correspond to the equivalent dc values,

$$E_{rms} = \frac{E_{ac}}{\sqrt{2}} \quad \text{and} \quad I_{rms} = \frac{I_{ac}}{\sqrt{2}} \quad (7\text{-}67)$$

For example, in a 120-V ac power circuit, $E_{rms} = 120$ V, then for an equivalent rate of power conversion

$$E_{dc} = 120 \text{ V} = E_{rms}$$

requires that

$$E_{ac} = \sqrt{2} \times 120 \text{ V} = 169.7 \text{ V peak}$$

The 120-V ac power circuits have the sine wave voltage $e(t) = 169.7 \sin \omega t$ even though the voltage is called by its rms value of 120 V.

SUMMARY

The effectiveness of Kirchhoff's laws in solving electrical circuits has been demonstrated for dc and ac circuits. The use of the rotational operator j makes the analysis of ac circuits nearly as easy as the analysis of dc circuits because the response of a circuit to any periodic voltage or current can be studied by using only sine waves.

The transient response of capacitance is governed by the requirement that the charge in a capacitance must change for the voltage across it to change. If a step in voltage is applied to one side of a capacitance, the step in voltage is transmitted immediately to the other side. Similarly, we saw that the current through an inductance cannot change instantaneously and that any rapid change generates a large voltage across the inductance, whose polarity is such as to oppose the change in current.

Thevenin's theorem states that any two-terminal network, element, or device can be represented by a voltage source and an internal impedance. The internal impedance of the two-terminal network is a resistance for dc and is frequency-dependent for ac systems. This idea relieves us of learning the details of complicated devices in order to consider them as circuit elements, and it suggests how their electrical characteristics can be obtained by simple measurements.

Circuits with both capacitance and inductance have the special property of being able to transfer stored energy between the capacitance and the inductance. They resonate at the rate of energy transfer, and a portion of the energy in each cycle is converted by ohm heating. If the energy loss is small, the quality of the resonance circuit is high and the resonance curve is sharp. In later chapters, we will encounter other kinds of circuits that also generate periodic waveforms.

PROBLEMS

7-1 The open-circuit voltage at circuit terminals is 8.3 V. With a 500-Ω resistance across the terminals, the voltage is 8.17 V.
  (a) What is the internal resistance of the circuit?
  (b) What will the power in a load (resistor) of 200 Ω be?
  (c) What is the power in the internal resistor?

7-2 Find the current through R_3 in Figure 7-5 by
  (a) Kirchhoff's laws
  (b) Thevenin's theorem

7-3 Show the equation for R_{int} in a Thevenin source when V_{oc}, V_0, and R_L are known. If $V_{oc} = 12.10$ V, $V_0 = 12.0$ V and $R_L = 50\ \Omega$, what is R_{int}?

7-4 Find the current I_5 to R_5 in Figure 7-4 by applying Thevenin's theorem.
  (a) Obtain the algebraic expression for I_5 in terms of R's and V.
  (b) Use the values in the text to find I_5.

7-5 Measure the voltage and current with two different resistances at the output of an ac power adapter for your hand-held calculator. Measure the voltage without any external load. Find the Thevenin equivalent source at 60 Hz. (Choose the resistor small enough to give measurable results but not so small that the power unit overheats. Usually a resistance that yields a current slightly in excess of the rated current is suitable.)

7-6 A logic gate is considered a Thevenin source. The measurements are $V_{OC} = 4.25$ V, $R_L = 1500\ \Omega$, and $V_O = 4.18$ V.
  (a) What are the Thevenin voltage source and the internal resistance?
  (b) What will the output voltage be if the load resistor is 100 Ω (the rated minimum load resistance)?
  (c) What is the current with a 100 Ω load?
  (d) Do you judge this to be a "quality" voltage source?

7-7 Use Euler's relation to show that
  (a) $\dfrac{A/\theta_1}{B/\theta_2}$ is $\dfrac{A}{B}/\theta_1 - \theta_2$
  (b) A/θ_1 times B/θ_2 is $AB/\theta_1 + \theta_2$
  (c) $A/\theta_1 + B/\theta_2$
  (d) $A/\theta_1 - B/\theta_2$
  (e) Use $A/\theta_1 = 3/25°$ and $B/\theta_2 = 2/-60°$ to calculate and plot each result to confirm the relations.

7-8 Find the current-voltage relation in Figure 7-23 by assuming that the current $i(t)$ is initially in the opposite direction than shown in the figure. Does the answer confirm the statement following Equation 7-36 that the current-voltage relation is given by the phase? Are these results and the one in the text in agreement? Use the numerical values in the text to calculate the result.

7-9 Find the root-mean-square value of the sawtooth wave.

$e(t) = Et \qquad 0 \le t < t_1$
$e(t) = E(1 - 3t) \qquad t_1 \le t < \tau$

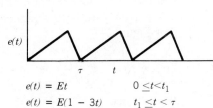

$e(t) = Et \qquad 0 \le t < t_1$
$e(t) = E(1 - 3t) \qquad t_1 \le t < \tau$

7-10 Show how the characteristics of the two-terminal Norton equivalent circuit of Figure 7-9 may be determined.

7-11 Solve for the currents in the circuit.

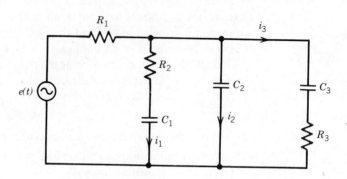

for $R_1 = 100\ \Omega$ $C_1 = 0.1\ \mu F$
$R_2 = 10^3$ $C_2 = 0.2$ $E(t) = 10 \sin \omega t$
$R_3 = 5 \times 10^3$ $C_3 = 0.3$

find i, i_1, i_2 and i_3 for $f = 10^4$ Hz.

7-12 Use the values $f = 5 \times 10^3$ Hz, $R_1 = 100\ \Omega$, $C_1 = 0.01\ \mu F$, $R_2 = 5\ \Omega$, and $C_2 = 0.05\ \mu F$ to find $Z/\underline{\theta}$, e_1 and e_2. Find $i_R(t)$ and $i_c(t)$ and $i(t)$ and confirm the node equation $i(t) - i_R(t) - i_c(t) = 0$ for Figure 7-29.

7-13 Using the $R = 10^3\ \Omega$, $C = 0.1\ \mu F$, and $f = 3 \times 10^3$ Hz, confirm the numerical example for the RC circuit by using Equation 7-42 and 7-43.

7-14 (a) Find the currents in Figure 7-29 for $E = 5$ V, $f = 5$ kHz, $R_1 = 10^3\ \Omega$, $R_2 = 0.5 \times 10^3\ \Omega$, $C_1 = 0.1\ \mu F$ and $C_2 = 0.05\ \mu F$.
(b) Find $e_1(t)$ and $e_2(t)$.
(c) Confirm the vector sums of the quantities.

7-15 Calculate and plot the frequency dependence of a high-pass filter, Figure 7-28, using Equations 7-43 and 7-44.

7-16 Show graphically and algebraically that the voltages E_R and E_L in the high-pass filter add to the supply voltage E at 3×10^3 Hz with $R = 10^3\ \Omega$, $L = 10^{-3}$ H.

7-17 For the series resonance circuit, Figure 7-30, plot $Z/\underline{\theta}$ (rather than $Z/\underline{\theta}/R$) versus frequency for both values of R by using the other parameters chosen to plot Figure 7-31.

7-18 Show that for $Z/\underline{\theta} = Z/\underline{0}°$ in a parallel resonance circuit the frequency difference $\Delta f = (f_0 - f_r)$ between the frequency f_0 for the purely resistive impedance and $f_r = (1/2\pi)(1/LC)^{1/2}$ is approximately $\Delta f \approx -f_r/(2Q_r^2)$ for high Q circuits.

7-19 Exchange the positions of the resistance and capacitance in Figure 7-23 and calculate the voltage across the resistor $E_r(t)$ as a function of frequency. Plot $A(f)$ and $\theta(f)$ for this arrangement. What would be an appropriate name for this arrangement?

7-20 Using Coulomb's law,

$$\mathbf{F} = \frac{q_1 q_2}{4\pi\varepsilon_0 r^2} \hat{r}$$

or

$$\mathbf{E} = \frac{\mathbf{F}}{q_{\text{test}}} = \frac{1}{4\pi\varepsilon_0} \frac{q}{r^2} \hat{r}$$

explain that there can be no accumulation of charge in a conductor in a circuit.

CHAPTER 8

SOLID-STATE ELECTRONIC DEVICES

Many materials, which are extremely poor conductors of electricity, are called *insulators*; others, such as metals, are extremely good conductors. The materials that form the foundation of modern electronics lie between these extremes of conduction and are called *semiconductors*. Silicon and germanium are the semiconducting materials most used in solid-state electronic devices.

All atoms are alike in that they have a number of positively charged protons in the nucleus and the same number of negatively charged electrons, so that the atom is electrically neutral. The number of nuclear charges or of negative electrons is the atomic number Z. The outermost electrons are called *valence* electrons; in solids they provide the molecular bonding with adjacent atoms by sharing their valence electrons. The periodic atomic structure of solids makes the behavior of the electrons different from that in an isolated atom or molecule. The electrons are restricted to certain states or ranges of energy called *energy bands*.

The lowest energy band is called the *valence band*, and at very low temperatures all the electrons in the solid remain in and fill the valence band. The *conduction band* occurs at an energy step higher than the valence band. Figure 8-1 shows the energy bands and the interval between them. The physics of electrons in solids allows electrons to exist in the valence band or the conduction band but not in the energy gap between them. An electron can move from the valence band to the

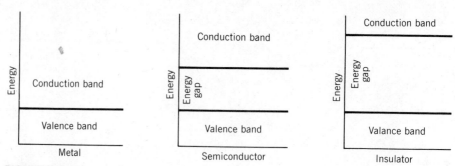

FIGURE 8-1

Energy Bands in the Periodic Atomic Structure of Solids

The magnitude of the energy gap determines the electrical conductivity of solids.

conduction band only if it has enough additional energy to overcome the energy gap. This energy may come from an increase in the temperature of the solid, but it is unlikely that an electron will make the jump if the energy gap is sizable.

The difference between insulators and semiconductors is the size of the energy gap. Table 8-1 shows the energy gap, the resistivity, and other properties of three materials at 300 K. Diamond is a very good insulator; it has a resistivity 10^9 to 10^{12} times greater than the other two materials because its energy gap is 5 to 8 times greater. These data are given in SI units. The sources are *Handbook of Chemistry and Physics* (1976) and G. L. Pearson and W. H. Brattain (1955).

Figure 8-2 shows the state of a semiconductor with some electrons in the conduction band and the vacancies (holes) that the electrons left in the valence band. One property of solids is that conduction (the transport of electrons) requires some electronic quantum states (positions) to be unoccupied. Accordingly an electron can move from its present state to an unoccupied one; this move leaves an unoccupied state. If all states are occupied, however, conduction cannot occur, according to the Pauli exclusion principle that no two electrons can occupy the

TABLE 8-1
Properties of Intrinsic Semiconductors

Property	Diamond (Carbon)	Silicon	Germanium
Atomic number Z	6	14	32
Energy gap at 300 K	5.4 eV	1.107 eV	0.67 eV
Melting point	4300 K	1685 K	1231 K
Dielectric constant	5.5	12	16
Intrinsic resistivity at 300 K	10^{12} Ω-m	2300 Ω-m	0.47 Ω-m
Electron mobility at 300 K	$0.18 \frac{m^2}{V\text{-sec}}$	$0.19 \frac{m^2}{V\text{-sec}}$	$0.38 \frac{m^2}{V\text{-sec}}$
Hole mobility at 300 K	$0.14 \frac{m^2}{V\text{-sec}}$	$0.05 \frac{m^2}{V\text{-sec}}$	$0.182 \frac{m^2}{V\text{-sec}}$

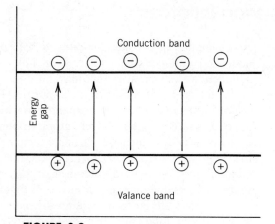

FIGURE 8-2

An Intrinsic Semiconductor

Holes ⊕ in the valence band are created by electrons ⊖ promoted to the conduction band by thermal energy.

same quantum state. If the states are filled, two electrons might exchange states (positions), but this would not constitute conduction.

The promotion of an electron from the valence band to the conduction band by thermal energy permits the conduction or motion of electrons in both bands. Electric voltage applied to the semiconductor will cause the electrons in the conduction band to drift, which constitutes a current. The voltage will cause electrons in the valence band to move in the same direction, but in the valence band, the only states open to accept the electrons are those left vacant by the promotion of other electrons to the conduction band. As an electron in the valence band moves in the direction of the electron drift to fill a hole, it leaves a hole behind it. This is equivalent to a hole moving in a direction opposite that of the electron drift.

The holes migrate in the same direction that a positive electronic charge would take under the influence of the voltage. For this reason holes are assumed to be like positive charges, and the behavior of holes in semiconductors is considered on the basis of this assumption. The differences in electron mobility and hole mobility given in Table 8-1 confirm that their drift mechanisms are different.

Materials that conduct in the manner just described are called *intrinsic semiconductors.* Their ability to conduct depends upon the promotion of electrons from the valence band to the conduction band by the thermal energy available to the material. As a result, conductivity depends strongly on temperature. (The intrinsic semiconductor elements used as temperature sensors to exploit this feature are called *thermistors.*) At ambient temperature (30°C) the fraction of electrons that are present in the conduction band because of thermal energy may be only 10^{-12} of those in the valence band.

EXTRINSIC SEMICONDUCTORS

A semiconductor becomes an *extrinsic semiconductor* when "impurity" atoms are added to enhance the electrons or holes that may participate in electrical conduction. To see how this may be done we will examine the nature of the semiconductor that is host to the impurities. Figure 8-3 shows the arrangement of silicon atoms in a solid and the manner in which the four valence electrons of each atom are shared among adjacent atoms in bonding the atoms together. Each atom of silicon has four valence electrons, which indicates that it belongs to column 4 of the chemical periodic chart. Germanium, the other important semiconductor host material, also belongs to column 4; it also has four valence electrons, which it shares in the same manner, as shown in Figure 8-3.

The host material from column 4 becomes an extrinsic semiconductor when material from columns 3 or 5 is added in very small amounts. To show how the addition of this material provides electrons for conduction or holes to affect conduction by positive charges, we place an atom of the impurity in the array of host atoms. Let us consider an atom from column 5, such as phosphorus (P) or antimony (Sb), in the silicon array. Figure 8-4 shows the result of the fifth electron in the array. In this case and in all others involving impurities, the material remains electrically neutral because the charge of the impurity nucleus is balanced by the electrons it has as a neutral atom.

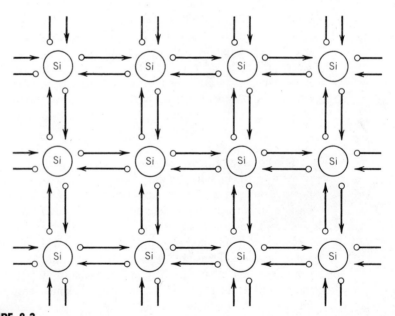

FIGURE 8-3

Silicon Atoms in a Crystal

The shared valence electrons bind atoms in a crystal.

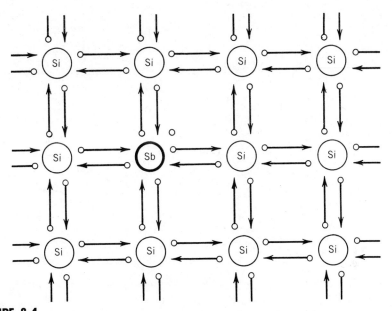

FIGURE 8-4
A Silicon Array with Antimony (Sb) Impurity
The antimony atom donates an electron to form an *n*-type extrinsic semiconductor.

  The dominant tetrahedral covalent bonding of the host material is satisfied by four of the five valence electrons, and in the presence of all the atoms the fifth valence electron of an antimony atom is tied only very weakly to its nucleus. Thus an electrical field produced by a voltage difference across the extrinsic semiconductor will cause the fifth electron to drift, which (like the drift of electrons in the conduction band of a semiconductor) constitutes a current.
  The impurity antimony atom in the host of silicon provides an electron for conduction independently of the thermal energy required to promote electrons into the conduction band of the intrinsic semiconductor host material. The extrinsic semiconductor made with one impurity atom for each hundred million host atoms will have several thousand times the conductivity of the intrinsic semiconductor at ambient temperature (30°C).
  Extrinsic semiconductors made with column 5 impurities are called *n-type* because the negative electron supplied by the impurity is free to contribute to conduction, but no hole conduction is associated with the impurity atom.
  An extrinsic semiconductor made with column 4 host materials and column 3 impurities is represented in Figure 8-5. A boron (B) or an indium (In) impurity atom has three valence electrons; thus it cannot complete the tetrahedral covalent bonding with the silicon atoms. The uncompleted bond may be viewed as a hole into which an electron from an adjacent atom may move under the influence of an electric field. In doing so it leaves a hole behind, so that the holes move opposite to

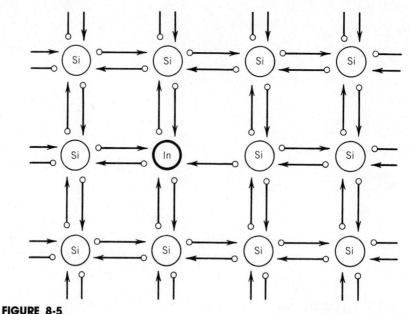

FIGURE 8-5

A Silicon Array with Indium (In) Impurity

The indium atom creates a hole to form *p*-type extrinsic semiconductor material.

the motion of electrons. This constitutes positive charge conduction, which is identical with the "holes" in intrinsic semiconductor material. The holes in the extrinsic material, however, exist independently of the thermal energy required to promote electrons into the conduction band.

Extrinsic semiconductors made with column 3 impurities are called *p-type* because the conduction is equivalent to the transport of positive charges. A very small fraction of impurity atoms (parts per million or less) makes the conduction thousands of times greater than that of an intrinsic semiconductor.

The process of adding impurities to make extrinsic semiconductor materials is called *doping*. When the host material, silicon or germanium, is grown as a crystal by being withdrawn slowly from a pool of melted material, an impurity may be added to the melt to make the extrinsic semiconductor. Another method of doping is to place pure host material in an oven containing the gas as an impurity and allowing the impurity to diffuse thermally into the host material. A mask placed photographically on the crystal will prevent diffusion into covered regions. Localized doping can be achieved by accelerating impurity atoms with a particle accelerator and driving them into the host material. The advances in masking, plating, controlled doping, and other techniques have made (and will continue to make) widespread integrated circuits that were unimaginable only a few years ago.

In *n*-type extrinsic semiconductors the impurity atoms are called *donors* because they donate electrons. The thermal promotion of electrons into the conduction band continues to produce holes in the host material, but the holes are filled in part by the donor electrons. Thus the material conducts predominantly by electrons, which are called the *majority carriers*. The remaining holes are the *minority carriers*.

In *p*-type material the column 3 impurity atoms are called *acceptors* because they accept the electrons. The conduction by holes makes the holes the majority carrier. Electrons are the minority carriers in *p*-type material.

In concluding the discussion of extrinsic semiconductors, we note that either *n*-type or *p*-type material is electrically neutral and that either will conduct when placed in an electric field. The electrical neutrality exists because of the balance between the positively charged protons of the nucleus and the negatively charged electrons associated with the atom. The impurity atoms have one more or less electron than the atoms of the host material, but the nucleus of the impurity atom has one more or one less nuclear charge.

The importance of extrinsic semiconductor materials lies in the phenomena that occur when *p*-type and *n*-type materials are placed in intimate contact.

PN **JUNCTION**

When *p*-type and *n*-type extrinsic semiconductor materials are placed in contact, a *pn junction* is formed. In the *pn* junction the majority carriers (electrons in *n*-type material and holes in *p*-type material) diffuse into the opposite material and combine with each other, which yields important results.

At the very narrow region of contact the electrons and holes combine, and no majority carriers are left uncombined. This region, about a micrometer thick, is called the *depletion region* or *depletion layer*.

The diffusion is not limited to the depletion layer, however. Electrons diffuse farther into the *p*-type material and combine with some of the holes, while the holes also diffuse into the *n*-type material and combine with the electrons. In this portion of the *n*-type material fewer electrons exist than before the diffusion; thus the material, which was electrically neutral before the diffusion, acquires a net positive charge. Similarly, the electrons diffusing into the *p*-type material diminish the number of holes, causing a net negative charge distribution. The charges in the regions adjacent to the depletion layer are called *uncovered charges*.

Because of the uncovered charges, the negatively charged region in the *p*-type material and the positively charged region in the *n*-type material create an electric field across the junction. The diffusion of majority carriers is arrested by the electric field, and an equilibrium in the charge distribution is established. Figure 8-6 shows the *pn* junction and its electrical properties.

The work done on a unit charge to transport it from one position to another in the presence of an electric field is called the *electrical potential*. The potential across the *pn* junction is shown in Figure 8-6.

The configuration of a *pn* junction, consisting of a positive charge and a negative charge separated by an intervening neutral space (the depletion region), is identical

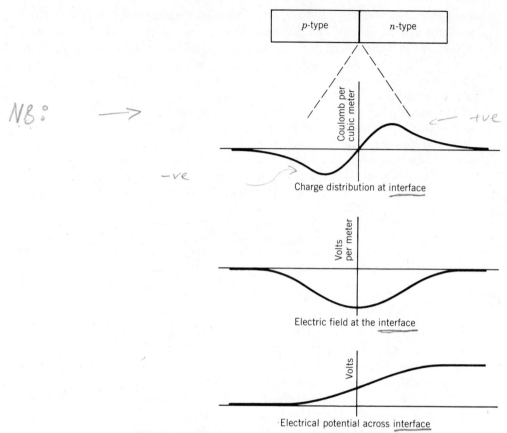

FIGURE 8-6

A *pn* Junction and its Electrical Properties

Electrons and holes diffuse into the narrow interface between the *p*-type and *n*-type materials. The depletion layer and the uncovered charges result in an electrically charged junction.

to a charged capacitor. Thus we can conclude that a *pn* junction has the properties of a capacitance. The magnitude of the capacitance depends upon the geometry of the junction and the nature of the semiconducting material and the doping.

THE DIODE: REVERSE-BIASED

The usefulness of the *pn* junction becomes apparent when it is placed in an electrical circuit. Figure 8-7 shows a *pn* junction connected to a voltage source, a battery. In this connection the negative potential is attached to the *p*-type material and the positive potential is attached to the *n*-type material. Positive charges flow

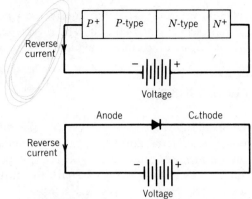

FIGURE 8-7

A *pn* Junction (Diode) with Reverse Bias

The polarity of the applied voltage separates the majority carriers, and no conduction occurs. The P^+ and N^+ regions are highly doped to allow semiconductor-to-metal ohmic connections.

from positive potential to negative potential. Figure 8-7 shows that the positive charges (holes) in the *p*-type material tend to move to the left and the negative charges (electrons) in the *n*-type material tend to move to the right. This motion results in more uncovered charges on both sides of the *pn* junction, which increase the electrical potential across the *pn* junction.

The majority carriers are farther apart than before, and conduction is essentially zero. With the voltage polarity shown in Figure 8-7, the *pn* junction is *reverse-biased*.

Two features of a reverse-biased *pn* junction must be noted. One is that the increase in the number of uncovered charges and the change in their separation will change the effective capacitance of the junction. Because the charge and voltage of a reverse-biased *pn* junction depend upon the electrical properties of the junction, it is useful to define the capacitance of a *pn* junction as $C = dQ/dV$, the ratio of the change in charge to the change in voltage. In *pn* junctions the capacitance may decrease from nominal values of tens of picofarads to a third of that magnitude as the reverse bias is increased from a few volts to many volts.

The second feature concerns the essentially zero current passing through a reverse-biased diode. The reverse-bias current, which is very small but not zero, arises from the minority carriers in both the *p*-type and the *n*-type materials. In semiconducting materials the thermal creation of electrons and holes is a continuous process. As mentioned above, one result of doping materials to make them predominately *p*-type or *n*-type is to reduce the number of minority carriers, because the impurity-induced majority carriers combine with some but not all of the thermally produced minority carriers. Because the migration of the minority

214 Solid-State Electronic Devices

carriers is opposite that of the majority carriers, minority carriers cross the junction and represent a small current. The reverse-bias current is 10^{-4} or less than the current through the diode when it is conducting.

THE DIODE: FORWARD-BIASED

Figure 8-8 shows a *pn* junction with the positive terminal of a battery connected to the *p*-type material and the negative terminal connected to the *n*-type material. This arrangement causes the holes in the *p*-type material to migrate to the right and the electrons in the *n*-type material to migrate to the left. If the applied voltage exceeds the electrical potential across the *pn* junction as generated by the uncovered charges, the holes and electrons migrate and combine, and a current exists through the junction. Figure 8-9 shows the current-voltage relation in a *pn* junction.

Figure 8-9 shows three significant features of a *pn* junction in an electrical circuit. First is the reverse current through the reverse-biased diode. This is a small value, limited by the number of minority carriers. The second feature is the small current for a small forward-applied voltage. The current remains very small until the applied voltage nearly equals the electrical potential formed in the *pn* junction. The third feature is the increasing forward current as the applied voltage exceeds the electrical potential of the *pn* junction. The current-voltage relation in Figure 8-9 shows clearly that the forward current through a *pn* junction depends in nonlinear fashion on the applied voltage.

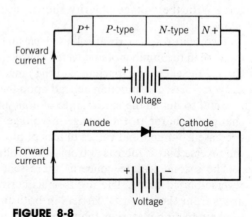

FIGURE 8-8

A *pn* Junction (Diode) with Forward Bias

The voltage polarity causes the flow of charges (electrons and holes) across the junction, which constitutes a forward current.

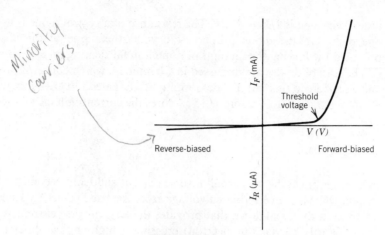

FIGURE 8-9

Voltage-Current Relations in a Diode

The forward current becomes large when the forward bias exceeds the threshold voltage generated by the uncovered charges. The reverse current of minority carriers is very small, approximately 10^{-4} of the forward current.

The applied voltage at which the current through a forward-biased *pn* junction begins to increase rapidly is called the *threshold* or *cut-in* voltage. This is approximately 0.3 V in *pn* junctions made by doping germanium and about 0.7 V for silicon-based *pn* junctions.

PRACTICAL DIODE

A *pn* junction is made by placing *p*-type and *n*-type material in contact, but to be used as an electrical device the junction must have conductors attached to it. Either *n*-type or *p*-type material in intimate contact with a metal (a *Schottky junction*) constitutes a diode; that is, it conducts readily when forward-biased and essentially not at all when reverse-biased. For a practical *pn* junction diode, the additional semiconductor-metal diode contacts are to be avoided, which is done by doping heavily each region in which the metal contacts are made. The result is an *ohmic junction*, one in which there is no rectification (diode action) but only a small resistance across the contact. The heavily doped regions, marked with N^+ and P^+, are shown in Figures 8-7 and 8-8.

In Figure 8-9 the current-voltage relation beyond the threshold or cut-in voltage is governed by the ohmic resistance of the *pn* junction diode, including the connections to the metal leads and the resistance of the semiconducting material itself. The slope is $1/R = dI/dV$. The resistance of a typical diode is approximately one ohm. Unique diode properties, beyond the scope of this discussion, are provided by varying the amount of doping in fabricating *pn* elements.

The uses of diodes are discussed in Chapter 14, which concerns power supplies, and elsewhere in this text. Diodes forming AND-gates and OR-gates were shown in Chapter 3. Laboratory Exercise 15 explores the current-voltage relations in diodes.

ZENER DIODE

In a *Zener diode* the very small inverse current suddenly becomes a large inverse current when the reverse-biased voltage exceeds a fixed value. Two mechanisms are involved in the breakdown that provides the large inverse current.

A minority carrier (an electron) crossing a high-voltage difference may gain sufficient kinetic energy to create an electron and a hole by colliding with an atom in the semiconductor material. The electron and the newly freed electron may both gain sufficient kinetic energy in the electric field to produce other pairs by collisions. The multiplication of the creation of carriers is called an *avalanche*.

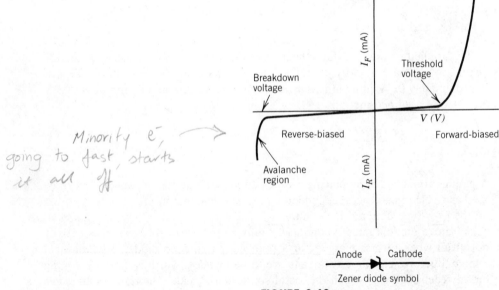

FIGURE 8-10

Voltage-Current Relations in a Zener Diode

The Zener diode has normal forward-bias properties. A reverse bias in excess of a value established in the fabrication of the diode suddenly initiates a strong reverse current.

2. The second mechanism is the ionization (the separation of an electron from its parent atom) by an intense electrical field. This process creates an electron and a hole, which participate in the inverse conduction. A few volts across a very thin depletion layer may represent a very high electrical field or voltage gradient. Typically, a voltage gradient of 10 million volts per meter will cause field ionization.

Figure 8-10 shows the conduction properties of a Zener diode. The symbol for a diode is modified as shown in Figure 8-10 to symbolize a Zener diode. The breakdown voltage, which is determined by the manufacturer, may range from two to several hundred volts. Zener diodes are used to provide a reference voltage (their breakdown voltage) for regulated power supplies and other applications. The applications of Zener diodes is presented in the chapter on power supplies and elsewhere later in the text.

THE TRANSISTOR

After years of semiconductor research (Pearson and Brattain, 1955) the transistor was invented in 1948 by John Bardeen, Walter Brattain, and William Shockley of the Bell Telephone Laboratories (Bardeen and Brattain, 1948). By placing two point contacts extremely close together on a semiconducting material, they demonstrated that a small amount of electrical energy could govern the flow of a larger amount. The point-contact transistor has evolved into various forms, some of which will be discussed below. The most widely used transistors are the *bipolar-junction transistor* and the *field-effect transistor*. We will study the bipolar-junction transistor first as a foundation for understanding other forms of the transistor.

The bipolar-junction transistor is formed by placing two diodes back to back with one element common to both. In a *PNP* bipolar junction transistor, a very thin layer of *n*-type semiconductor material is sandwiched between two pieces of *p*-type semiconductor material. In contrast, the *NPN* bipolar-junction transistor is made by using *p*-type material as the common element.

Figure 8-11 shows a *PNP* bipolar transistor in schematic form, with voltage sources attached to it. The symbol for the *PNP* transistor in Figure 8-11 has an arrow on the emitter lead pointing toward the base. The arrow indicates the direction of conventional current (flow of positive charges) from the emitter to the base. We recognize that the voltage polarity causes the emitter-base junction to conduct as a forward-biased diode.

The key to the operation of the transistor is that only a small portion of the current in the emitter-base junction leaves the transistor by the base lead. Perhaps 100 times as much current penetrates through the very thin base, enters the collector, and is carried by the collector lead. The total current in the transistor, however, is controlled by the emitter-base current, governed by the voltage across the emitter-base diode. The much larger emitter-collector current is controlled by controlling the emitter-base current; in this way electrical signals are amplified by the transistor.

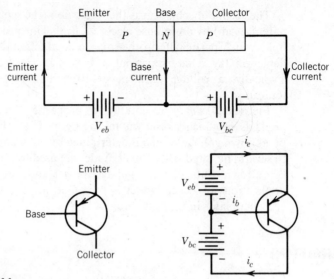

FIGURE 8-11

A *PNP* Bipolar Junction Transistor with Supply Voltages

The transistor emitter-base junction is forward-biased, and it conducts. Most of the current leaks through the reverse-biased base-collector junction and passes through the collector. The arrow on the symbol shows the forward-bias polarity and the direction of the current. The arrow indicates that the transistor is *PNP*.

Figure 8-12 shows the configuration for the *NPN* bipolar-junction transistor with the voltage sources. We note again that the voltage polarity forward-biases the emitter-base diode. The arrow on the emitter lead shows the direction of conventional current; the direction of this arrow is opposite to the direction shown in Figure 8-11 for the *PNP* transistor. (In the symbol for the transistor, the arrow indicates the type, *PNP* or *NPN*.)

It is important to note that in both figures the current from the base to the collector is across a reverse-biased diode. As stated above, the current passes from the emitter to the collector through the base. After the current reaches the collector, it is conducted through the semiconducting material. A thin base facilitates the passage of the current.

Because the arrow on the emitter shows the direction of conventional current, it also indicates the voltage polarity required to make the transistor operate. The arrow, a useful indicator, corresponds to the voltage polarity from positive to negative across the transistor.

In Chapter 10 we discuss the techniques for establishing the voltages and other circuit parameters for proper transistor operation in digital and analog applications.

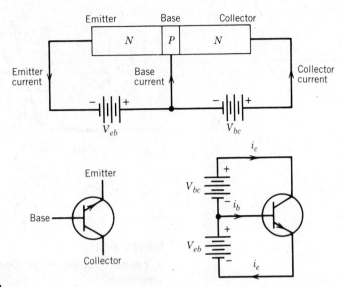

FIGURE 8-12

An *NPN* Bipolar Junction Transistor with Supply Voltages

The emitter-base junction is forward-biased, causing a current that passes mostly through the collector. Note the reverse polarity of the voltages compared to those in Figure 8-11. The direction of the arrow shows that it is an *NPN* transistor.

CURRENT RELATIONS IN A TRANSISTOR

In a bipolar-junction transistor, from 0.5 to 5.0 percent of the current from the emitter leaves the transistor by the base lead; the remaining current leaks through the base and passes through the collector. The ratio of the collector current to the emitter current is called *alpha*.

$$\alpha = \frac{i_{collector}}{i_{emitter}} \quad (8\text{-}1)$$

Alpha, which is one measure of the characteristics of a transistor, lies typically between 0.95 and 0.995.

The voltages shown in Figures 8-11 and 8-12 must be selected to allow the transistor to operate in a reasonable range of currents. These values or conditions are called *quiescent values* or *quiescent operating points*. (The techniques for choosing and establishing quiescent points, E_Q and I_Q, are discussed in Chapter 10.) The time-varying departures from the quiescent points are the signals, which are amplified by the transistor. The time-varying quantities are designated by lowercase letters. Figure 8-13 shows the quiescent points and the signals for the

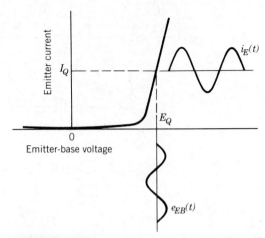

FIGURE 8-13

A Emitter Current in a Bipolar-Junction Transistor

A voltage signal $e_{EB}(t)$ varying about a quiescent voltage E_Q causes the emitter current $i_E(t)$ to vary about a quiescent value I_Q.

forward-biased junction formed by the emitter and the base in a transistor. (The details of a simple bipolar-junction transistor amplifier are given in Chapter 10.) The signals are assumed to be sine wave variations about the quiescent points.

Because the emitter current $i_E(t)$ is divided between the base current $i_B(t)$ and the collector current $i_C(t)$, we have

$$i_E = i_B + i_C \tag{8-2}$$

and from Equation 8-1, $i_C = \alpha i_E$, we find

$$i_C = \frac{\alpha}{(1-\alpha)} i_B = \beta i_B \tag{8-3}$$

We note that

$$\beta = \frac{\alpha}{(1-\alpha)} \approx \frac{1}{(1-\alpha)} \tag{8-4}$$

where the approximation is possible because α is so close to unity. Because the denominator $(1 - \alpha)$ is small, β is large. In commonly used transistors β may range from 50 to several hundred; β may differ by 50 percent or more among transistors of the same kind made by the same manufacturer. The proper design of circuits with transistors allows for variations in the properties of the transistors.

We use Equations 8-2 and 8-3 to find

$$i_E = \frac{1}{(1-\alpha)} i_B \approx \beta i_B \qquad (8\text{-}5)$$

which confirms our earlier statement that the emitter and collector currents differ only slightly because the base current is such a small part of the emitter current. For many purposes the emitter and collector currents may be assumed to be equal.

To conclude the discussion of bipolar-junction transistors, we note that they are also made with different configurations from those shown in Figures 8-11 and 8-12. In all configurations, however, the emitter and the collector are separated by a thin base of the opposite doping.

The generic or typical characteristics of junction transistors are shown in Figure 8-14. The curves show how the collector current depends upon the base current. For a given base current, for example, say 40 μA, which is governed by the forward-biased voltage across the emitter-base junction (diode), the collector current changes only slightly, while the voltage across the transistor, V_{CE}, changes greatly. That is, the current that leaks through the base to the collector depends essentially upon the base current alone.

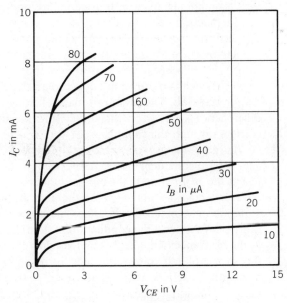

FIGURE 8-14

Generic Characteristics of Bipolar-Junction Transistor

The base current I_B determines the relation between the collector current I_C and the voltage across the transistor V_{CE}.

The decision whether to select a *PNP* transistor or an *NPN* transistor for a circuit frequently depends upon the voltage relations in the circuit that allow the quiescent points to be achieved. Many circuits use both *PNP* and *NPN* transistors. Chapter 10 illustrates the reasons for the selection of transistor types.

JUNCTION FIELD-EFFECT TRANSISTOR (JFET)

Many forms of transistors have been developed, and other forms are being created. Two of the more important basic types of field-effect transistors are the *junction field-effect transistor* (JFET) and the *metal-oxide-semiconductor field-effect transistor* (MOSFET). Their operating mechanisms differ from that of the bipolar-junction transistor, but the basic components are still *p*-type and *n*-type semiconductor materials. The interactions between the *p*-type and the *n*-type, particularly the features that relate to the depletion layer and the uncovered charges, also govern the performance of these two forms.

The JFET and the MOSFET are fabricated by photomasking arrays. Very large-scale integrated circuits such as very high-density arrays are made with FETs. The FET has an advantage over the bipolar-junction transistor because its input resistance is much larger than that of the bipolar-junction transistor. Bipolar-junction transistors, however, provide more voltage amplification than FETs; therefore bipolar-junction transistors are preferred in simple amplifier circuits.

The junction field-effect transistor, is shown in schematic forms in Figure 8-15. Because the mechanism of controlling conduction is different from that in the bipolar-junction transistor, the parts are renamed and the symbol is different. The *source* and the *drain* are at opposite ends of a semiconductor path called the *channel*, which is more lightly doped than the bipolar-junction transistor semiconductor materials. The channel is made of either *p*-type or *n*-type material. A voltage difference, V_{DS}, between the source and the drain causes a current of majority carriers through the channel.

The gate, made with the opposite type of semiconductor material from the channel, causes a depletion layer to occur at the interface of the gate and channel, just as the *pn* junction in the bipolar junction transistor generates a depletion layer. A reverse-biased voltage across the gate and the channel of the semiconductor materials that make the interface increases the depletion layer, which, in turn, decreases (narrows) the conduction path in the channel from source to drain. That is, the resistance of the FET is increased by the signal (voltage) on the gate because the reverse-biased voltage thickens the depletion layer and diminishes the number of majority carriers in the channel. In contrast to the semiconductor doping in bipolar-junction transistors, the lightly doped semiconductor materials in the JFET facilitate the growth of the depletion layer when the interface is reverse-biased.

The term *field effect* is used to describe the manner in which the electrical field controls the current path. In the FET the conduction is determined by the gate

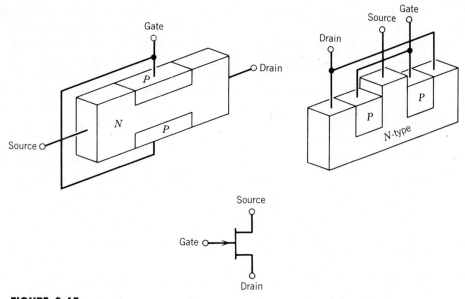

FIGURE 8-15

Forms of an *n*-Channel Junction Field-Effect Transistor, JFET

A channel exists from source to drain. The symbol contains the arrow that shows the forward bias between the gate and the channel *pn* junction. The gate is always reverse-biased in operation.

voltage, not by the current into the gate. In this respect, the FET differs from a bipolar-junction transistor, which uses the current into the base as a means of controlling the current from the emitter to the collector. In proper use the FET gate is reverse-biased at all times with respect to the channel. A JFET is designated as *p*-channel or *n*-channel, depending upon the semiconductor material used for the channel. The JFET is bidirectional; either "end" of the channel may be considered the source and the other the drain, because the direction of the current through the channel depends only upon the voltage polarity.

In the JFET the voltage drop along the semiconductor material of the channel between the source and the drain causes the reverse-biased voltage to vary from point to point along the gate. The position-varying voltage increases the reverse bias between the channel and the gate. The channel becomes narrower as one moves along the gate in the direction of the current. Figure 8-16 shows schematically the growth in the depletion region along the gate in the direction of the current. For small currents ($i = V_{DS}/R_{semiconductor}$) because V_{DS} is small, channel narrowing along the gate is negligible and the behavior of the JFET follows Ohm's law. As the current grows because of increasing V_{DS}, the channel narrows and the current becomes nearly independent of V_{DS}. This condition is called *current saturation*. Finally, if V_{DS} is increased further, the JFET will break down in a manner similar to a Zener diode. This breakdown occurs when the voltage between the source and drain is large enough to induce an avalanche mode.

The lower drawing in Figure 8-16 shows the voltage drop across the JFET with different source-drain voltages. The low-current condition (curve *a*) represents the ohmic condition. The shape is formed by three resistors: one from the source to the gate with the resistivity of the semiconductor material; a second under the gate

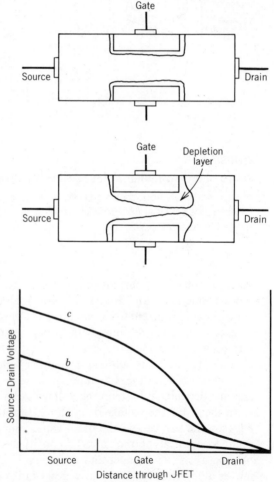

FIGURE 8-16

Channel Narrowing in a Junction Field-Effect Transistor

As the source drain voltage is increased, the depletion layer is enlarged to reduce the channel. Curve *a* (bottom figure) corresponds to the nominal depletion area (top figure) with ohmic conduction. Curves *b* and *c* represent voltage drops through the depleted channel, corresponding to a constant current and the two values of V_{DS}.

with a smaller cross-sectional area but with the same resistivity; and the third from the gate to the drain, similar to the first portion. Curves *b* and *c* represent higher but identical currents. The voltage drops in the semiconductor material before and after the gate are the same and have identical slopes, but the narrowing of the channel is greater for curve *c* than for curve *b*; therefore the voltage drop under the gate is greater and nonlinear and reflects the narrowing of the channel.

The relations among the drain current, I_D, and the drain-source voltage, V_{DS}, are shown in Figure 8-17. The Ohm's law region (the linearly rising curve), the saturation region (the constant current or flat portion of the curve), and the breakdown region are indicated in the figure. The leakage current through the reverse-biased gate due to minority carriers is extremely small, which results in the very large input resistance observed earlier.

It is worth noting that a JFET with its gate tied to its source may be used as a load resistor in a circuit with other JFETs as active transistors. Thus in the construction of integrated circuits the photomasking fabrication builds both kinds

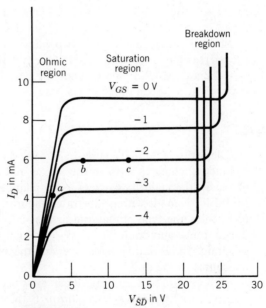

FIGURE 8-17

Characteristics of a Junction Field-Effect Transistor

Drain current has three distinct features as the source-drain voltage is increased. Points marked *a*, *b*, and *c* correspond to the curves in Figure 8-16. At high V_{SD} an avalanche produces rapid current increase.

of components as well as their interconnections and avoids the difficult task of fabricating actual resistors in integrated circuits.

The arrow on the gate lead in the symbol for the JFET signifies the forward-biased direction of the *pn* junction formed by the gate and channel, but we must remember that the JFET is never forward-biased in use. The symbol in Figure 8-15 designates an *n*-channel JFET.

METAL-OXIDE-SEMICONDUCTOR FIELD-EFFECT TRANSISTOR (MOSFET)

The metal-oxide-semiconductor field-effect transistor is so named because the current is controlled by the electric field between the gate and channel, as in the FET discussed above, but the configuration has been modified by placing an insulating cover (the oxide layer) over the channel and a metal plate on the insulating layer. The initials MOS describe the sandwiched arrangement of a *m*etal electrode separated by a very thin insulating layer of silicon *o*xide from the *s*emiconducting material. Another name for the MOSFET is the *insulated gate FET*.

MOSFETs are made with two basically different arrangements of extrinsic (doped) semiconducting materials. One type is called a *depletion MOS*, the other an *enhancement MOS*. In both types the gate is insulated from the semiconductor material by a silicon dioxide (glass) layer.

Depletion MOSFET

The depletion MOSFET (Figure 8-18) is very similar to a JFET because it has a conducting channel between the source and the drain. The gate voltage modifies (depletes) the majority carriers in the channel to control the current. A depletion MOSFET is made with a substrate, a source, a drain, and a channel lightly doped to be the same type of material as the source and the drain between the heavily doped source and drain. The heavy doping is noted with the N^+ on the source and the drain. An insulating oxide layer and the metal electrode of the gate extend from the source to the drain.

With no gate voltage, current can exist between the source and the drain by the transport of majority carriers through the three commonly doped elements. A positive gate voltage (in the case of an *n*-type channel on a *p*-type substrate) induces electrons (the majority carriers) into the doped region, which increases the conduction. A negative gate voltage depletes the majority carriers and reduces the conduction. Figure 8-19 shows the generic characteristics of a depletion MOSFET and illustrates that it may be operated either with the majority carriers depleted or with the majority carriers increased by the gate voltage. Both positive and negative gate voltages are allowed because the silicon dioxide insulation does not allow conduction through the gate.

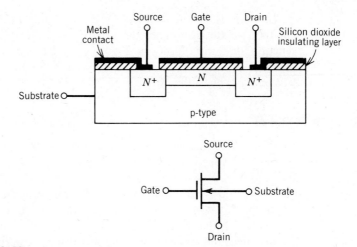

FIGURE 8-18

An n-Channel Depletion Metal-Oxide-Semiconductor Field Effect Transistor

The silicon oxide layer prevents conduction to the gate. The N^+ regions are heavily doped to allow metal ohmic contacts. The MOSFET is a four-terminal element.

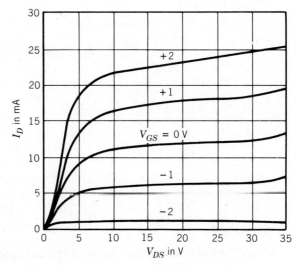

FIGURE 8-19

Generic Characteristics of n-Channel Depletion MOSFET

Conduction through the channel occurs at zero gate voltage and may be increased or decreased by the gate voltage. The insulated gate does not draw current.

The symbol for the depletion MOSFET indicates the channel-substrate semiconductor materials by showing an arrow in the direction of current through the *pn* junction that they form. In Figure 8-18 the arrow pointing toward the gate indicates that a forward-biased *pn* junction would have current in the direction from the *p*-type to the *n*-type.

Enhancement MOSFET

On the other hand, in the enhancement MOSFET (Figure 8-20) the channel between the source and the drain is omitted. No conduction exists because of the diode formed by the *pn* junctions in the element. A voltage on the gate, however, "draws" majority carriers from the substrate into the region between source and drain so that conduction can occur; that is, conduction is made possible or enhanced by the gate voltage. The physics of providing majority carriers is complicated and not simply a matter of attracting them. Figure 8-20 shows a schematic of an *n*-channel enhancement MOS made with *p*-type substrate, as indicated by the arrow. The separated segments indicate that no fabricated channel exists between the source and the drain. The MOSFET does not conduct because of the *pn* junctions between the substrate and the source and the drain. When a positive voltage is placed on the gate, the positive charges accumulate on the gate (the gate-insulator-substrate is a small capacitor) and induce negative carriers (electrons) from the substrate into the region just under the gate. Conduction by

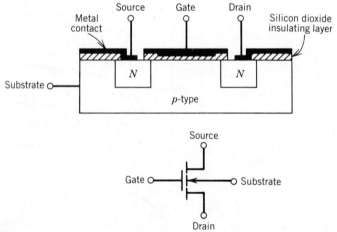

FIGURE 8-20

An *n*-Channel Enhancement Metal-Oxide-Semiconductor Field Effect Transistor

The gate voltage induces majority carriers into the region between the source and the drain to provide conduction. The elements in the symbol are broken into segments to signify that the channel was not fabricated in the transistor. The arrow shows the forward-biased current direction.

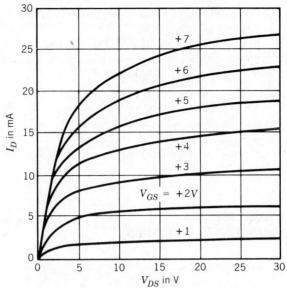

FIGURE 8-21

Generic Characteristics of n-Channel Enhancement MOSFET

The positive gate voltage generates majority carriers in the region between the source and drain for conduction.

electrons occurs between the source and the drain. Figure 8-21 gives the generic characteristics of an enhancement MOSFET.

In all MOSFETs the *pn* junction between the source or the drain and the substrate must be kept at zero voltage or be reverse-biased. Therefore a lead to the substrate is needed, making the MOS a four-terminal device. (The proper connection for the substrate of a MOSFET in a circuit is shown in Chapter 10.) Frequently the substrate is connected to the source, if that arrangement will ensure that it has zero bias relative to the source and is reverse-biased with respect to the drain and the channel.

By photomasking, doping, etching, and the evaporation of oxide and metal in the proper sequence, and finally etching to separate the source, drain, and gate-metal contact, MOSFETs can be formed in dense arrays. Because the oxide insulation does not cover the source and the drain, metal contacts are made on them, but a very thin layer of the excellent insulator silicon oxide (10^{10} to 10^{15} Ω-m) separates the substrate from the gate electrode, which extends from the source to the drain.

The input resistance to the gate signal is extremely high, which is an advantage in many applications. A hazard exists, however, because of the extremely thin (0.5-μm) silicon oxide insulating layer between the gate electrode and the substrate. If static charge from the person handling a MOSFET reaches the gate, it can cause a very large voltage across the insulating layer (recall that $V = Q/C$); the insulating layer may rupture, and the MOSFET will be destroyed. Precautions are necessary in handling and storing MOSFETs to prevent static charges; as a rule, the pins of DIPs are stuck into conducting plastic foam or kept in conducting plastic sleeves during storage and handling.

Either p-type or n-type material may be used to fabricate an enhancement MOSFET. n-channel MOSFETs are preferred because higher-density arrays in integrated circuits can be made and faster switching times achieved than with p-channel MOSFETs. The channel refers to the type of semiconductor majority carriers, n-type (electrons) or p-type (holes), induced to cause conduction.

COMPLEMENTARY METAL-OXIDE SEMICONDUCTOR (CMOS)

A variation of the MOSFET widely used in integrated circuits for digital applications is the fabrication of two MOSFETs together, a p-channel enhancement MOSFET and an n-channel enhancement MOSFET with their drains connected and with a common connection to both gates. Figure 8-22 shows the arrangement of a *complementary metal-oxide semiconductor* (*CMOS*). The advantage of the complementary interconnected transistors is that they draw significant power only when they switch between states. (CMOS circuits are discussed in Chapter 10.)

A number of innovations in semiconductor devices display special properties. One recent and possibly important innovation is a solid-state device that has both high-current and high-voltage properties and is analogous in many ways to the gas-filled tube known as a thyratron (Kapoor and Henderson, 1980). This device consists of a very lightly doped piece of extrinsic semiconductor material between

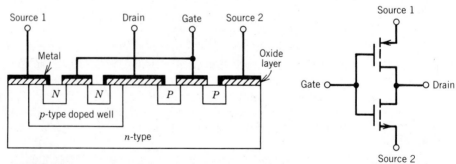

FIGURE 8-22

A Complementary Metal-Oxide-Semiconductor (CMOS) Configuration

Two transistors with opposite substrates are fabricated as a single element.

two plate electrodes. Each electrode has central needle
conduction is initiated when the voltage across the
material reaches a predetermined value. The conducti
semiconducting material as a solid-state plasmalike d
quired to initiate conduction depends upon the separat
the semiconductor wafer, while the current capacity is
the plates. Currents of many amperes with voltages
breakdown suggest that the element will be useful in electrical pow..

SUMMARY

The intrinsic semiconductors, germanium and silicon, can be doped to become either p-type or n-type extrinsic semiconductors. The majority carriers, holes in p-type and electrons in n-type material, move under the influence of an electric field to constitute a current in semiconductors. Noteworthy and useful properties arise when p-type and n-type materials make a pn junction. (A single pn junction is a diode.) Holes and electrons diffuse to establish a depletion region and uncovered charges. The latter generate a potential difference across the junction.

Two basically different kinds of transistors are made with semiconducting materials. The bipolar-junction transistor is formed by two pn junctions back to back, enclosing a very thin common element (the base). The current from the emitter to the base governs the much larger emitter-collector current, which leaks through the base. The field-effect transistor has a channel from source to drain, in which majority carriers move. In the junction field-effect transistor (JFET), the electric field from the gate modifies the depletion region in the pn junction between the gate and channel to control the current through the channel. In metal-oxide-semiconductor field-effect transistors (MOSFETs), the gate is insulated from the channel carrying the majority carriers. The depletion MOSFETs govern the current by depleting the channel of majority carriers. In enhancement MOSFETs the electric field from the insulated gate induces majority carriers into the region between the source and the drain to provide the current.

Transistors in analog amplifiers, logic gates, digital memory elements, and other circuits are presented in Chapter 10.

PROBLEMS

8-1 On logarithmic graph paper, plot the resistivity versus the energy gap for the materials in Table 8-1. Determine an approximate power relation between them, $\rho = \rho_0(\Delta E)^a$.

8-2 The following data relate the forward-current and the forward-biased voltage in a *pn* junction diode. Fit the data to $I = I_0 V^n$ for the curve beyond the threshold to find I_0 and n (Figure 8-9).

V_{volts}	I_{ma}
0	0
0.25	0.003
0.50	0.005
0.56	0.008
0.58	0.015
0.60	0.025
0.62	0.040
0.64	0.060
0.66	0.103
0.68	0.213
0.70	0.431
0.71	0.710
0.72	1.04
0.73	1.42
0.74	2.10
0.75	3.13
0.76	4.55

8-3 Calculate the electric field in volts per meter, across a 10^{-6} m *pn* junction when the reverse voltage is 5 V. What is the kinetic energy of an electron crossing the *pn* junction in this electric field? If the ionization energy of an atom in the semiconductor is 0.5 eV, how far must an electron travel to acquire energy enough to produce an electron and hole by collision?

8-4 Field ionization requires a voltage difference across the atom equal to or exceeding the binding energy of an electron to its atom. What voltage is required if the atomic dimensions are 3×10^{-10} m and the binding energy is 0.1 eV.

8-5 Using Figure 8-16 to deduce the resistance versus path length from source to drain in both curves *a* and *b* in the lower drawing:

(a) Show that for curve *a* the following relation applies:

$$i = \frac{V}{R} \quad \text{(Ohm's law)}$$

(b) Show how $i = $ constant can result when the depletion area grows with each increase in V_{DS}.

(c) Show how the area under the curves in the lower drawing must depend upon V_{DS} to yield $i = $ constant.

REFERENCES

1948   Bardeen, J., and Brattain, W. H., (1948) "The Transistor, a Semiconductor Triode," *Physical Review*, **74**, 230.

1955   Pearson, G. L., and Brattain, W. H., (1955) "History of Semiconductor Research," *Proceedings of IRE*, **43**, 1794.

1976   *Handbook of Chemistry and Physics*, 57th Edition, CRC Press, Cleveland, Ohio.

1980   Kapoor, Ashok K., and Henderson, H. Thurman, "A New Planar Injection-Gate Bulk Switching Device Base Upon Deep Impurity Trapping," *IEEE Transactions on Electron Devices*, **ED27**, 1268 (1980).

CHAPTER 9

AMPLIFIERS AND FEEDBACK

Earlier in this book (in Chapter 7) we learned that an electrical signal source can be represented as a signal generator and a series impedance. This information is summarized as Thevenin's theorem. We will now examine the behavior of analog circuits that use the output of the signal source. These circuits may be passive circuits, which are networks of resistors, capacitors, and inductors or, as we emphasize in this chapter, they may be active circuits, which contain amplifiers and other elements that respond to the input signals and may draw energy from power sources in doing so.

In particular, we are interested in the behavior of circuits when a portion of their output is returned to their input by one means or another. This is known as *feedback*, and it may occur either by design or inadvertently. Feedback may improve the performance of the circuits in some ways, or it may be very detrimental. In this chapter we establish the parameters that determine which will be the case.

In our work with logic gates we found that returning output states to the inputs had certain beneficial effects. Flip-flops and other digital circuits with connections of that kind could perform data storage, counting, and other valuable functions. In the *JK* FF, however (shown in Figure 4-6), the connections from the outputs to the inputs when *J* and *K* are high (1) cause the output to switch repeatedly from high to low and back again as long as the clock pulse is present. The feedback causes an indeterminate state, so that one does not know whether Q or $\bar{Q}$ will be high when the clock pulse ends. The undefined state is a detrimental effect of feedback.

Analog circuits, in which the signals vary continuously, use feedback in particular ways to improve their performance. Unwanted instability may arise, however, unless one is careful to avoid the circumstances that cause it.

ANALOG AMPLIFIER CHARACTERISTICS

In order to address feedback in a general way, we establish the general features of amplifying circuits. Without discussing the details of the internal electrical circuit elements and their interconnections, we note in particular the relations between input and output signals. Figure 9-1 shows an amplifier with an input from a Thevenin-equivalent source.

In order to form a basis for discussing the general characteristics of amplifiers, we consider that the amplifier with gain A, shown in Figure 9-1, is an audio amplifier. An audio amplifier amplifies signals whose frequencies lie within the range of human hearing.

Figure 9-2 shows the amplification and the phase angle between the signal source and the amplified output of the amplifier. The amplification or gain is the ratio of the amplitudes of the output signal, e_o, and the input signal, e_i, that is, $A = e_o/e_i$. The phase is the difference in angle between identical points on the two waves that constitute the input and output signals. The amplifier has a gain of 50 over a wide band of audio signals, the midfrequency band, from about 100 Hz to about 20,000 Hz, but the input signal is not amplified so much if its frequency is lower or higher than those frequencies. The phase angle in the midfrequency band is close to zero degrees, but it increases at low and high frequencies. This behavior is typical for ac amplifiers.

Circuits that amplify sine waves signals use coupling capacitors to transmit the ac signals between elements and to block the dc components. (The capacitor

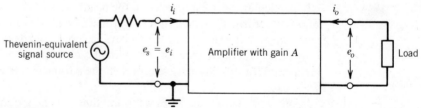

FIGURE 9-1

An Analog Amplifier

The capacitance-coupled amplifier has a frequency-dependent gain for periodic signals.

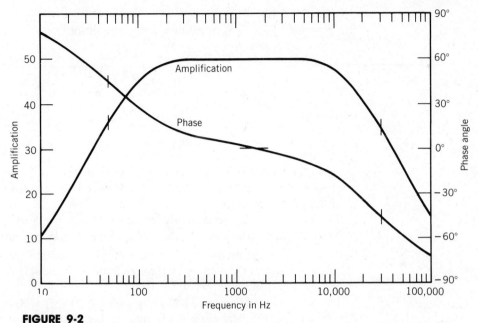

FIGURE 9-2

Amplification and Phase versus Frequency

The performance characteristics of an audio amplifier are displayed. The constant midfrequency gain of 50 falls off at both high and low frequencies. The curves are marked at the half-power points.

between the input and the transistor in the figure for Laboratory Exercise 16 serves this function.) Because the reactance of the capacitor, $X_c = 1/(2\pi f C)$, becomes very large for low frequencies, the low-frequency signals are attenuated; at higher frequencies the X_c is small, and the signals are transmitted without attenuation. As a result, the overall gain of the capacity-coupled amplifier is zero for dc, small for low frequencies, and relatively large and constant for midfrequencies.

At high frequencies the gain diminishes again because of capacitance; the problem, however, arises not from the coupling capacitors but from the stray capacitance in the circuit and between the elements in the circuit and ground. The capacitive reactance is initially very large because the stray capacitance is quite small, but as the frequency increases, X_c diminishes. The gain of the amplifier falls as the signal is bypassed to ground by the small capacitive reactance. At both low and high frequencies the capacitances cause a difference in phase between the input and the output signals.

The reduction in the amplitude or gain of the output signal reduces the power that the signal can provide. As shown in Chapter 7, the power carried by a voltage signal is proportional to the square of the signal amplitude

$$P \propto E^2$$

Therefore one can define the *half-power point* as the frequency at which the signal has dropped to $1/\sqrt{2}$ of its midfrequency amplitude,

$$E_{1/2} = \frac{1}{\sqrt{2}} E_{mid} = 0.707 E_{mid}$$

or

$$P_{1/2} \propto E_{1/2}^2 = 0.5 E_{mid}^2$$

Because the reduction in power occurs at both low and high frequencies, the frequency band between the half-power points is a measure of the frequency response of the amplifier.

The audio amplifier whose performance is shown in Figure 9-2 has half-power points at 50 and 30,000 Hz. At these frequencies the amplification is $35.36/50 = 0.707 = 1/\sqrt{2}$ of the midfrequency values of 50. The output (amplified) voltage is 45° ahead of the signal source for the low frequency and 45° behind for the high frequency. The band width is 50 to 30,000 Hz.

The magnitude of the amplification or *gain* is $e_o(t)/e_i(t) = A$, but the gain itself is a complex quantity because it is the ratio of the output and input voltages. The voltages themselves are complex quantities, each with a magnitude and a phase angle. The gain magnitude is constant over a wide range of middle frequencies, but the gain is less at both lower and higher frequencies. The relative phase angle between the output and input signals varies with frequency and passes through zero degrees (or 180°) in the midfrequency range. The phase shifts referred to in this chapter are the *changes* from the midfrequency phase shifts.

Parenthetically we note that a single-stage amplifier has a phase difference of 180° between the output and input voltage signals, while a two-stage amplifier has twice that amount, or an equivalent midfrequency zero-degree phase shift. A 180° midfrequency phase shift means that the output signal is a reflection of the input signal. For example, as a point on the input sine wave moves above the time base (abscissa), the same point on the output sine wave is moving downward. In the next chapter, when we study the one-stage transistor amplifier, we will see this relation.

Figure 9-3 gives the universal amplification and phase curves for capacitance-coupled amplifier stages. The symmetry between the low-frequency and the high-frequency response is emphasized by the presentation of the frequency scale as a ratio of the frequency to the frequency at the half-power point. The half-power point is the high frequency, f_H, or low frequency, f_L, at which the voltage amplification has fallen to 0.707 of the midfrequency amplification. The relative amplification is the amplification or gain at any frequency, divided by the amplification at the midfrequency,

$$\frac{A(f)}{A(\text{midfrequency})}$$

Analog Amplifier Characteristics **239**

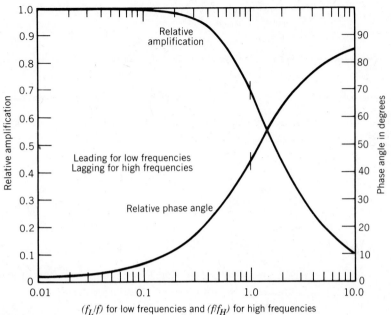

FIGURE 9-3

Universal Amplification and Phase

The general characteristics of capacitance-coupled amplifiers are given by the curves, which emphasize the symmetry of low-frequency and high-frequency behavior. The low-frequency f_L and high-frequency f_H half-power points are parameters for evaluating the gain at any frequency relative to the midfrequency gain.

In Figure 9-3, f is the frequency, f_L is the half-power *low* frequency, and f_H is the half-power *high* frequency. If we apply the universal curves to low frequencies, the relative frequency is f_L/f; for the high frequency the ratio f/f_H applies.

The amplification at a frequency f may be expressed in terms of the midfrequency amplification, which has a phase angle of zero, or 180 degrees. The angle $\underline{/\theta}$ is the change relative to the midfrequency phase angle.

$$A(f)\underline{/\theta} = \frac{A(\text{midfreq})}{[1 - jf_L/f]} \tag{9-1}$$

for frequencies lower than the midfrequency. This expression may be put into more useful form if we multiply both numerator and denominator by the complex conjugate of the denominator, $[1 + jf_L/f]$, to obtain

$$A(f)\underline{/\theta} = \frac{A(\text{midfreq})[1 + jf_L/f]}{[1 + (f_L/f)^2]}$$

$$= A(f)\exp(j\theta) \tag{9-2}$$

The phase angle θ is given by

$$\theta = \tan^{-1}(f_L/f) \tag{9-3}$$

When the frequency f is equal to the half-power frequency f_L, the ratio $(f_L/f) = 1$ and

$$A(f_L)\underline{/\theta} = 0.707\, A(\text{midfreq})\underline{/45°}$$

For frequencies below the midfrequency the output voltage leads the input signal.
For high frequencies

$$A(f)\underline{/\theta} = \frac{A(\text{midfreq})}{[1 + jf/f_H]}$$

$$= \frac{A(\text{midfreq})[1 - jf/f_H]}{[1 + (f/f_H)^2]}$$

$$= A(f)\exp(-j\theta) \tag{9-4}$$

This expression shows that for frequencies above the midfrequency the output voltage lags the input signal. The phase angle is

$$\theta = -\tan^{-1}\left(\frac{f}{f_H}\right) \tag{9-5}$$

For many purposes a phase difference between the amplified signal and the input signal is not significant in the use of the amplified signal. If a portion of the output signal is returned to the input of the amplifier, however, the phase relation becomes important.

DECIBELS

A common way to state the gain of an amplifier is to express it in units called *decibels* (dB). The relation between two power levels (or intensities of sound or light) is defined in terms of a common logarithm by

$$\text{Gain} = 10\log_{10}\left(\frac{P_2}{P_1}\right) \quad \text{dB} \tag{9-6}$$

where P_1 is the power (intensity) at one state, and P_2 is the power (intensity) at the second state.

Because the power or intensity is proportional to the square of the voltage amplitude, the voltage gain of an amplifier may be expressed in decibels by

$$\text{Gain} = 10\log_{10}\left(\frac{e_{\text{out}}^2}{e_{\text{in}}^2}\right) = 20\log_{10}\left(\frac{e_{\text{out}}}{e_{\text{in}}}\right) \quad \text{dB} \tag{9-7}$$

In calculating values in decibels, we must keep in mind the distinction between voltage amplitude and electrical power.

Common logarithms are based on the powers of 10 and are defined by

$$\log_{10} x = y$$

where y is the exponent in $10^y = x$. For example, $\log_{10} 100 = 2$ because $10^2 = 100$; thus, if $e_{\text{out}}^2/e_{\text{in}}^2 = 100$, the gain is 20 dB.

In the previous section the half-power points corresponded to the low or high frequencies at which the power output of the amplifier was one-half of the output at the midfrequency, $P(f_H) = \frac{1}{2} P(f_{\text{mid}})$. Thus we find

$$\text{Gain} = 10 \log_{10}(\tfrac{1}{2}) = 10(-0.301) = -3.01 \text{ dB}$$

The half-power point is said to be "down by 3 dB" from the midfrequency power level. At the half-power point the voltage at f_H is 0.707 of the midfrequency voltage; thus we find

$$\text{Gain} = 20 \log_{10}[e(f_H)/e(f_{\text{mid}})] = 20 \log_{10}(0.707)$$
$$= 20(-0.15) = -3.0 \text{ dB}$$

which corresponds to the value found when we use the power levels. If a voltage at a particular frequency is one-half the midfrequency voltage, however, then the value of the relative gain is

$$\text{Gain} = 20 \log_{10}(\tfrac{1}{2}) = 20(-3.01) = -6.0 \text{ dB}$$

which corresponds to a decrease in power level to one-fourth the midfrequency level.

Decibels are useful in describing amplifier performance because human senses are approximately logarithmic in sensitivity. For example, a change in sound intensity must be large enough to be recognized with certainty. A one-decibel change is approximately the degree of change that nearly everyone in a large group of people will recognize when it occurs. One dB corresponds to $P_2/P_1 = 1.26$, or sound-intensity change of about 26 percent. Humans can hear sounds that range from a very low detectable sound level of 10^{-12} W/m² to very intense levels. The intensity of sound may be expressed on an absolute scale by designating 10^{-12} W/m² $= P_0$ as the reference level expressing all other sound intensities in decibels. A sound intensity of $P = 90$ dB, for example, corresponds to $P = 10^{-3}$ W/m² or 1 mW/m², about the loudness of a pneumatic hammer 6 m away.

FEEDBACK IN ANALOG AMPLIFIERS

Feedback is provided by returning some of the amplified signal to the input of the amplifier. *Negative* or *degenerative feedback* occurs when the input signal is diminished by the portion of the output signal that is returned; *positive feedback* uses the output signal to reinforce the input signal so that the amplification is

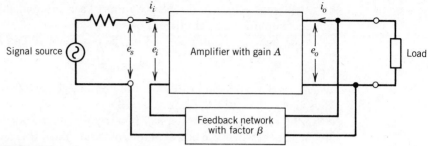

FIGURE 9-4

An Analog Amplifier with Feedback

The addition of a circuit that returns a portion of the output signal to the input greatly modifies the amplifier performance. The feedback may be advantageous or detrimental.

increased. Positive feedback is also called *regenerative feedback*. If positive or regenerative feedback is strong enough, the amplifier becomes an oscillator, which may or may not be the result intended by the designer. Most of the practical advantages of feedback are derived from negative or degenerative feedback.

Figure 9-4 shows an amplifier with a feedback circuit whose properties are designated by β. β may be a complex quantity that is frequency-dependent, or it may describe a resistor network with simpler properties. For the general case we set

$$\beta(f)\underline{/\theta_\beta} = \mathcal{R}e\,\beta + j\,\mathcal{I}m\,\beta = \beta(f)\exp(j\theta_\beta) \tag{9-8}$$

where θ_β is the phase angle generated in the feedback network itself. $\mathcal{R}e$ refers to the real part and $\mathcal{I}m$ refers to the imaginary part of β. The amplifier also has a phase angle, which it generates between output and input.

The amplification of an amplifier without the feedback network is called the *open-circuit gain*. We designate the open-circuit gain by $A\underline{/\theta}$ as it was used in the previous sections (see Figure 9-1), by the ratio

$$A\underline{/\theta} = \frac{e_o}{e_s}\underline{/\theta} = \frac{e_o}{e_i}\underline{/\theta} \tag{9-9}$$

The amplification with feedback is

$$A_f\underline{/\theta_f} = \frac{e_o}{e_s}\underline{/\theta_f} \tag{9-10}$$

where θ_f is the combined phase shift due to the amplifier and the feedback network. From Figure 9-4 we find the relation

$$e_i = e_s + \beta e_o \tag{9-11}$$

MIDFREQUENCY FEEDBACK

Because in the midfrequency range, the phase angles of $A\underline{/\theta}$ and $\underline{/\theta_\beta}$ are small, we assume for the present that they are zero. Combining Equations 9-10 and 9-11, we find that the relation between the amplifications with and without feedback for the midfrequency range is

$$A_f = \frac{e_o}{e_i - \beta e_o} = \frac{A}{1 - \beta A} \qquad (9\text{-}12)$$

Two possibilities exist with regard to the term βA. If βA is positive, then positive or regenerative feedback exists. If $0 \leq \beta A \leq 1$, the amplification with positive feedback is larger than the open-circuit amplification because the denominator is less than unity. As the magnitude of $1 - \beta A$ approaches zero, the amplification with positive feedback grows to large values, and the output signal no longer resembles the input signal. The amplifier may break into oscillations and generate an output signal even if the input signal is removed.

For positive feedback to be useful the term $1 - \beta A$ must remain relatively close to unity; that is, βA must be one-tenth or less.

The other possibility is negative feedback, in which the product βA is negative. The term $1 - (-\beta A)$ in the denominator of Equation 9-12 is larger than unity, and the amplification with feedback is less than the open-circuit gain. Amplifiers designed to use negative feedback usually have large open-circuit gains so that the reduced gains from negative feedback meet the amplification requirements of the system.

ADVANTAGES OF NEGATIVE FEEDBACK

1. We can now enumerate several advantages of negative feedback. Let us consider an amplifier in which the negative feedback is provided by a network of resistors; resistors remain very stable over long periods of time. If a fraction β, say one-tenth of the output signal, is returned to the input and if the open-circuit gain is large, perhaps 100 or more, the magnitude of the term $-\beta A$ will certainly be very much larger than unity. We may write Equation 9-12 as

$$A_f = \frac{A}{1 - (-\beta A)} \approx \frac{A}{\beta A} = \frac{1}{\beta} \qquad (9\text{-}13)$$

Under the conditions of large open-circuit gain and negative feedback we have an amplifier whose gain is established by the feedback network itself.

The advantage becomes apparent when we note that components, such as transistors of a particular kind made by a single manufacturer, may differ by as much as 50 percent in performance. If the gain is determined by the feedback network, however, no special selection of transistors is required, and transistors can be replaced with little effect on the overall performance of the system.

2. A second advantage of negative feedback is the ability to ensure amplifier performance against deterioration of its active components. To illustrate this advantage we may consider an amplifier with an initial open-circuit gain of 1000, which deteriorates to 500. With a feedback factor $\beta = -0.05$ the gain with negative feedback is

$$A_f = \frac{1000}{1 + 50} = 19.61$$

with the initial amplification and

$$A_f = \frac{500}{1 + 25} = 19.23$$

with the later amplification. The performance of the feedback amplifier declines only 2 percent (-0.17 dB) when the open-circuit gain drops to one-half of its initial value (-6.02 dB). Electronic instruments are made with feedback amplifiers so that the instruments can virtually "ignore" changes in active component performance due to electrical-line voltage variations, changes in ambient temperature, or other occurrences, which may alter the open-circuit amplification by as much as several percent.

3. A third advantage of feedback is the reduction in distortion. Distortion arises when the signal amplitude is so large that it extends into the nonlinear range of the amplifier. In the case of a sine wave signal the peak amplitudes will not be amplified at the same ratio as the smaller portions of the waveform. The result is a flat-topped wave. In a multistage amplifier the distortion arises in the final stage, where the signal amplitudes are large.

Figure 9-5 shows a feedback amplifier with the two amplifying stages. Because the signal amplitudes in the first stage are small, the distortion due to large amplitudes arises only in the second stage. We represent the distortion as a signal source e_d, which is added to the signal output of the first stage e_1 even though e_d is generated within the second stage. The source e_d might be similar to the third harmonic in Figure 7-17, which flattens the fundamental.

The output signal

$$e_o = A_2(e_1 + e_d)$$

is returned partially to the input by the feedback network. The input signal is

$$e_i = e_s + \beta e_o$$

Thus we may write

$$e_1 = A_1(e_s + \beta e_o)$$

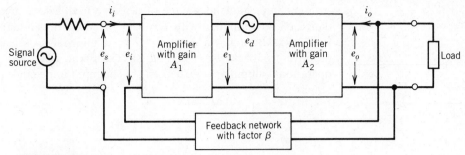

FIGURE 9-5

A Amplifier with Distortion in the Final Amplifying Stage

The contributions of negative feedback in reducing distortion introduced in the final stage is found by adding a distortion signal, e_d, to the amplified output of earlier stages, e_1.

and eliminate e_1 from the two equations. A little algebraic calculation gives us

$$e_o = \frac{A_1 A_2 e_s}{1 - \beta A_1 A_2} + \frac{A_2 e_d}{1 - \beta A_1 A_2} \qquad (9\text{-}14)$$

The total open-circuit gain $A = A_1 A_2$ is applied to the input signal e_s, but only A_2 amplifies the distortion signal e_d. In the multistage amplifiers the voltage gain of the final (power) stage is usually modest; $A_1 = 300$ and $A_2 = 10$ might be typical. Using these values with $\beta = -0.05$, we find for the feedback amplifier

$$e_o = 19.87 e_s + 0.07 e_d$$

The advantage of reducing distortion by negative feedback becomes evident when the distortion is introduced internally after the first stage. A distorted input signal is not improved, and if distortion arises in the first stage because of improper operating conditions, it is not corrected.

4. When amplifiers amplify signals, noise may intrude. Noise is any contaminating electrical signal, and in audio amplifiers it may be a hum or hiss that arises from thermal sources or poor junctions. It may be represented as a signal source, as distortion is, and the advantages of negative feedback can be found by an identical analysis. The reduction in noise generated internally after the first stage is

$$N_f = \frac{A_2 N}{1 - \beta A_1 A_2} \qquad (9\text{-}15)$$

where N is the noise signal and N_f is the noise magnitude with feedback. It is important to note that the amplifier cannot distinguish between the desired signal and noise in the initial signal into the amplifier; feedback treats both types of signal identically.

5. Another important advantage of negative feedback is that the input impedance of an amplifier with feedback may be much higher than without feedback. Although an increase in input impedance is usually achieved by negative feedback, we will find an exception when we discuss inverting operational amplifiers in Chapter 11. The input impedance of an open-circuit amplifier is

$$Z_i = \frac{e_i}{i_i}$$

Thus with feedback $e_i = e_s + \beta e_o$, and we find that the input impedance of the amplifier with feedback is

$$Z_{if} = \frac{e_s}{i_i} = \frac{e_i - \beta e_o}{i_i} = \frac{e_i - \beta A e_i}{i_i}$$

$$= Z_i(1 - \beta A) \qquad (9\text{-}16)$$

When βA itself is negative and larger than one, the input impedance is greatly increased.

6. The output impedance may be diminished by feedback. The analysis that shows this requires us to consider the nature of the output of an amplifier. Figure 9-6a shows an open-circuit amplifier with no input signal; that is, the input terminals are tied together so that the quiescent (no-signal) output voltage and current prevail. The output impedance of the amplifier is

$$Z_o = \frac{e_o}{i_o}$$

Figure 9-6b shows the amplifier with the feedback network and again with no input signal. The feedback network, however, causes a voltage $-\beta e_o$ at the input. As shown in Figure 9-6b, this voltage is amplified and makes a Thevenin-theory equivalent signal source at the output. We find that the output current is

$$i_{of} = \frac{e_o - \beta A e_o}{Z_o} = \frac{e_o(1 - \beta A)}{Z_o}$$

The output impedance with feedback is

$$Z_{of} = \frac{e_o}{i_{of}} = \frac{Z_o}{(1 - \beta A)} \qquad (9\text{-}17)$$

Thus negative feedback may reduce the output impedance and increase the input impedance of the amplifier. The reduction in output impedance is important in providing fan-out, that is, allowing an amplifier to drive many of the circuits that proceed from it.

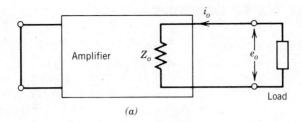

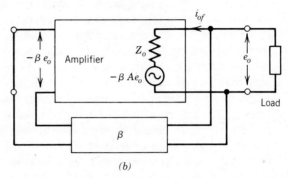

FIGURE 9-6

An Output Impedance of Amplifiers

The open-circuit output impedance, $Z_o = e_o/i_o$ is modified by feedback. The feedback network creates a Thevenin-like internal signal source.

7. A final advantage of negative feedback is the expansion of the frequency interval between half-power points. The *bandwidth* of the amplifier is extended while the midfrequency amplification is reduced relative to the open-circuit gain of the amplifier. The advantage of increased bandwidth is illustrated in the following sections.

A word about amplification is in order. The reduction in amplification by negative feedback may appear to be a great sacrifice, but in fact the power gain may be substantial if the voltage amplitude is increased only modestly. A crystal cartridge on a phonograph, for example, may produce a rms voltage of 100 mV, so that a voltage gain of about 200 is required to drive a 50-W, 8-Ω speaker system ($E = 10^{-1} \times 200 = 20$ V, $P = E^2/R = (20)^2/8 = 50$ W). An amplifier with a midfrequency open-circuit gain $A_1 A_2 = 10{,}000$ and a feedback factor $\beta = 0.005$ will suffice. An amplifier with modest voltage gain as the last stage will match the power requirements of the speakers. In many applications an increase in bandwidth, freedom from changes in the performance of active elements, and the reduction in distortion and noise are more important than high gain.

FEEDBACK AT ALL FREQUENCIES

In the discussion of midfrequency feedback and its advantages we said nothing about the phase angles of the amplifier and the feedback network but assumed that they were zero degrees. Now let us consider the general case in which frequencies beyond the midfrequency cause the phase angles of the amplifier and the feedback network to be significant in the operation of the feedback amplifier.

As the frequency of the signal source increases well beyond the midfrequency, the gain drops and the lagging phase angle grows. The reason for this is that the capacitive reactance from both the small intrinsic capacitance of the electronic elements and from the small stray capacitance between conductors and components acts as a low-impedance path to ground. This inevitable result can be shifted somewhat to higher frequencies by the choice of elements and by care in circuit configuration. Technical advancements in electronic components have extended (and may be expected to extend) their high-frequency response by reducing their intrinsic capacitance, but ultimately a high-frequency half-power point will be reached. The low-pass filter in Figure 7-23, whose response is given in Figure 7-27, and the amplifier, whose response is given in Figures 9-2 and 9-3, have identical shapes of high-frequency relative amplitude and phase angle curves for the same reason: the capacitive reactance to ground. Only the frequency scales are different.

In the discussion above, we disregarded the phase angles, but now that we are considering them, we can obtain a more general analysis of feedback by writing Equation 9-12 with $A(f)/\underline{\theta}$ and $\beta(f)/\underline{\theta_\beta}$ from Equations 9-9 and 9-8, respectively.

$$A_f/\underline{\theta_f} = \frac{A \exp(j\theta)}{1 - \beta A \exp(j(\theta_\beta + \theta))} = \frac{A(f)/\underline{\theta}}{1 - \beta(f)A(f)/\underline{\theta + \theta_\beta}}$$

$$= \frac{A}{\exp(-j\theta) - \beta A \exp(j\theta_\beta)} \qquad (9\text{-}18)$$

Using the relation $\exp(j\theta) = \cos\theta + j\sin\theta$, we can write the amplification with feedback as

$$A_f(f)/\underline{\theta_f} = \frac{A(\cos\theta - \beta A \cos\theta_\beta) + jA(\sin\theta + \beta A \sin\theta_\beta)}{1 + \beta^2 A^2 - 2\beta A \cos(\theta + \theta_\beta)} \qquad (9\text{-}19)$$

with the phase angle

$$\theta_f = \tan^{-1} \frac{\sin\theta + \beta A \sin\theta_\beta}{\cos\theta - \beta A \cos\theta_\beta} \qquad (9\text{-}20)$$

The difference between the open-circuit amplifier and the same amplifier with negative feedback is illustrated in Figure 9-7. The open-circuit amplification was taken to be $A(\text{midfreq}) = 100/\underline{0°}$ (40 dB); the feedback factor is

$$\beta/\underline{\theta} = \beta/\underline{180°} = -\beta = -0.05$$

When one part in 20 of the output signal is returned to the input, the midfrequency

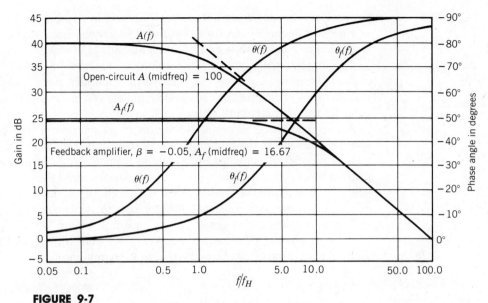

FIGURE 9-7
A Bode Plot of Amplifier Characteristics

A Bode plot uses the gain in decibels versus the logarithm of the frequency; that is, a log-log plot of gain versus frequency. The simplified Bode plot extends the midfrequency gain horizontally and extrapolates the asymptoic gain falloff back to the point of intersection. The dashed lines represent this construction. The lines meet at the high-frequency half-power point.

gain is $A_f(\text{midfreq}) = 16.67 \underline{/0°}$ (24.44 dB). The curves in Figure 9-7 for the feedback amplifier were calculated with these values; Equations 9-19 and 9-20 were calculated with $A(f)$ and $\theta(f)$ from Equations 9-4 and 9-5. In calculating the material plotted in Figure 9-7, we assumed the feedback factor $\beta/\theta_\beta = -0.05$ to be frequency-independent. This is essentially the case if the feedback network is a resistor network and its stray capacitance is small in relation to the intrinsic capacitance of the amplifier itself.

The amplifications in Figure 9-7 are stated in decibels, so that the ordinate of Figure 9-7 is really a logarithmic scale. Amplification in decibels and the logarithmic scale for frequency provide a very useful presentation of the data; this kind of presentation is often called a *Bode plot.*

BODE PLOT

An examination of Figure 9-7 shows that the open-circuit amplifier has a half-power point, gain down by 3 dB, at $f/f_H = 1$ as defined. The gain with negative feedback, however, drops by 3 dB only when $f/f_H = 6$. If the open-circuit high-frequency half-power point is 10 kH, then the half-power point with negative feedback is extended to 60 kH, a great increase in bandwidth. A similar relative extension of frequency response occurs at low frequencies. As observed above, the expansion of bandwidth is an important advantage of negative feedback.

Far from the midfrequency the falloff in gain reaches -6 dB per octave—that is, each doubling of frequency—or -20 dB for each factor of 10 in frequency. The values characterize resistance-coupled amplifiers and low-pass filters and represent the expected amplifier performance.

As an expedient, the essential features of an amplifier's performance can be found by measuring the midfrequency amplification and the asymptotic falloff at high frequencies. A *simplified Bode plot* can be made by drawing the midfrequency gain in dB as a horizontal line to high frequencies and extending the asymptotic falloff back to where the lines intersect. The point of intersection is the half-power point. In Figure 9-7 the dashed lines represent this construction.

The phase angles $\theta(f)$ and $\theta_f(f)$ are shown in Figure 9-7. Each has the value $-45°$ at its respective half-power point, $f/f_H = 1$ and $f/f_H = 6$, and they approach each other asymptotically in the high-frequency region.

STABILITY OF ANALOG AMPLIFIERS WITH FEEDBACK

The stability of an analog amplifier is determined by the amplitude of the feedback signal and its phase relation to the input signal. When the returned signal is added to the initial signal—that is, when the feedback is positive or regenerative—the gain of the amplifier increases. The denominator of Equation 9-12 is reduced by positive feedback, and A_f becomes larger than A. As the denominator becomes smaller, the stability of the amplifier becomes uncertain. Two criteria exist for evaluating the stability of analog amplifiers with feedback; these are presented below.

If the denominator of Equation 9-12 approaches zero, the amplifier becomes unstable and a free-running oscillation begins. The amplitude of the oscillation is bounded by the power limitations and nonlinearities of the electronic elements. Sometimes, if the voltage extremes block off the operation of an element, the instability results in an intermittent signal. This condition is called *motorboating* because the sound in an unstable audio amplifier resembles that of a motorboat.

An amplifier with negative feedback in the midfrequency range may also be unstable because the feedback becomes positive or regenerative at high frequencies. As the frequency of the input signal increases, the output signal becomes farther out of phase with the input signal, as seen in Figure 9-7. If the feedback network also shifts the phase of the feedback voltage, the product $\beta A \underline{/\theta + \theta_\beta}$ can play a role at high frequencies. Noise and other sources that have very high-frequency components may initiate the instability. An amplifier will be stable only if excessive positive feedback is prevented at all frequencies. The falloff in gain at frequency extremes is important in avoiding unwanted positive feedback.

Sometimes regenerative feedback arises from coupling through the power sources or from factors that are not actually part of the feedback amplifier. One example of an external factor is the regenerative feedback that occurs when a microphone picks up its own amplified sound from a loudspeaker and a shrill tone occurs.

NYQUIST CRITERION FOR STABILITY

As mentioned earlier, two criteria exist for judging the stability of a feedback amplifier system. The Nyquist criterion requires an analysis or measurement of the complex term $\beta A / \theta + \theta_\beta$. It is evaluated for all frequencies, and the real and imaginary parts, $\mathcal{R}e\,\beta A + j\,\mathcal{I}m\,\beta A$, are plotted on the complex plane. Figure 9-8 illustrates the plot of the real and imaginary parts. Unlike the scales on the other axes, the scale of the negative real axis in Figure 9-8 differs by a factor of 10. The midfrequency value of βA is -26.0; the curve is plotted point by point as the frequency varies. The portion of the curve with large negative real values (left side of Figure 9-8) corresponds to the midfrequency behavior of the amplifier.

The Nyquist criterion states that if the point $(1 + j0)$ on the complex plane is encircled, the amplifying system is unstable. The value $\beta A = (1 + j0)$ makes the denominator of Equation 9-12 zero, and the amplification with feedback becomes infinite.

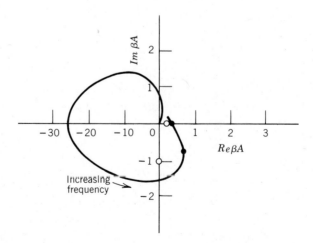

FIGURE 9-8

A Nyquist Plot of $\mathcal{R}e\,\beta A + j\,\mathcal{I}m\,\beta A$

The real and imaginary parts of the term βA are plotted for all frequencies. The left real axis is a factor of 10, unlike the other axes. The open circles represent values taken from Figure 9-9. The solid dots correspond to the minimum requirements for amplifier stability.

BODE CRITERION FOR STABILITY

The Bode criterion is based upon a Bode plot of the amplitude and phase of the factor βA. The values to be plotted—the amplitude in dB and the phase in degrees versus the logarithm of frequency—are obtained by calculation or by driving the feedback amplifier with a variable-frequency signal generator and measuring the relative amplitude and phase between the feedback signal and the input signal. Figure 9-9 shows a plot of βA for a stable feedback amplifier.

Stability is judged by the values of βA and the phase angles at two conditions. The first one corresponds to the point at which the magnitude of βA is unity; $\beta A = 1$ (0 dB). The phase angle is determined at this point. The second significant

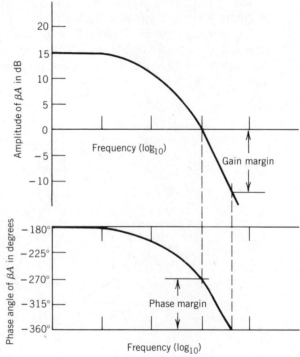

FIGURE 9-9

A Bode Plot of A

The feedback term βA in decibels is plotted versus the logarithm of frequency. The gain margin and the phase margin are found to judge the stability of the feedback amplifier.

value is the magnitude of βA where the phase angle of βA is equal to $-360°$. The magnitude of βA is evaluated in decibels, that is, $20\log_{10}\beta A$. This value is called the *gain margin*. The difference between the phase angle found at zero decibel and $-360°$ is called the *phase margin*.

A feedback amplifier is judged to be stable if the gain margin is $-10\,\text{dB}$ or a greater negative amount and if the phase margin is at least $-50°$. The gain margin and the phase margin are shown in Figure 9-9.

The two criteria for stability use different displays of the feedback term βA to judge the performance of the amplifier, but they are equivalent techniques. In applying either, we must examine the entire frequency spectrum. The high frequencies are most likely to contribute to the instability because of the phase shifts that arise from the stray capacitances, which are very difficult to know. The drop in gain at the high frequencies reduces the term βA; both criteria recognize that a rapid decrease in magnitude of βA contributes to stability.

The similarity between the Bode and the Nyquist criteria for stability is illustrated by plotting the values from the Bode curve, Figure 9-9, onto Figure 9-8. One open circle corresponds to the value $\beta A = 1$ (0 dB), where the phase angle is $-270°$. This value is plotted at $1/\underline{-270°} = 0 - j1.0$. The second open circle corresponds to the value of βA at $-360°$, which, as we learned from Figure 9-9, is $\beta A = -12.4\,\text{dB} = 20\log_{10}\beta A$. Therefore, $\beta A = \log_{10}^{-1}(-12.4/20) = 10^{-0.62} = 0.24$; $A/\underline{-360°} = 0.24 + j0$.

The minimum requirements for amplifier stability are plotted as solid dots on Figure 9-8. They represent the gain margin ($-10\,\text{dB}$ at $360°$) and the phase margin ($-50°$ less than $-360°$ where $A = 0\,\text{dB}$). The values for these limits are $\beta A/\theta = 1/\underline{-310°} = 0.64 - j77$, plotted as one solid dot, and $\beta A/\underline{-360} = 0.32\,\underline{0°} = 0.32 + j0$, the second solid dot. When the open circles are compared with the solid dots, the amplifier whose βA characteristics versus frequency are plotted in Figure 9-9 appears to have more acceptable values for its gain margin and phase margin than do the minimum stability requirements.

The applications of feedback are illustrated in Chapter 11, which deals with operational amplifiers, and elsewhere in this book. The concepts presented here are encountered in control theory, where the problems of correcting or directing a device entail returning a sample of the output to the input. In addition, the time delays in control theory add complications not suggested here.

SUMMARY

Amplifiers for periodic signals have capacitive coupling to isolate the dc voltages and currents. The capacitive coupling affects the low-frequency gain and the phase angle of the output signal relative to the input signal. As the frequency increases, the gain also increases, and the relative phase angle decreases from $+90°$ at zero frequency. For midfrequencies the gain is practically constant and the phase angle is near zero. At higher frequencies the gain and the phase angle both decrease, not because of the capacitive coupling, but because of stray capacitance in the elements and in the circuit that bypass the signals to ground. This behavior is so general that the relative gain and phase angles can be presented by a universal curve (Figure 9-3). The low and high frequencies at which the voltage gain is 0.707 or $1/\sqrt{2}$ of the midfrequency gain are called the half-power points. The frequency range between half-power points is the frequency bandwidth of the amplifier.

The behavior of the amplifier is modified greatly when a portion of the output is returned to the input. The feedback can be negative, or it can be positive (regenerative); if positive, it can increase the gain or lead to instability. Negative feedback has a number of salutary effects, the most important of which are the expansion of the frequency bandwidth and the decrease of the amplifier's sensitivity to changes in circuit components.

With calculations or measurements, the stability of feedback amplifiers can be judged in two ways, by the Nyquist and the Bode criteria. The entire frequency range must be considered in applying these criteria.

PROBLEMS

9-1 An instrument with a negative-feedback amplifier suffers a line voltage change that reduces the open-circuit gain of the amplifier by 3 percent. If the open-circuit gain is 10,000 and the feedback factor is 0.05, find the change in the instrument response in percentage and in decibels.

9-2 Find the values in decibels for the following voltage gains: 1.26, 2, 5, 10, 20, 50, 100, 2000, 20,000, and 200,000.

9-3 The intensity of the sound near a jet aircraft engine during takeoff is approximately 1.0 W/m^2. What is the value in decibels of this painful sound intensity, which can permanently damage the ability to hear? The sound intensity reference P_0 is 10^{-12} W/m^2.

9-4 An amplifier with an open-circuit gain of 500 has a high-frequency half-power point at 8000 Hz. How much feedback is required to move the half-power point to 18,000 Hz? What is the gain with this feedback? If the initial low-frequency half-power point is at 80 Hz, what is the bandwidth after the feedback? Make Bode plots of the amplifier response before and with feedback.

CHAPTER 10

ELEMENTS OF TRANSISTOR CIRCUITS

The theme of this text is the use of integrated circuits as logic gates, digital elements, operational amplifiers, and linear devices rather than the construction of electronic circuits with elementary components. Yet the frequent use of individual transistors in digital and analog circuits to provide power gain and other advantages suggests that an understanding of them is useful and, indeed, necessary. Furthermore, a knowledge of the basics of transistor use contribute to understanding the nature and use of digital and analog devices.

In the spirit of this text we limit our attention to a few particularly useful transistor circuits and to the design considerations that are essential to satisfactory transistor operation. The essentials are (1) the choice of the transistor and its quiescent operating point, (2) the choice of resistors and capacitors to give the desired amplification, impedance match, frequency range, and other features, and (3) protection against thermal runaway. The bibliography lists several good sources that discuss transistor amplifier design in great detail.

The principles of bipolar-junction transistors were developed in Chapter 8. As you may recall, the three elements—emitter, base, and collector—require the proper voltage polarity and values to operate in a useful mode. For analog amplifiers the quiescent conduction must be in a linear portion of the operating range. For digital use the two allowed states, "on" or "off," must result in a low output or a high output, each of which can be distinguished clearly from the other.

...ng the characteristic data of transistors provided by the manufacturer, we will ...hat both analog and digital applications follow from the same design analysis. ...a reminder, note that the arrow on the emitter in the symbol for a bipolar-...n transistor indicates the direction of conventional (positive) current; for ...junction transistor operation, it indicates the direction from positive to ... voltage.

COMMON-EMITTER AMPLIFIER

The *common-emitter amplifier* is the configuration that represents the conventional voltage amplifier with voltage gain, current gain, and power gain. (Power gain is the product of the voltage gain and the current gain.) The configuration is called a common-emitter amplifier because the emitter is the element of the transistor common to both the input circuit and the output circuit. Figure 10-1 shows the circuit with which a *NPN* bipolar-junction transistor operates as a one-stage common-emitter ac analog amplifier.

The common-emitter amplifier is symbolized ideally in Figure 1-1, where the output voltage is the portion of the supply voltage that occurs across the resistor R and is given by

$$v_o = V_{cc} \frac{R}{R + R_L} \qquad (10\text{-}1)$$

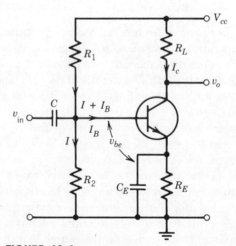

FIGURE 10-1

A Common-Emitter Amplifier

A *NPN* junction bipolar transistor amplifies ac input signals as a typical low-power voltage amplifier. The values of the components are given in the design example.

In the case of the common-emitter amplifier the transistor corresponds to the variable resistor R. Here we consider the resistance values of R and R_L similar in magnitude that correspond to an analog amplifier. Later we examine the digital case, in which R is much larger or much smaller than R_L.

In the common-emitter amplifier the conduction through the transistor is determined by the base current; the base current varies according to the input signal $v_{in}(t)$. Conduction is the reciprocal of resistance; thus a high conductance is a low resistance. A high base current from a high input signal causes a smaller portion of the supply voltage, V_{cc}, to appear across the transistor as the output voltage, $v_o(t)$. When the input signal reduces the base current, the transistor conduction drops and the effective resistance of the transistor increases, providing a larger portion of the supply voltage for the output. The base current is dependent upon the forward bias of the emitter-base pn junction. An increase in v_{be} (emitter-base ac voltage) to the amplifier circuit, as shown in Figure 10-1, increases the conduction of the transistor and thus reduces the output voltage. We will show later how the quiescent emitter-base voltage, V_{BE}, is established so that the ac input signal, $v_{in}(t)$, properly affects the base current. An important property of this one-stage amplifier is that an increase of the input voltage, $v_{in}(t)$, reduces the output voltage, $v_o(t)$. A 180° phase difference exists between the input and output signals in the midfrequency range.

The ratio of the collector current to the base current is large. Because β is usually between 50 and 400 for low-power bipolar-junction transistors, a small change in base current results in a large change in conduction through the transistor,

$$\beta = \frac{i_c}{i_b} \tag{10-2}$$

In the manufacturers' manuals, which give transistor characteristics, the quantities labeled h_{FE} (forward-current transfer ratio) and h_{fe} (small signal forward-current transfer ratio) correspond effectively to β. (It is customary to use capital letters in subscripts for direct current or quiescent values and lowercase subscripts for alternating-current values.) The quantity h_{FE} is given by $h_{FE} = I_c/I_b$, the dc ratio of the collector current, I_c, to the base current, I_b; thus h_{FE} may be equated with β for dc conditions. The quantity h_{fe} is measured at a particular steady, dc, voltage across the transistor, collector to emitter, V_{ce}. The ratio of the change in the collector current, ΔI_c, to the change in the base current, ΔI_b, gives $h_{fe} = \Delta I_c/\Delta I_b$ at V_{ce}. h_{fe} depends upon the point at which it is measured and differs slightly from h_{FE}. β and h_{fe} are frequently assumed to be the same. The quantity h_{fe} is given in the specification sheet for specific values of V_{CE} (collector-emitter quiescent voltage) and I_C (quiescent-collector current) at a signal frequency of 1 kHz. Because β is not well controlled in manufacture, minimum and maximum values of h_{FE} and h_{fe} may be quoted in transistor specification sheets. The actual value of β is not very important, except that it must be reasonably large.

Next we give an example of the design for a simple common-emitter transistor voltage amplifier to show how the quiescent points and circuit elements of the transistor amplifier are determined.

DESIGN EXAMPLE

To start the design of the one-stage common-emitter analog voltage amplifier, we choose a transistor and its quiescent operating point. As the example shows, that as the resistors and capacitors are chosen to yield the quiescent operating point, precise values are not required because the currents adjust by a small percentage as the transistor seeks its operating point. A number of arbitrary choices based upon experience are involved but many parameters, such as the supply voltage and the gain, are frequently dictated by available power sources and the specific application.

Table 10-1 gives values taken from a manufacturer's specification sheet for a typical low-power *NPN* transistor (Texas Instruments, 1973).

Because the power supply voltage, V_{cc}, sets the voltage ranges in the circuit, an early step in the design is the selection of the quiescent collector current, I_C, consistent with the voltage available. The maximum ratings from the specification sheet are not very useful for this purpose because a much smaller quiescent current is suitable. Experience or examples are better guides. We choose $V_{cc} = 12$ V and $I_C = 4$ mA as a starting point. Figure 8-14 suggests that these are reasonable values. The maximum values are limits that should not damage the transistor, but they do not suggest operating values in amplifiers.

Because the amplifier is to be an analog voltage amplifier, we let R (the effective quiescent-transistor resistance) equal $R_L + R_E$ (the effective load resistance); thus approximately half of the supply voltage will occur across the transistor at the

TABLE 10-1
Data Selected from a Specification Sheet for a Silicon *NPN* Transistor 2N4123

Absolute maximum ratings at 25°C free-air temperature		
Collector-base voltage		40 V
Collector-emitter voltage		30 V
Emitter-base voltage		5 V
Continuous collector current		200 mA
Continuous device dissipation		625 mW
Electrical characteristics at 25°C free-air temperature	Min	Max
$h_{FE}(V_{CE} = 10$ V, $I_C = 2$ mA)	50	150
$h_{fe}(V_{CE} = 10$ V, $I_C = 2$ mA, $F = 1$ kHz)	50	200

quiescent point. We use $I_E = I_C$ to find

$$\frac{V_{cc}}{I_C} = R + R_L + R_E = 3.0 \text{ k}\Omega$$

With $R = R_E + R_L = 1.5 \text{ k}\Omega$, we let $R_E = R_L/3$ and find that $R_L = 1.1 \text{ k}\Omega$ and $R_E = 0.4 \text{ k}\Omega$. The ratio between R_L and R_E is a matter of arbitrary choice. R_E should be at least $R_L/10$, but our choice of a larger ratio is consistent with providing thermal runaway protection, as discussed below.

At the quiescent operating current the emitter voltage, V_E, will be $V_E = I_C R_E = 1.6$ V above ground. The quiescent base current can be estimated from I_C and β. Unless β is known with certainty, it is frequently taken as 50 or 60 percent of the maximum value of h_{fe} that is given in the specification sheet. We take $\beta = 100$ and find $I_B = I_C/\beta = 40 \, \mu\text{A}$.

The voltage divider, R_1 and R_2, which establishes the quiescent base voltage, is $V_B = V_{BE} + V_E$. If we assume that V_{BE} is 0.8 V, slightly more than the cut-in voltage of the silicon pn junction, $V_B = V_{BE} + V_E = 0.8 + 1.6 = 2.4$ V. Figure 8-13 illustrates that the choice of a V_{BE} slightly larger than the cut-in voltage will yield a reasonable collector (emitter) current. The base is held at the voltage established by the voltage divider, which is formed by R_1 and R_2. A small change in V_{BE} causes a large change in I_C (because of the steep slope of I_C versus E_{EB}; see Figure 8-13) and the shift in the quiescent emitter voltage, $V_E = I_E R_E = I_C R_E$, establishes an equilibrium or quiescent operating point very near $V_{BE} = 0.8$ V.

Let I be the current through R_2; thus $I + I_B$ is the quiescent current through R_1; the voltage relations are

$$V_{cc} = R_1(I + I_B) + R_2 I = 12 \text{ V}$$

and

$$V_B = R_2 I = V_{BE} + V_E = 0.8 + 1.6 = 2.4 \text{ V}.$$

The current I should be much larger than the base current, I_B, so that a change in base current will not significantly shift the voltage at the base. We choose $I = 10 I_B$ to obtain

$$R_1 = 22 \text{ k}\Omega$$

and

$$R_2 - 6 \text{ k}\Omega$$

The voltage, V_E, caused by the current, I_E, through R_E should also remain nearly unchanged when an ac signal is amplified. This condition is ensured by placing a capacitor across R_E. A large time constant, $R_E C_E$, sustains V_E during the ac variations. At low frequencies the V_E begins to follow the signal and the gain is diminished. Using 50 Hz as an arbitrary choice for the half-power point, we find that $C_E = 8 \, \mu\text{F}$, so that a 10-μF capacitor with at least a 5-V rating can be used to ensure a constant-emitter voltage.

R_1, R_2, and R_E also serve another important function, which is to prevent thermal runaway in the transistor.

A transistor will be destroyed by overheating when maximum ratings are exceeded or by thermal runaway. Thermal runaway occurs when the temperature of the transistor becomes too high; then the transistor quickly grows even hotter. As we noted in Chapter 8, there exist some minority carriers in extrinsic semiconductor material that are not neutralized by the donor or acceptor impurities added to the intrinsic semiconductor material to form extrinsic semiconductor material. The residual minority carriers were noted as the source of the very small inverse current in a *pn* diode. We also noted that minority carriers are produced by the thermal promotion of electrons from the valence band to the conduction band in both intrinsic and extrinsic semiconductor materials.

As the temperature of the transistor increases, the number of minority carriers grows exponentially, and the inverse collector current across the reverse-biased collector-base *pn* junction increases substantially. This increase leads to greater power dissipation, which heats the transistor even more. Because of the strong thermal dependence in the creation of minority carriers, the self-regenerative process raises the transistor temperature to destructive levels.

Thermal runaway can be prevented by assuring that the thermally generated base current causes a change in the base-emitter voltage to reduce the current through the transistor, an example of negative feedback. An examination of Figure 10-1 shows that the thermally generated base current can reach ground through either R_1 (and the power supply) or R_2. R_1 and R_2 are parallel resistors with respect to the thermally generated base current. The current-coupled feedback network of the parallel resistors, $R_1 R_2/(R_1 + R_2)$, and the emitter resistor, R_E, will suppress thermal runaway if the ratio

$$\frac{R_1 R_2}{R_1 + R_2} \bigg/ R_E$$

is no greater than approximately 10. A small ratio provides greater protection against thermal runaway but diminishes the gain of the amplifier.

In checking our design example, we find the ratio

$$\frac{R_1 R_2}{R_1 + R_2} \bigg/ R_E = 11.8$$

which is beyond our criterion. By reducing R_1 and R_2 proportionally, we choose $R_1 = 18\ \text{k}\Omega$ and $R_2 = 5\ \text{k}\Omega$, the ratio becomes 9.8, which is acceptable for thermal runaway protection and does not significantly diminish the gain.

The input capacitor, C, isolates the amplifier from dc voltages in the signal source. The reactance, $X_C = 1/(2\pi f C)$, should be approximately the same magnitude as the resistance R_2 at the frequency used to choose C_E. We find that at 50 Hz $C = 0.6\ \mu f$.

Table 10-2 lists the values in the design example and the actual values used in making a single-stage common-emitter amplifier to verify the design. The self-

TABLE 10-2
Common-Emitter Amplifier with NPN 2N4123

	Design Values	Actual Values
R_L	1.1 kΩ	1.0 kΩ
R_E	0.4 kΩ	0.39 kΩ
C_E	8 μF	10 μF
R_1	18 kΩ	19 kΩ
R_2	5 kΩ	4.7 kΩ
C	0.6 μF	0.15 μF
β	100	81
I_C	4.0 mA	3.75 mA
I_B	40 μA	46 μA
V_E	1.6 V	1.5 V
V_C	7.6 V	7.3 V
V_{BE}	0.8 V	0.71 V
Gain	100	>100

biasing provided by R_E and C_E allows the emitter voltage, V_E, to grow to just V_{BE} less than the quiescent base voltage established by the voltage divider circuit of R_1 and R_2, and a quiescent operating point close to the design value is achieved as expected. Laboratory Exercise 16 examines a similar transistor amplifier.

EMITTER-FOLLOWER AMPLIFIER

In our limited examination of transistor amplifiers, we will now examine an *emitter follower*, which is widely used with integrated circuits. The configuration has the dual advantages of high input impedance and low output impedance. These properties make the amplifier very useful as a buffer amplifier. It is called an "emitter follower" in analogy to the "cathode follower" of vacuum tube technology, which has the same advantages.

Figure 10-2 shows the circuit for an emitter follower. The quiescent base voltage E_B is supplied by the voltage drop in R_1, which carries the base current, $E_B = V_{cc} - R_1 I_B$. The emitter voltage, which is established by the emitter current through R_L, is slightly below the base voltage for an *NPN* transistor; the difference is the amount of the voltage drop across the forward-biased base-emitter *pn* junction. As an input signal raises or lowers the base voltage, the emitter voltage follows. The reader will recognize that the relation between the input signal and the output signal represents negative feedback and explains the dual advantages mentioned above. Because no phase reversal exists between the input and the output signals, the term "buffer" is appropriate.

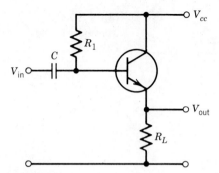

FIGURE 10-2
A Emitter-Follower Amplifier
The high input impedance and the low output impedance make this amplifier useful for load matching.

The output impedance of the emitter follower made with a low-power junction transistor may be as low as 10 Ω; the input impedance is approximately βR_L, which can be hundreds of kilohms. The voltage gain of the emitter follower is very close to unity, 0.99; the current gain can be 200 or more. The current gain is an important advantage, which is illustrated in the discussion of output limitations of operational amplifiers and linear devices.

SIMPLE REGULATING CIRCUITS

Two applications of simple transistor amplifiers are the regulation of voltage and the maintenance of constant current. The term "voltage regulation" means the property of maintaining a constant voltage when the supply voltage varies, when the load current changes, or when both occur. Constant current regulation is important in stabilizing difference amplifiers, in the production of constant magnetic fields for experimental purposes, and in other applications. These simple circuits are discussed to present the principles of regulation. More sophisticated circuits use these principles as well as greater amplification to provide excellent regulation.

In Figure 10-3a, a series voltage regulation is accomplished by causing the collector-emitter voltage of the series transistor (the voltage drop across the transistor) to change in such a manner as to keep the output voltage constant. If the voltage at the load increases because of load changes, the emitter voltage becomes higher, while the base voltage remains at a fixed value maintained by the Zener diode. (The breakdown voltage of the reverse-biased Zener diode, see Figure 8-10, becomes the fixed voltage.) Thus the relative emitter-base voltage is reduced and the transistor develops a higher voltage drop across itself, and the output voltage falls nearly to its proper value. The error or difference voltage is very small because

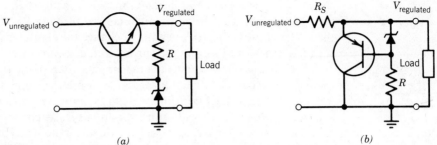

FIGURE 10-3

Voltage Regulators

The circuits are the bases for the providing a constant output voltage independent of a change in the voltage source or the load current. The series regulator in (a) uses the bipolar junction transistor gain and the Zener reference voltage to modify the voltage drop across it. In (b) the shunt regulator in parallel with the load bypasses current so that the voltage drop in the series resistor, R_s, compensates for voltage variations.

of the amplifying property of the transistor circuit. In a series regulator the transistor is endangered if the output is shorted.

Figure 10-3b shows an arrangement called a shunt voltage regulator. In this arrangement the voltage drop in R_S varies according to the sum of the load current and the current bypassed (shunted) through the transistor. If the output voltage increases, the transistor conducts more current because of the emitter-base voltage change. The increased voltage drop across R_S counteracts the increase in the output voltage. In the shunt regulator, the transistor is not damaged if the output is shorted, but an energy loss does occur in R_S.

A nearly constant current can be maintained by the transistor circuit of Figure 10-4. The voltage divider, R_1 and R_2, maintains the base voltage; this changes only

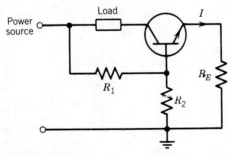

FIGURE 10-4

A Constant Current Regulator

The base-emitter voltage, given by IR_E, insures that the current I does not change within the amplification factor of the circuit.

slightly if the very small base current through R_1 shifts slightly, whereas the emitter voltage depends upon the much larger current through R_E. As the current through the load and the transistor attempts to change the emitter-base voltage changes in such a manner as to sustain the current at a constant value.

It is clear that these circuits use the amplification of the transistor and negative feedback to accomplish their purposes. They are the simplest (and least effective) arrangements, but their performance can be improved many times over by using active negative feedback with transistor amplifiers as part of the feedback circuits. Integrated circuit regulators based on these simple concepts are available with excellent properties arising from internal transistor-amplifying circuits. The task of regulation will be discussed in Chapter 14.

DIFFERENCE AMPLIFIER

A *difference amplifier*, also called a *differential amplifier*, can be constructed with two matched bipolar junction transistors in a common-emitter configuration. Figure 10-5 shows the basic arrangement of a difference amplifier. Because this amplifier does not use capacitor coupling, it is a dc as well as an ac amplifier.

An ideal difference amplifier has equal currents in identical transistors with identical collector resistors. Identical signals at inputs A and B cause identical voltages at outputs A and B, and the amplified difference voltage is zero. For zero

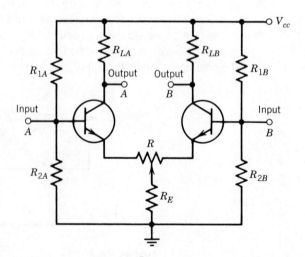

FIGURE 10-5

A Difference Amplifier

The difference amplifier accepts dc and ac signals. The output difference voltage is the difference between the input voltages amplified by the circuit. The difference amplifier is the foundation of operational amplifiers.

difference in voltage inputs there is zero difference in voltage outputs, none of which is at zero volts. Ideal difference amplifiers never exist, however, because components are never identical. Carefully matched components and elements, such as the small adjustable resistor R used to balance the currents, can make useful difference amplifiers, but a better solution is to use ICs.

Identical inputs are called *common-mode* inputs. In contrast to ideal difference amplifiers, the common-mode input of an actual difference amplifier, in which both inputs are connected to the same signal, results in a small but not a zero output-difference voltage. The quality of a difference amplifier is judged in part by the smallness of the common-mode output-voltage difference. We designate the ratio of ac voltages $v_{\text{out}}/v_{\text{in}}$, applying v_{in} simultaneously to each input (common-mode input) as the common-mode voltage gain, A_v (common-mode),

$$A_v(\text{common-mode}) = \frac{v_{\text{out}}(\text{common-mode})}{v_{\text{in}}(\text{common-mode})} \qquad (10\text{-}3)$$

For different input signals, however, different input voltages on A and on B cause the respective transistors to conduct at different values. Because the collector voltages differ, an amplified difference voltage occurs between the outputs A and B. The transistors share a common emitter resistor; thus, the current in either transistor affects the conduction of both. Let input B, for example, be held at a fixed voltage and input A increased slightly above the fixed voltage. The conduction in transistor A increases, causing the output voltage at A to fall and the emitter voltage of both transistors to rise. The rise in emitter voltage makes transistor B conduct less, which reduces the voltage drop in R_{LB}, and raises the output B. Most important, the reduced conduction diminishes the change in the emitter voltage shared by both transistors. With matched transistors operating in a linear range the emitter voltage changes very little, but a difference voltage exists between outputs A and B. The output-difference voltage is the difference between the input voltages amplified by the system with a phase difference of $180°$.

The ac voltage amplification with different input voltages is

$$A_v(\text{difference}) = \frac{v_{\text{out}}(\text{difference})}{v_{\text{in}}(\text{difference})} \qquad (10\text{-}4)$$

The ratio of the difference voltage gain to the common-mode voltage gain is called the common-mode rejection ratio (CMRR),

$$\text{CMRR} = \frac{A_v(\text{difference})}{A_v(\text{common-mode})} \qquad (10\text{-}5)$$

A good difference amplifier has a large common-mode rejection ratio; the importance of a large CMRR can be shown in an example. Frequently, when a very small difference voltage, such as the voltage across a nerve fiber, must be measured, the leads to the instrument pick up a much larger common-mode input from the electrical noise of the power circuits in the building. The common-mode signal will overwhelm the difference signal of interest unless the common-mode input is rejected.

Operational amplifiers, available as integrated circuits, are essentially high-gain difference amplifiers that are compensated internally to give excellent response and a very high common-mode rejection ratio. Difference amplifiers will be illustrated in the discussion of operational amplifiers in Chapter 11.

TRANSISTOR-TRANSISTOR-LOGIC GATES

Transistor-transistor-logic (TTL) is the basis for the digital-gate integrated circuits of the 7400 series. (The 5400 series are the same elements, but with military specifications.) Because digital elements have only two states, on and off, they differ in two respects from the integrated circuits used in analog amplifiers and linear devices. First, digital circuits are not capacitor-coupled because signals may persist for indefinitely long periods. With capacitive coupling, constant voltages would not be sustained because of resistance-capacitance voltage decay (discussed in Chapter 7). Second, digital elements operate in the extremes of conduction and cutoff rather than in a linear range.

All digital circuits, both combinational and sequential, are formed with elementary logic gates. They are the AND-gate (or NAND-gate), the OR-gate (or NOR-gate), and the inverter. An understanding of these three gates will allow us to understand the features of all digital circuits.

Suitable digital logic gates must have the following properties. The input and output voltages (states) must be compatible so that the cascading of gates is possible. Input states must require little current; output states must provide sufficient current to drive several, perhaps 10, following gates. This property is called Fan-Out. The two states, high and low, must be unambiguously different voltages. Finally, gates must change states very quickly.

The finite time in which the gate changes state depends upon the capacitance (the charge-storage capacity) of both the gate's transistors and the circuit that is being driven (i.e., the next stage) and on the resistance of the charging and discharging paths. That is, the time constant of the configuration governs the rate of change from a low-voltage state to a high-voltage state and vice versa. For a state to change from low to high, the effective capacitance must be charged through the effective resistance of the gate output stage until the voltage corresponding to a high state is reached. An interval of several time constants, $r_{\text{effective}} C_{\text{effective}} = \tau$ sec, is involved in completing the change of state. The time interval is stated as the *propagation time delay*.

Table 10-3 shows the input and output characteristics of the 7400-series TTL gates as stated in the manufacturer's specifications.

The simple inverter in Figure 10-6 is a single-stage common-emitter amplifier without capacitance coupling. This kind of RTL (resistor-transistor-logic) is not

TABLE 10-3
TTL-gate Input and Output Characteristics

Quantity	Value		
	Min	Nominal	Max
High-level input voltage	2.0 v		
Low-level input voltage			0.8 v
High-level output voltage	2.4 v	3.4 v	
Low-level output voltage		0.2 v	0.4 v
Propagation time delay (nanoseconds)			
Low-to-high level output		11 ns	22 ns
High-to-low level output		7 ns	15 ns

Source: TTL Data Book, 1976.

suitable for very fast digital circuits because the transfer to the high state requires too much time. The circuit is limited by the time to charge the effective capacitance, C (50 pf or more), of the next stage through R_L. The time constant $R_L C$ may typically be 20 to 30 nsec; thus, a hundred or more nanoseconds may be needed for the state to change completely from low to high. The transfer from the high state to the low state, however, occurs when the transistor conducts; it happens relatively quickly because the discharging path has the very low resistance of a conducting transistor.

Transistor-transistor-logic (TTL) overcomes the charging time by providing less resistance in the charging path. Figure 10-7 shows how the several interconnected transistors provide fast transfer between states in the 7404 TTL inverter. The two transistors at the output, T_3 and T_4, are complementary electronic switches; that is,

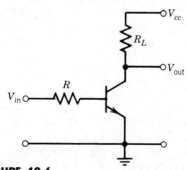

FIGURE 10-6

A RTL Inverter

The resistor-transistor-logic inverter is an elementary digital gate with the logic function $A \rightarrow \bar{A}$.

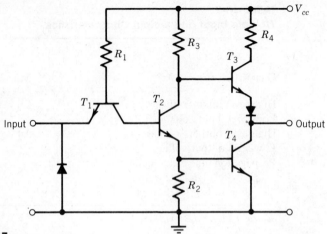

FIGURE 10-7

A TTL Inverter (7404)

The transistor-transistor-logic inverter is a fast-responding digital gate. The output stage is used through the TTL family of logic gates. Its logic function is $A \to \bar{A}$.

one is conducting while the other is cut off. In this way the capacitive load is both charged and discharged through the low resistance of a conducting transistor. R_4, a small resistance of 100 ohms or less, is only one-tenth of R_L in Figure 10-6.

The operation of the inverter can be understood if we assume an input state and then follow the consequence of the input state through the circuit. High input signal, for instance, cuts off T_1, which raises the base-emitter voltage of T_2 and causes it to conduct. The voltage drop in R_3 lowers the base-emitter voltage of T_3, cutting it off. The current through T_2 simultaneously raises the voltage across R_2. This voltage provides the base-emitter voltage to make T_4 conduct. The cutoff T_3 and the conducting T_4 form the voltage divider from which the output is taken. A

TABLE 10-4
Various TTL Inverters

Inverter	Designation	Propagation Delay	Power at 50% Duty Cycle
Standard gate	7404	10 ns	10 mW
Low-power gate	74L04	32 ns	1 mW
High-speed gate	74H04	6 ns	22.5 mW
Schottky gate	74S04	3 ns	28.8 mW
Low-power Schottky gate	74LS04	10 ns	2 mW

Source: TTL Data Book, 1976.

small fraction of V_{cc} exists across T_4 to make the output state low, the complement of the high-input state. As the reader may recognize, a low input will reverse the voltage relations throughout the inverter.

An examination of the TTL data book will reveal other TTL inverters with different configurations to minimize propagation delays and power requirements. The different logic gates are labeled to indicate their features. Table 10-4 shows the designations on TTL inverters and the properties that the designations imply (*TTL Data Book*, 1976).

Schottky gates employ Schottky diodes (metal-semiconductor diodes that were mentioned when practical diodes were treated in Chapter 8) to prevent saturation currents in the transistors of the gate. Unsaturated transistors respond more quickly than those whose currents are at maximum levels. Low-power Schottky gates have replaced the high-speed and low-power gates in most applications.

NAND-gate

The TTL NAND-gate is identical to the inverter except for the input stage. The input signals are brought to two or more emitters of the input transistor, which is constructed with two or more emitters. Two or more *NPN* transistors with their bases and collectors tied together are equivalent to the input transistor in Figure 10-8. A low signal on any input line or low signals on several or all input lines will make the input transistor conduct. The response in the rest of the circuit is exactly as described for the inverter, and a high output results. If all inputs are high,

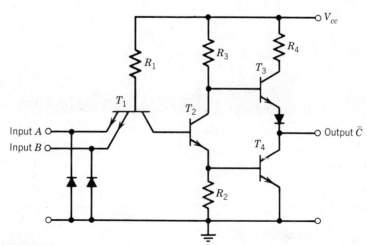

FIGURE 10-8

A TTL Nand-gate (7400)

The transistor-transistor-logic NAND-gate satisfies the logic function $AB = \overline{C}$.

however, the input transistor T_1 will not conduct and the output is low. Thus the NAND-gate logic $AB = \bar{C}$ is satisfied. Where several emitter leads are used as inputs, we note that a high input on one emitter lead causes a reverse bias between that emitter and the base. Because of this reverse bias, simultaneous low inputs on one input emitter or the other are isolated electrically from the high input, and both high and low input states in any combination are acceptable.

NOR-gate

The TTL NOR-gate has the same output stage as the TTL inverter and the NAND-gate, but the input stage differs to provide the NOR-gate logic $A + B = \bar{C}$. The NOR-gate configuration is shown in Figure 10-9. To analyze the operation of the NOR-gate, let us assume that input A is high and input B is low. T_1 is cut off while T_2 conducts. The base of T_3 is high relative to its emitter; thus it conducts with current through the two resistors (R_3 and R_4) that it shares with T_4. The base of T_6 (the output transistor) rises above the grounded emitter of T_6, making T_6 conduct. The voltage on the base of T_5, which depends on the current in the upper load resistor, R_3, cuts T_5 off. The output voltage across the voltage divider, formed by the high resistance of T_5 and the low resistance of T_6, is low. Whether the signals between inputs A and B are exchanged or whether both A and B are high, the states described above are not altered.

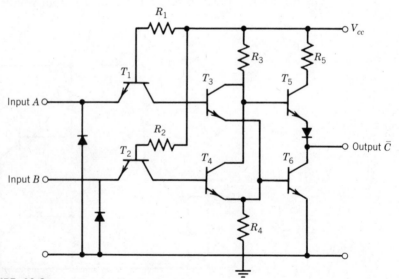

FIGURE 10-9

A TTL NOR-gate (7402)

The transistor-transistor-logic NOR-gate satisfies the logic function $A + B = \bar{C}$.

If both inputs A and B are low, however, the input transistors T_1 and T_2 both conduct. The voltage drop in their respective load resistors places the base voltage below the emitter voltage in both T_3 and T_4, cutting them off. With no current in the resistors R_3 and R_4, the base of T_5 is at a higher voltage than its emitter; it conducts while the base of T_6 is at ground and T_6 is cut off. The output is high.

The arrangement of the gates in Figures 10-7, 10-8, and 10-9, in which one transistor is placed above the other, is called a *totem-pole output*.

OPEN-COLLECTOR GATES

In the TTL gates just discussed, the output voltage-dividing circuit was made with a resistor, two transistors and a diode. In open-collector gates, however, the output circuit is made only with a transistor, as shown in Figure 10-10. The circuit must be completed by a resistor connecting the output terminal with the supply voltage V_{cc}. The resistance value must be large enough to limit the current to the rated value of the internal transistor.

A typical "load" resistor of an open-collector gate is one segment of a seven-segment LED (light-emitting diode) in series with a current-limiting resistor. Most decoder/drivers for LED displays are open-collector devices with an output pin for each of the seven segments from the seven internal gates. Each gate must have its collector connected independently to $+V_{cc}$ through a resistor and its segment.

Because the output of an open-collector gate is an RTL (resistor-transistor-logic) element, its propagation delay is greater than that of the totem-pole gates, but this is not normally a disadvantage in the applications of open-circuit gates.

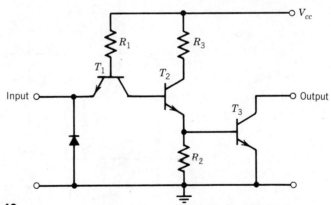

FIGURE 10-10

An open-Collector Inverter (7405)

The output of the open-collector inverter may be tied to other open-collector devices without damage. An external resistor must be placed between V_{cc} and output for the gate to function. Its logic function is $A \rightarrow \bar{A}$.

TRISTATE GATES

To conclude the discussion of TTL logic gates, let us examine a tristate gate. In digital logic there are only two states, high and low or 1 and 0, but in many applications, gates should not indicate their states to the system. Data (bits) on a data bus, for example, may originate from many different sources such as memory arrays and shift registers; each of these is connected physically to the bus, but only one of the elements may be the data source at any one time. It is impractical at any moment to disconnect physically all the elements except the one providing data, but it is possible to disconnect gates in effect by means of an electrical signal.

The third state of these tristate gates is the state of being electrically disconnected or isolated. In this state, the gate output has no specific binary output state, high or low. The gate that provides data on the same conductor of the bus has a definite state, high or low (1 or 0). The high impedance of all the other tristate gates, explained below, removes them as loads on the gate that drives the bus.

The third state—being isolated—is achieved with an enable signal to the gate. Figure 10-11 shows a tristate inverter. Its operation is the same as the inverter examined above when the enable signal is high. A low enable sign, however, causes the voltage of each base to be lower than its emitter voltage in both output transistors. Both T_3 and T_4 are cut off. There is a very high impedance from the output line to $+V_{cc}$ through the cut off T_3. A similar isolation to ground occurs

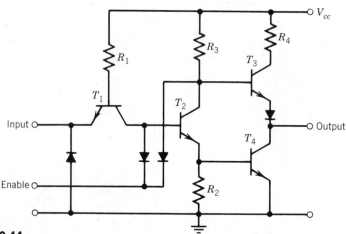

FIGURE 10-11

A tristate Inverter

The tristate inverter output may have values 0, 1, or "electrically disconnected." A low (0) on enable cuts off both output transistors, so that the impedance from the output is very large; in effect, it is removed electrically from the circuit to which the output is physically attached.

through the cut off T_4. The voltage that can be measure
state of the enabled gate that is using the bus.

JUNCTION FIELD-EFFECT TRANSISTORS

A *junction field-effect transistor* (JFET) differs from the b
whose typical circuits we examined above. Both *p*-type and *n*-type semiconducting materials are used in JFETs, but the conduction through JFETs is not controlled by a conducting *pn* diode junction as it is for emitter and base in a bipolar-junction transistor. In field-effect transistors, conduction from the source to the drain depends upon the voltage between them and upon the condition of the semiconducting path or channel from the source to the drain. The conductivity of the channel is affected by the electric field placed on the channel by the gate.

In semiconductors the majority carriers (holes in *p*-type material and electrons in *n*-type) move under the influence of electric fields and constitute the current from the source to the drain. At the junction between the gate and channel, however, the two kinds of majority carriers combine and the number of remaining majority carriers is diminished. This is called the *depletion region*. When the *pn* junction is reverse-biased, the depth of the depletion region increases and the conduction drops. By increasing the volume of the channel, which is depleted of majority carriers, the field-effect transistor uses the reverse bias provided by the voltage on the gate relative to the source-to-drain channel to modify the current in the channel. Figure 8-15 shows schematic forms of the JFET, and Figure 10-12 shows

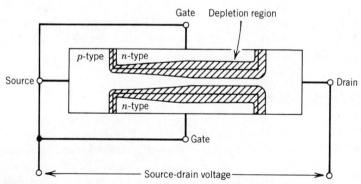

FIGURE 10-12

Schematic of a JFET-Equivalent Resistor

The voltage between source and drain increases the depletion region when the gate of the junction field-effect transistor is tied to the source. A long, narrow JFET is a readily produced equivalent resistor in large-scale intergrated circuits.

the depletion region in a JFET with the gate tied to the source and a voltage across the channel.

We emphasize that the conduction in a JFET from source to drain does not involve a diode junction, since the channel from source to drain is simply one continuous path of a single type of semiconducting material. We note also that because of the voltage difference between the source and drain, the gate at a fixed (source) voltage has a reverse bias with regard to the channel, which increases along the channel. This increase causes the depth of the depletion region to increase along the channel as one goes from source to drain.

This feature is used in JFETs that are made long and narrow (relative to other JFETs) and whose gates are attached electrically to the source. As the drain is approached, the narrowing conduction in the channel indicates an effective resistance much higher than the resistance of the bulk semiconducting material of the channel. In large-scale integrated (LSI) circuits the use of JFETs as equivalent resistors avoids the very difficult task of making actual resistors.

Figure 10-13 shows a schematic of a read only memory (ROM) array prepared by a manufacturer's photo mask. The active JFETs are formed in the pattern specified by the user, and long, narrow JFETs are used as load resistors. The pattern of this figure matches Figure 5-11 exactly, but Figure 10-13 shows that the open-collector inverters in Figure 5-11 may actually be JFETs.

A similar relation exists between Figures 5-12 and 10-14. Bipolar-junction transistors are shown in Figure 10-14. The PROM, however, does not contain the load resistors; they are added externally as required on the bit lines.

Because the gate of a field-effect transistor is reverse-biased, it draws only the very small leakage current of a reverse-biased *pn* junction. Thus the input resistance of a JFET is very high, perhaps greater than $10^9 \, \Omega$. JFETs have this distinct advantage over bipolar-junction transistors.

Another property of JFETs is that a vast difference in resistance exists between the source and the drain when the JFET is conducting and when it is biased off. The low resistance of $100 \, \Omega$ or less that exists when the JFET is conducting fully can become $10^8 \, \Omega$ or more when the JFET is cut off. The JFET is an excellent voltage-activated switch.

MOSFET AMPLIFIERS

In the construction of MOSFETs, the gate electrode is insulated from the channel by an oxide layer. The name is derived from the arrangement of a *m*etal gate on an *o*xide insulator on a *s*emiconductor. The gate and the semiconductor material with the oxide between them form a capacitor, which is charged by the voltage placed on the gate. The two types of MOSFETs—depletion and enhancement—differ in the presence or absence, respectively, of a conducting channel fabricated during manufacture between the source and drain.

Depletion MOSFETs behave similarly to the FETs discussed above with this advantage: the oxide layer provides high resistance between the gate and the rest of the element. Figure 8-18 shows the configuration of a depletion MOSFET; Figure

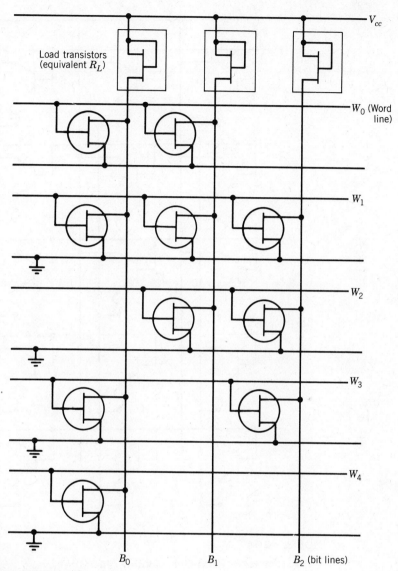

FIGURE 10-13

A Read Only Memory (ROM) Array

The cells are fabricated by the manufacturer in accord with the bit pattern supplied by the user. The ROM is unalterable and nonvolatile.

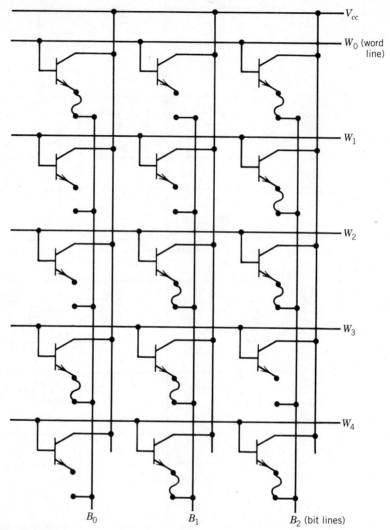

FIGURE 10-14

A Programmable Read-Only Memory (PROM) Array

All cells are fabricated, and each includes a fuse that allows the user to form a nonvolatile bit pattern by deleting the unwanted cells.

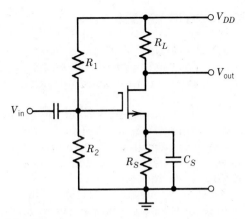

FIGURE 10-15

A Depletion MOSFET Analog Amplifier

The metal-oxide-semiconductor field-effect transistor is used in an ac amplifier similar to the one in Figure 10-1. Depletion MOSFETs have a conducting channel between source and drain, as indicated by the solid bar between source and drain. The insulated gate provides a very high input impedance.

8-19 shows that the gate can be made positive or negative in relation to the more lightly doped channel between the source and the drain. The conduction between the source and the drain can be increased or diminished by the charge on the gate, which increases or depletes the majority carriers in the channel under the gate.

Figure 10-15 shows a simple amplifier made with a depletion MOSFET. The voltage divider formed by R_1 and R_2 sets the gate voltage relative to the source, which is self-biased by R_s and C_s. Because of the good insulating properties of the oxide layer, the gate current is truly negligible. The arrow on the source indicates the direction in which positive charge would flow from gate to channel if no insulating layer were present and if the junction were forward-biased. The arrow designates the MOSFET as a N-channel (NMOS) transistor.

The other type of MOSFET, an enhancement MOSFET will not conduct until the charge on the insulated gate induces (enhances) carriers like those in the doped source and drain into the region under the gate. Then conduction can occur between the source and the drain in what has become a channel of the same type of semiconductor as the source and the drain. As shown in Figure 8-21, the enhancement MOSFET will not conduct without a gate voltage to induce carriers in the channel. Figure 8-20 shows the configuration on an enhancement MOSFET and the symbol for a N-channel MOSFET (NMOS). In practice, the substrate or base is connected to the source to ensure that it will not be forward-biased.

As we will see shortly, MOSFETs are used widely in logic gates, where the high density in which they are formed enables thousands to be placed in medium (MSI), large (LSI), or very large-scale (VLSI) integrated circuits.

MOS AND CMOS LOGIC GATES

Figures 10-16 and 10-17 show that MOSFET NAND-gates and NOR-gates are made with a load and active MOSFETs. By choosing input states and deducing the output states, the truth tables or logic functions generated confirm the operation of the arrangements.

MOSFET inverters are shown in Figure 10-18. In the MOSFET circuits with the load transistor (for a depletion MOSFET, see Figures 10-16, 10-17, and 10-18), power is consumed when the output state is low. In very dense arrays the total power usage can heat the chip, because it requires a substantial power supply to ensure satisfactory operation.

The complementary CMOSFET inverter in Figure 10-19 modifies the power requirements because the transistors conduct only during the instant of transition. The form of the CMOSFET is presented in Figure 8-22. The voltage supply range of 15 V or more allowed with CMOSFETs is considerably larger than 5-V TTL gates. The output state is closer to ground (0 V), or V_{ss} (5 V, for example) in CMOS gates than in TTL gates, where the voltage range is typically 0.2 V for a low and 3.4 V for a high (see Table 10-3).

CMOSFET logic gates are made by combining similar circuit elements to yield the logic function (Pawloski et al., 1983). Figure 10-20 shows a two-input NAND-gate and a two-input NOR-gate, each requiring two CMOSFETs, for a total of four transistors. By comparison, only three N-channel MOSFETs (Figures 10-16

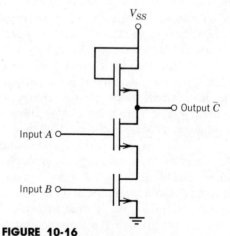

FIGURE 10-16

A MOSFET NAND-gate

The NAND-gate includes an equivalent resistor and two depletion MOSFETs in series to yield the logic function $AB = \overline{C}$.

MOS and CMOS Logic Gates **279**

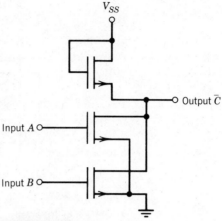

FIGURE 10-17

A MOSFET NOR-gate

The NOR-gate includes an equivalent resistor and two depletion MOSFETs in parallel to yield the logic function $A + B = \bar{C}$.

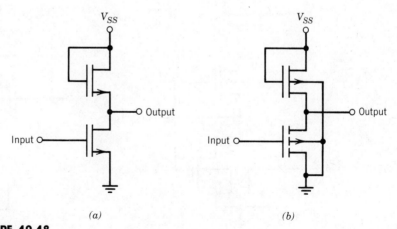

FIGURE 10-18

MOSFET Inverters

Two types of MOSFETs are used to make inverters. (*a*) Depletion MOSFETs are used for the equivalent resistor and the active element. (*b*) The equivalent resistor is a depletion MOSFET, and the active element is an enhancement MOSFET, in which no channel has been formed between source and drain in manufacture. Both (*a*) and (*b*) yield $A \rightarrow \bar{A}$.

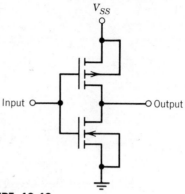

FIGURE 10-19

A CMOSFET Inverter

The complementary MOSFET elements alternate in conduction as the input signal is low (0) or high (1). The logic function is $A \rightarrow \bar{A}$.

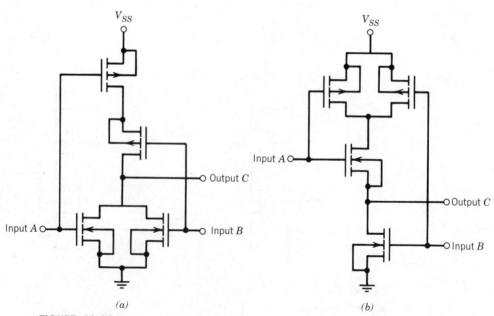

FIGURE 10-20

A CMOSFET NOR-gate and a NAND-gate

Complementary MOSFETs are used in pairs to yield the logic functions, (a) $A + B = \bar{C}$ and (b) $AB = \bar{C}$.

and 10-17) are required to make the comparable two-input gates. In the production of VLSIs, manufacturers weigh the lower power needs of CMOSFETs against the higher gate density possible with MOSFETs.

DYNAMIC SHIFT REGISTER

The shift registers discussed in Chapter 5 are bistable elements tied together so that each clock pulse transfers the bit pattern one element in the direction of shift. Figures 5-1 and 5-4 show the *RS* flip-flops and data flip-flops used to make shift registers. These are static elements, which can remain indefinitely in one state or another without attention. Dynamic shift registers, however, use a different method of storing and shifting bits. The bit is stored as a charge in the intrinsic capacity of the transistor. The intrinsic capacity is quite small, approximately one picofarad, 10^{-12}F, and the charge leaks off the capacitor in a fraction of a second. The bit can be retained only by refreshing the charge a thousand or more times a second. The charge is refreshed by transferring it continuously along the dynamic shift register in a closed loop. The bit pattern is written in and read out serially at one point in the loop. The absence of a charge can represent a zero; the presence of a charge, a one.

The high density of MOSFETs on silicon chips makes dynamic shift registers attractive for many applications. It is likely that the display of your hand-held calculator is activated by a dynamic shift register with 1024 bits in a continuous loop. Figure 10-21 shows a register cell in the loop and provides a means of describing the transfer-refresh process. The MOSFETs are driven with two clock pulses; one high governs one set of elements and the second high, out of phase with the first, governs the second set. In this sense they are master-slave units, in which the bit and its complement are transferred in a two-step process. The intrinsic capacitors are shown in dotted lines in the figure.

We follow the sequence by assuming that there is a high bit (1) at the transfer point 1 when clock 1 places a high (1) on the gate of transistor A. This transistor and transistor E are called transmission gates; they act as switches to charge or discharge the intrinsic capacitors. Capacitor C_A is charged through transistor A in accordance with the high bit at transfer point 1. The inverter formed by transistors B and D is driven low (0) by the high from charged C_A on the gate of transistor D. After clock 1 goes low, clock 2 goes high, and the intrinsic capacitor C_B discharges through the conducting transistors E and D. The second inverter, formed by transistors F and G, has a low (0), because of the discharged C_B, on the gate of transistor G, cutting it off, so that the voltage at the transfer point 2 is high (1). A sequence of two clock pulses has transferred the high (1) input state across the dynamic register cell and refreshed the bit. The reader should confirm that a low (0) at the input would be replicated at the output by the response to the two clock pulses.

Transistor A' is the transmission gate of the next memory cell. This exact sequence is used in one form of a dynamic RAM, except that the bit is not

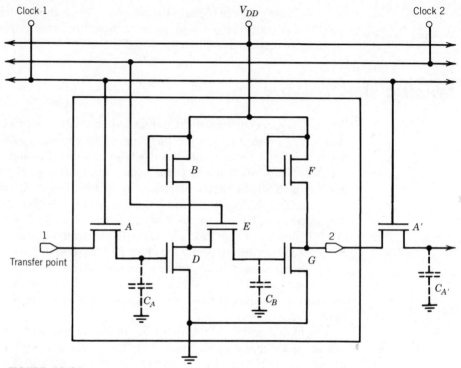

FIGURE 10-21

A Cell of a Dynamic-Shift Register

One cell of a dynamic-shift register in a large loop (perhaps 1024) of similar cells is shown. The bit at transfer point 1 is shifted to transfer point 2 in two clock pulses, during which the stored charge is restored.

transferred as it is refreshed. When we examine the three-transistor dynamic RAM, we find that the bit storage cell is like the first inverter of the dynamic shift register and that the refresh amplifier is analogous to the inverter driven by clock 2.

DYNAMIC RAMs

Dynamic random access memories are made in dense arrays of memory cells; two types of cells will be discussed here. One type is closely related to the elements in the dynamic shift register, both in the manner of bit storage and in the refresh process. These dynamic RAM cells have three transistors each, and the refresh process uses the complement of the stored bit to implement its refresh. The other type of dynamic RAM uses only one transistor per cell, but its refresh method is more complicated than the three-transistor form mentioned above. A high (1) bit stored in the single-transistor memory cell is destroyed as it is read, and the read process includes the restoration of the high to the cell.

The dynamic RAMs are formed in columns of 64 or more cells. Each column has its own refresh circuitry and column-address lines. The cells in adjacent columns are connected by a row-address line at each level. There are as many row-address lines as there are cells in the column. Any cell can be addressed individually by selecting its column and row; that is, by placing a high on both lines.

Figure 10-22 shows a column of dynamic memory cells with three transistors in each cell. Transistors A and D and the intrinsic capacitor C_A have the same functions as transistors A and D and intrinsic capacitor C_A in the dynamic-shift register in Figure 10-21. Transistor B, which corresponds to the load transistor B in Figure 10-21, is not in the cell; here, it is a load transistor for all cells in the column, though it acts as the load transistor only for the cell that is addressed at any instant. The read signal (a high) activates transistor B and transistor J, whose drain voltage is provided by the horizontal address line. During the read cycle this line is high (1). Transistor H is a transmission gate, which places the output of the inverter formed by transistors B and D on the data-out line. The data-out signal is the complement of the stored bit. A high on the column-address line causes the transistor L to transmit the datum. The stored bit is not lost in the read process, but it must be refreshed approximately once every millisecond, along with all of the other bits in the column.

A bit is read into the cell by raising the horizontal row and the vertical column-address lines to high (1) when the bit is on the data-in line. A high (1) on the write line causes transistor K to transmit the bit to the cell through transistor A, which is activated by the horizontal row-address line. A high on the data-in line will be stored as a high on capacitor C_A, and a low (0) will discharge the capacitor to store a low (0) in the cell.

The refresh circuit in each column is similar to the second stage in the dynamic shift register. Figure 10-23 shows the refresh circuit, which senses the complement of the stored bit on the data-out line when the horizontal row-address line is raised to high. The column-address line is not addressed; that is, it remains low, thus preventing any signals from reaching the data-in or data-out lines. When the read line is raised, the capacitor C_B acquires a charge complementary to the stored bit in the cell being refreshed. This charge travels through the transmission gate E, which is activated by the refresh signal. The state of the inverter formed by transistors F and G, which corresponds to the state of the cell, is transmitted to the cell by the write transistor K. The refresh process requires that the highs (1s) be placed on the horizontal address line, the read line, the write line, and the refresh line, and that the column-address line be kept low (0). The sequence is repeated by raising sequentially each horizontal address line until each cell in the column is refreshed. The time required to refresh a column is the product of the number of cells in the column and the time required to refresh a single cell.

It is clear that the charge on the intrinsic capacitor is sufficient to determine the state of a memory cell. Figure 10-24 shows a dynamic RAM array made with only one transistor per memory cell. The figure shows two cells of 64 on the left of the bit-sense amplifier and one reference cell (also called the dummy cell) on the right. In a 16K dynamic RAM, there are 128 rows of 64 cells each in the left column and

284 Elements of Transistor Circuits

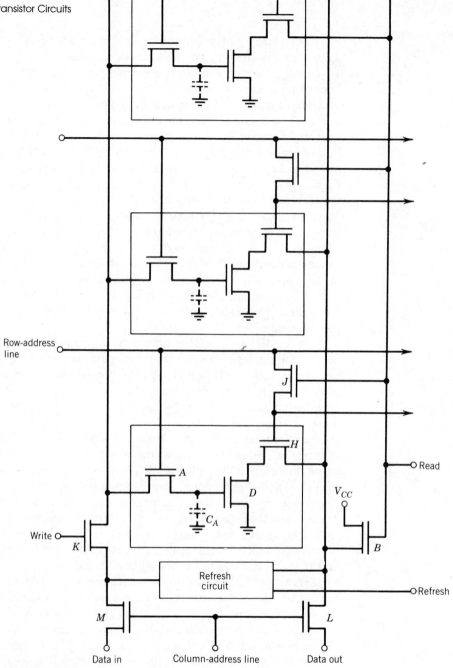

FIGURE 10-22

One Column of a Dynamic RAM

Three-transistor cells in the dynamic RAM are formed in columns of many cells (perhaps 128) per column. The transistors are labeled the same as in Figure 10-21 to identify their similar functions.

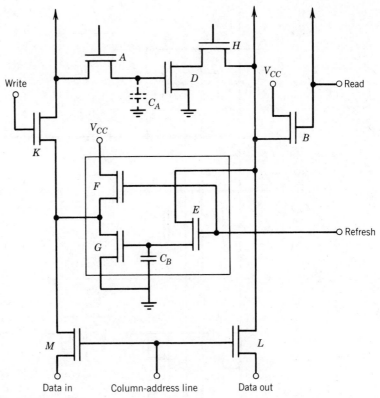

FIGURE 10-23

A Refresh Circuit for Dynamic RAM

The refresh process for each cell in the dynamic RAM of Figure 10-22 is accomplished by this circuit. The transistors are labeled the same as in Figure 10-21 to identify their similar functions.

in the right column. The reference cells for the right column, which are not shown, are on the far left. These columns are mirror images of each other, with one on each side of the column of bit-sense amplifiers.

When one of the select lines is raised high, the voltage of the capacitor that represents the stored bit is transmitted to the bit-sense amplifier. At the same time the voltage on the reference cell is sensed. The reference cell is charged by the precharge clock to a value midway between that of a high (1) bit and a low (0) bit in the memory cell. The bit-sense amplifier, which is a bistable element (flip-flop), is set in one state or the other, depending on whether the stored bit presents a higher or lower voltage than that of the reference cell. The value that was stored in the selected cell is read at the bit-sense amplifier; if it is high (1), it is restored by charging the capacitor in the cell. The refresh process addresses, senses, and refreshes the 128 cells on one select line at one time.

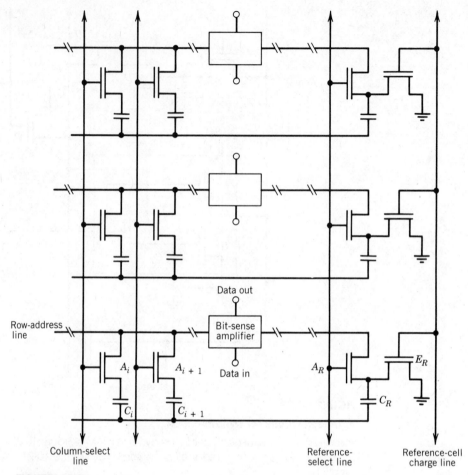

FIGURE 10-24
A Portion of Single-Transistor Dynamic RAM
An array of cells is addressed by row lines and column-select lines; the bit content is determined by the bit-sense amplifier. The reference cell compares the stored charge to determine the stored bit and to restore and refresh the cell.

A bit can be written into a cell by setting its bit-sense amplifier with the data-in signal and activating its select line. (The details of the read, write, and refresh processes vary with the manufacturer; they will not be explored here.) The goals in developing dynamic RAMs are to achieve the greatest possible cell density, to have as little power as possible drawn by the RAM, to minimize the refresh time, and to provide reliable bit storage. Dynamic RAMs with a megabit of storage capacity have been announced.

EPROMs

We conclude the discussion of transistor-based circuits by considering erasable programmable read-only memories (EPROMs). We examine only the cell because we know how the row- and column-address lines may be arranged to read the cell, and in this case there is no refresh process to be performed every millisecond. In fact, the decay time of the stored charge in the EPROM is several years unless it is subjected to conditions that erase the bit pattern.

In Figure 10-25, which shows a cell of the memory, the transistor has a floating gate under the gate in the MOS transistor. The method of programming the EPROM is to address the cell and apply a high-voltage pulse (about 25 V) between the drain and the control gate. The high-energy electrons are injected onto the floating gate and trapped on the floating gate by the isolating oxide. The charge can be induced to leak off the floating gate by exposing the memory-cell array to intense ultraviolet light for approximately one half-hour. About one month of sunlight is needed to erase the programmed cells.

The electrons on the floating gate mask the transistor from the gate signal when it is applied to read the state of the cell. An unprogrammed cell—that is, one with no charge on its floating gate—will conduct by an induced N-channel between the source and the drain when the gate signal is applied, but a programmed cell with negative charge on the floating gate, which induces a P-region, requires a larger gate signal to invert the region to an N-channel. When the floating gate has a sufficient negative charge, the typical 5-V signal on the gate will not cause the EPROM cell to conduct; in this way it is recognized as a high stored bit.

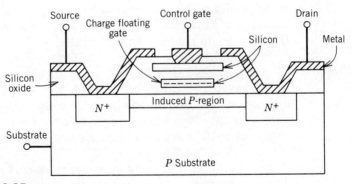

FIGURE 10-25

Schematic of EPROM Cell

An erasable programmable read-only memory stores a bit by electrons on the insulated floating gate. The induced p-region is present in the figure only because of the stored negative charge. The cell is erased by ultraviolet light, which allows the stored electrons to leak off the floating gate. The response to a signal on the control gate is modified by charge stored on the floating gate.

Other types of EPROMs are available and new ones are being developed. EEPROMs, for example, which are electrically erasable programmable read-only memories, have an advantage over the ultraviolet-erasable EPROMs because their cells can be reprogrammed individually without requiring that the entire bit pattern be erased to modify the PROM.

SUMMARY

Bipolar-junction transistors are the basis of analog amplifiers and logic gates. As gates, the advantage of their rapid response time is offset because of their power usage and because of the low input impedance of the current-conducting emitter-base junction. The simple configurations of bipolar-junction transistors in voltage- and current-regulating circuits are the foundations of the highly useful integrated circuit regulators. The amazing operational amplifiers are similar extensions of the simple difference amplifier.

The gate logic is acquired by the interconnections of transistors, in which the output stage always has the form of an inverter. Transistor-transistor logic is superior to resistor-transistor logic because it requires less time to charge the capacity of the load on the transition to a high state. The time, called propagation delay, is an important feature of logic gates.

The tristate gate can be electrically disconnected from others because it has the ability to place both of the two output transistors that form the inverter into a nonconducting state. This important advantage permits many interconnections to be made with a limited number of bus lines. Only selected elements are sources of signals at any one time, and the high impedances of the electrically disconnected tristate gates do not load the active gates.

Junction field-effect transistors (JFETs) and metal-oxide-semiconductor field-effect transistors (MOSFETs) have particular value in large-scale integrated circuits, where they are made in high-density arrays. They are important in other applications because of their high input impedance and because of the high impedance they have when they are gated not to conduct; that is, when they are used as electrically activated switches. Complementary MOSFETs provide alternative ways of making logic gates.

A small charge stored in a transistor represents a stored bit. The rapid loss of charge requires that the stored charge be refreshed approximately every millisecond. The charge is refreshed as a pair of inverters sense and restore the charge. dynamic shift registers do this by transferring the bit; dynamic RAMs do it in a similar way by addressing each cell, sensing its state, and refreshing the charge systematically throughout the RAM.

As the title of the chapter indicates, only a sampling of the current technology of transistor circuits is considered here. Binary states can be created and manipulated electrically in other ingenious electronic systems, such as charge-coupled devices (CCD) (Beynon and Lamb, 1980). Only the future can tell which elements will be selected for general use and what new possibilities will arise.

PROBLEMS

10-1 Explain the function of the diodes in Figure 10-8.

10-2 An unconnected input on a TTL gate assumes a high input condition. Explain how this happens in the circuit of Figure 10-8.

10-3 In Figure 10-14, resistors (not shown) on the bit lines are connected to ground. What is the nature of the data when a word line is addressed? What signal is required to read a word?

10-4 In a dynamic shift register the state (1 or 0) is determined by a charge that sustains the state effectively for about 1 msec as it leaks away through the reverse-biased junction formed by the drain and the substrate of the MOSFET. Could similar dynamic shift registers be made with bipolar-junction transistors?

REFERENCES

1973 Bulletin No. DL-S 7311471, November 1971, revised March 1973, Texas Instruments, Inc., Dallas, Tex.

1980 Beynon, J. D. E., and Lamb, D. R., *Charge-Coupled Devices and Their Applications*, McGraw-Hill, New York.

1983 Pawloski, M. B., Moroyan, T., and Altnether, J., "Inside CMOS Technology," *Byte*, 8, No. 9, 94, (1983).

CHAPTER 11

OPERATIONAL AMPLIFIERS

Operational amplifiers are high-gain difference amplifiers, which were perfected during World War II. They became the foundation of analog computers; at one time analog computers were called "differential analyzers" because they are used to solve differential equations. Operational amplifiers are also the basis of many important instruments.

The analog amplifier, originally a tube-type and presently a solid-state device, consists of a basic difference amplifier, implemented by feedback and other compensating amplifying circuits to give linear response, stability, freedom from drift, and other desirable properties. The complexity is required because operational amplifiers amplify dc as well as ac signals; capacitive coupling between amplifying stages is not permitted. Thus it is more difficult to isolate the long-term changes that arise from variations in temperature and power-supply voltage and from other effects that cause the output voltage to drift.

In the second decade after the invention of the transistor, solid-state operational amplifiers were introduced as integrated circuits. Now operational amplifiers are used to make high-quality, low-power analog amplifiers, and it is possible to avoid designing individual transistor amplifier stages for many applications. For most amplifying purposes and for many measuring and control applications, simple arrangements of operational amplifiers with feedback circuits will meet the designer's needs. The availability of operational amplifiers as integrated circuits in the form of dual in-line packages (DIPs) or in other compact forms makes the solution of analog signal problems analogous in many respects to the solution of digital logic problems, that is, through the interconnection of integrated circuits.

This chapter does not discuss the details of the internal operational amplifier circuits of the IC packages, but concentrates on the features that characterize operational amplifiers and circumscribe their use.

OPERATIONAL AMPLIFIER

Figure 11-1 shows the symbol for an operational amplifier. There are two inputs; the one marked with a plus sign is the *noninverting input*, and the one marked with a minus sign is the *inverting input*. The voltage amplified by the operational amplifier is the voltage difference between the two inputs. The open-circuit gain is so large, 10^5 to 10^6, that a voltage difference of only a few microvolts will give an appreciable output. Because the operational amplifier is a difference amplifier, connections must always be made to both input terminals for proper operation.

If two different positive voltages are applied separately to the two inputs, the output of the operational amplifier will be at the maximum (saturated) value if the voltage at the noninverting input is larger than the voltage at the inverting input. If the voltages exchange their relative values—that is, if the inverting input exceeds the noninverting input—the operational amplifier will switch to a minimum output, which is the saturated negative voltage. The switching occurs very quickly whenever the relative values of the voltages change. We will see that an operational amplifier is an excellent voltage comparator because it produces a large voltage swing at the instant of crossing, when an earlier relation in voltage magnitudes is reversed by only a few microvolts.

If the operational amplifier is to operate in its linear range, a very small voltage difference must exist between the inverting and the noninverting inputs. In

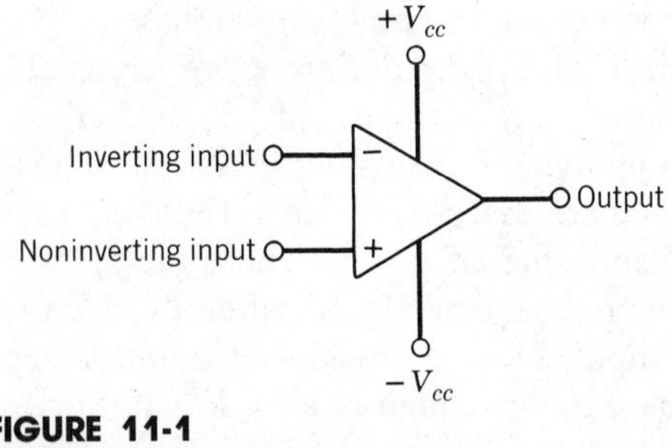

FIGURE 11-1

A Symbol for an Operational Amplifier

The $(-)$ inverting and $(+)$ noninverting properties of the inputs are designated.

analyzing operational amplifying circuits, we assume, because of the large open-circuit gain, that the voltage difference across the inputs is negligible. In some cases we will set the voltage to zero. When a portion of the output signal is returned to the input as negative feedback, the voltage difference cannot grow large because the output changes to minimize the voltage difference between the inputs.

Two additional properties are present in operational amplifiers. They are a very high input impedance, approximately $10^6\,\Omega$, and a low output impedance, approximately $100\,\Omega$. These properties contribute to the usefulness of operational amplifiers by allowing signal sources with small current capabilities to drive operational amplifiers directly. In turn, operational amplifiers may drive devices that have severe signal requirements.

To summarize, a good operational amplifier is characterized by a very large open-circuit gain of a million or so, an extremely small difference voltage between the inverting and the noninverting input terminals, a very high input impedance, and a small output impedance. An ideal operational amplifier, however, would have an infinite gain, a zero difference voltage between inputs, an infinite input impedance, and a zero output impedance.

Many operational amplifiers require two equal but opposite supply voltages, one positive and one negative. Typical values are $\pm 12\,\text{V}$ or $\pm 15\,\text{V}$. Other operational amplifiers may be operated with a single-ended power source, such as $+15\,\text{V}$. The useful range of the output voltage in an operational amplifier is approximately 80 percent of the supply voltages. With a two-sided power supply of $\pm 15\,\text{V}$, the output signal is restricted to about $\pm 12\,\text{V}$.

Many operational amplifiers have limitations; two of these need to be mentioned. One is that the gain decreases rapidly as the frequency increases. At small frequencies—as low as 10 Hz—the voltage gain begins to drop. (Power begins to drop at -20 dB per decade.) The gain-versus-frequency limitation is offset by the use of negative feedback to expand the frequency band. The other limitation is the rate at which most operational amplifiers can respond to a step change in the input signal. Compared to digital gates, operational amplifiers are poor in this respect; common operational amplifiers change at a rate of approximately $1\,\text{V}/\mu\text{sec}$. A TTL digital gate changes from one state to the other nearly 500 times faster. Newly developed operational amplifiers, however, have reduced this difference.

The importance of operational amplifiers lies in the advantages of using negative feedback networks. We will discuss several feedback arrangements that make operational amplifiers excellent amplifiers for analog signals.

INVERTING AMPLIFIER

The simplest arrangement of an operational amplifier is as an *inverting amplifier*. Figure 11-2 shows the arrangement in which the noninverting input is grounded. Because operational amplifiers with feedback have the special property of keeping the voltage across the input terminals essentially at zero, the inverting terminal is also essentially grounded. This relation is called *virtual ground*. The meaning of

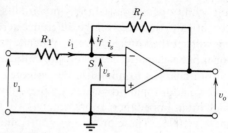

FIGURE 11-2

An Inverting Amplifier

In an inverting amplifier, a 180° phase difference exists between input and output signals, and a gain is established by the ratio of the resistors, $-R_f/R_1$.

virtual ground is very useful in understanding inverting operational amplifier circuits because the inverting input can depart from the noninverting input by only a few microvolts when the noninverting input is grounded.

Using Figure 11-2, which shows an inverting amplifier with the currents and voltages identified, we write the Kirchhoff loop and node equations for the circuit and find that

$$v_1 - i_1 R_1 - v_s = 0$$
$$v_s - i_f R_f - v_o = 0$$
$$i_1 - i_f + i_s = 0 \tag{11-1}$$

Here v_s is the voltage difference betweeen the inverting and the noninverting inputs. As pointed out above, v_s is a very small voltage.

We use the relation

$$v_o = -A_{oc} v_s \tag{11-2}$$

in which A_{oc} is the *open-circuit gain* (about 10^5) and the minus sign pertains to the inverting input, to obtain

$$\frac{v_1}{R_1} = -\frac{v_o}{R_f} - \frac{v_o}{A_{oc}}\left(\frac{1}{R_1} + \frac{1}{R_f}\right) - i_s \tag{11-3}$$

Equation 11-3 can be reduced to Equation 11-4 by discarding the terms containing $v_s = v_o/A_{oc}$ and i_s, which are negligible when compared to the two remaining terms. For an ideal operational amplifier the two discarded terms are identically zero. The voltage gain of the ideal inverting amplifier is

$$A_f = \frac{v_o}{v_1} = -\frac{R_f}{R_1} \tag{11-4}$$

The gain of the inverting amplifier is established by two stable resistors and not by the gain of the operational amplifier. The output signal is the factor $-R_f/R_1$ times the input signal; it does not matter whether the signals are dc, ac, or a combination of the two, as long as the output signal amplitude does not exceed the limits governed by the supply voltages.

If we retain the term divided by A_{oc}, which we discarded from Equation 11-3 to obtain Equation 11-4, the gain equation applies to a nonideal inverting amplifier. A little algebra shows that the gain of the inverting amplifier with finite open-circuit gain can be written

$$A_f = \frac{V_o}{V_1} = -\frac{R_f}{R_1}\left(\frac{1}{1 + \dfrac{R_1 + R_f}{R_1 A_{oc}}}\right) \tag{11-5}$$

The gains, A_f, given by Equations 11-4 and 11-5, differ little from each other; for example, if $R_1 = R_f$ and $A_{oc} = 10^5$, the term in parentheses differs from unity by 0.002 percent. Even with a gain of -500, which requires $R_f = 500\ R_1$, the parenthetical term differs from unity by only one-half percent. This conclusion verifies that Equation 11-4 is generally useful.

Negative feedback with amplifiers may be expected to provide the dual advantages of increased input impedance and decreased output impedance. For the inverting amplifier, however, this is not the case. We use Figure 11-3 and

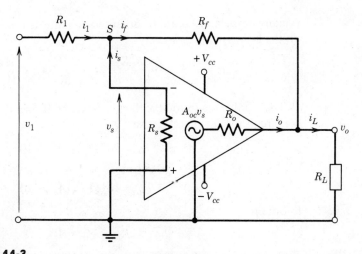

FIGURE 11-3

An Inverting Amplifier with a Thevenin Source

The input and output resistances are found by identifiying the internal components of the operational amplifier and calculating the feedback network equations.

Kirchhoff's laws to calculate the input resistance, R_{in}. The loop and node equations are

$$v_1 - i_1 R_1 + i_s R_s = 0$$
$$v_1 - i_1 R_1 - i_f R_f - v_o = 0$$
$$A_{oc} v_s - i_o R_o - v_o = 0$$
$$i_1 + i_s - i_f = 0$$
$$i_o + i_f - i_L = 0 \qquad (11\text{-}6)$$

Algebraic calculation yields

$$R_{in} = \frac{v_1}{i_1} = R_1 + \frac{\dfrac{R_s R_f}{R_s + R_f}}{\left[1 + \dfrac{R_s}{R_f + R_s} A_{oc} - \dfrac{R_s R_o}{R_f (R_f + R_s)}\right]} + f(v_o, i_1, R, A_{oc}) \qquad (11\text{-}7)$$

The second term on the right pertains to the internal input resistance of the operational amplifier itself. We note that the numerator

$$\frac{R_s R_f}{R_s + R_f}$$

is the parallel combination of the feedback resistor and the internal resistance between the inputs of the operational amplifier. The dominant term in the denominator

$$\frac{R_s}{R_f + R_s} A_{oc}$$

is many times larger than the other terms in the denominator; thus the second term on the right is approximately

$$\frac{R_s R_f}{R_s + R_f} \bigg/ \frac{R_s}{R_s + R_f} A_{oc} = \frac{R_f}{A_{oc}}$$

The relative magnitude of the terms can be shown by choosing typical values for the circuit parameters. We use

$$R_o = 200\ \Omega \qquad R_s = 10^5\ \Omega$$
$$R_f = 10^3\ \Omega$$
$$R_L = 10^3\ \Omega \qquad A_{oc} = 10^5$$

to obtain

$$R_{in} = R_1 + 10^{-2}\ \Omega$$

The term $f(v_o, i_1, R, A_{oc})$ in Equation 11-7 is approximately $\pm 10^{-4}\ \Omega$.

The extremely small effective resistance between the input terminals of the operational amplifier, approximately R_f/A_{oc}, justifies the claim that the virtual ground is a feature of the circuit. It must be understood, however, that the input current does not reach ground through the input terminals; rather, the operational amplifier generates the virtual ground. Because R_s, R_f, and A_{oc} are all much greater than 1, Equation 11-7 can be expressed in the limit as

$$R_{in} = R_1 + \frac{R_f}{A_{oc}} \tag{11-8}$$

The inverting amplifier is truly an unusual feedback amplifier whose effective input impedance is determined by the resistance R_1. An examination of Figure 11-2 shows that R_1 stands between the input signal and the virtual ground. Even though the resistance R_s between the noninverting and the inverting inputs is extremely large, the virtual ground exists because of the feedback properties of the amplifier.

In cases where R_1 is made large to increase the input resistance, R_f must be increased to retain the same amplification factor A_f. If the magnitude of $R_f = A_f R_1$ exceeds 10 MΩ or so, current may leak around the high resistance because of dirt or high humidity and cause trouble in the circuit. A better means of providing a high impedance input is shown below, when the noninverting amplifier is discussed.

In Chapter 9 we discussed the reduction of the output impedance of an amplifier by negative feedback. Although the input impedance of the inverting amplifier does not increase as expected, the output impedance is modified as expected, and we may use the equations found in Chapter 9 to give the output impedance.

Because of the negative feedback, the output resistance of the inverting amplifier, R_{of}, is considerably less than the intrinsic output resistance of the operational amplifier itself. We can rewrite Equation 9-17 to obtain

$$R_{of} = \frac{R_o}{1 - \beta A_{oc}} \tag{11-9}$$

We can multiply top and bottom with A_{oc} and use Equation 9-12 to find

$$R_{of} = \frac{R_o}{A_{oc}} \left(\frac{A_{oc}}{1 - \beta A_{oc}} \right)$$

$$= R_o \left(\frac{A_f}{A_{oc}} \right) \tag{11-10}$$

R_o is the intrinsic output resistance of the operational amplifier. Because the open-circuit gain, A_{oc}, is typically a 1000 times greater than the gain of the inverting amplifier, A_f, the output resistance has been diminished appreciably by the feedback. An output resistance of 1 Ω or less is typical.

The reduced output impedance can be an important advantage in many applications, but the operational amplifier has only a limited ability to supply current to the load on the output. Even though the amplifier is a Thevenin-like voltage source with a minuscule internal resistance, it cannot supply large currents because of internal current limitations of the operational amplifiers. Some operational amplifiers are protected internally from damage if the output terminals are shorted or if excessive current is required. Buffer amplifiers are used to offset the current limitations; a simple arrangement with a transistor emitter follower as the buffer amplifier is discussed at the end of the chapter.

NONINVERTING AMPLIFIER

An alternative to the inverting amplifier is made by placing the input signal on the positive or noninverting input terminal. Figure 11-4 shows the configuration of the *noninverting amplifier*. We note that the feedback network still involves the inverting input to ensure that the feedback is negative; that is, that the signal returned to the input from the output diminishes the output.

We write the Kirchhoff loop and node equations to obtain

$$v_{in} + v_d - i_1 R_1 = 0$$
$$v_{in} + v_d + i_f R_f - v_o = 0$$
$$i_f + i_s - i_1 = 0 \qquad (11\text{-}11)$$

If v_d and i_s are set equal to zero to correspond to an ideal operational amplifier, the equation formed from Equation 11-11 becomes

$$A_f = \frac{v_o}{v_{in}} = \frac{R_1 + R_f}{R_1} = 1 + \frac{R_f}{R_1} \qquad (11\text{-}12)$$

In the equations we use v_d rather than v_s to analyze the inverting amplifier because v_d, the difference voltage between the terminals in the noninverting amplifier, is not measured relative to ground.

The gain of the noninverting amplifier given by Equation 11-12 is 1 more than the gain of the inverting amplifier with the same resistors, R_1 and R_f, given by Equation 11-4, and the amplified signal has the same polarity as the input signal.

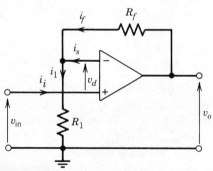

FIGURE 11-4

A Noninverting Amplifier

In the noninverting amplifier the input and output voltages are in phase. The gain is established by the resistor ratio, $1 + (R_f/R_1)$.

We can identify β, the feedback factor for the noninverting amplifier, by using the equation for amplification with feedback, Equation 9-13, to obtain

$$\beta = -\frac{R_1}{R_1 + R_f} \qquad (11\text{-}13)$$

An examination of Figure 11-4 shows that this is exactly the fraction of the output voltage provided at the inverting input by the voltage divider formed from R_1 and R_f, which corresponds to the definition of the feedback factor.

In the case of the noninverting amplifier, R_1 does not determine the input resistance; it may be set as small as desired within the limits of the output current of the operational amplifier. The input impedance of an amplifier with feedback is given by Equation 9-16. In accordance with this equation, the input resistance with feedback, R_{if}, is related to the intrinsic input resistance of the operational amplifier, R_i, by

$$R_{if} = R_i(1 - \beta A_{oc}) = R_i \frac{A_{oc}(1 - \beta A_{oc})}{A_{oc}}$$

$$= R_i \frac{A_{oc}}{A_f} \qquad (11\text{-}14)$$

The input resistance R_i is approximately R_s shown in Figure 11-3 between the input terminals of the operational amplifier, but because resistances exist within the operational amplifier from the terminals to ground in parallel with R_s, R_i is somewhat smaller than R_s. Because the intrinsic input resistances R_i of bipolar transistor operational amplifiers are approximately 2 $M\Omega$, the noninverting amplifier has a very high input resistance. Operational amplifiers with FET (field-effect transistor) inputs have intrinsic input resistances of 10^{12} Ω and are selected for this reason. In both types of amplifier the large ratio of the open-circuit gain to the closed-circuit gain increases the input resistance thousands of times.

The output resistance of the noninverting amplifier is reduced in the same manner as shown for the inverting amplifier. Equations 11-9 and 11-10 apply to noninverting amplifiers as well; both types of amplifiers exhibit output resistances of about 1 Ω.

The noninverting amplifier approaches the ideal analog amplifier. This amplifier, constructed with high-quality operational amplifiers, has the important advantages of negative-feedback amplifiers: high input impedance, low output impedance, and gain that is independent of variations in the amplifying elements.

VOLTAGE FOLLOWER

A variation of the noninverting amplifier occurs when the feedback resistor is made very small; that is, when it is replaced by a conducting wire. The closed-circuit gain becomes

$$A_f = 1 + \frac{0}{R_1} = 1 \qquad (11\text{-}15)$$

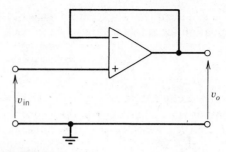

FIGURE 11-5

A Voltage Follower

The voltage follower is a noninverting amplifier with $R_f = 0$ and $R_1 = \infty$ to yield a ratio of one between input and output signals.

If R_1 is made very large (i.e., if it is removed to correspond to an infinite resistance), the noninverting amplifier retains the gain of one and the output voltage corresponds in magnitude and phase to the input voltage.

Figure 11-5 shows the voltage follower, which is actually a buffer amplifier with very high input resistance, very low output resistance, and faithful reproduction of the input signal. A direct connection from the output to the inverting input is all that is needed to provide the feedback. The signal into the voltage follower is replicated at the output as long as it is about 80 percent of the supply voltages $\pm V_{cc}$.

For voltage measurements in digital curcuits, an inexpensive voltmeter can be converted to an instrument with very high input resistance by placing a voltage follower between it and the measuring probe. A voltage follower placed before R_1 in Figure 11-2 transmits the input signal v_{in} unchanged, while providing an extremely high input impedance to the signal source. This arrangement provides the solution to the problem of a high input impedance for an inverting amplifier, as promised above.

FREQUENCY DEPENDENCE

Because of the frequency limitations of operational amplifiers we now turn to the frequency response of analog amplifiers made with them. Operational amplifiers are dc as well as ac amplifiers; thus the gain does not drop as low frequencies are used, unlike the capacitance-coupled amplifiers discussed in Chapter 9. The open-circuit frequency response, however, is limited seriously by the direct coupling of signals in operational amplifiers and the internal amplifiers that ensure the stability and linearity of the elements.

Figure 11-6 is a Bode plot of the gain in decibels versus the logarithm of the frequency for a commonly used operational amplifier. Other existing operational amplifiers, as well as those that may be expected to be available in the future, may have better frequency responses than the amplifier shown in Figure 11-6, but a half-

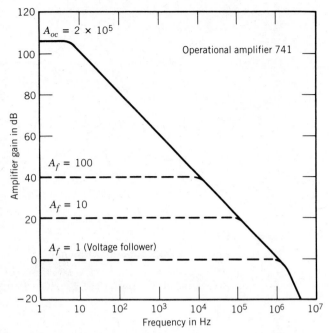

FIGURE 11-6

Voltage Amplification versus Frequency

The expansion of the frequency band is a result of increasing the feedback and reducing the gain (Texas Instruments, 1980).

power point at low frequencies will be a common feature. As we recall, a general feature of Bode plots is that the falloff at high frequencies is the same, -6 dB per octave. The half-power point of -3 dB, however, occurs at the frequency at which a line extended horizontally from the constant gain intersects the falling gain curve.

The frequency range of an analog amplifier becomes quite adequate for control and other applications with a modest closed-circuit gain from negative feedback. Figure 11-6 shows the frequency bands for three amplification values, with the simplified Bode construction mentioned in Chapter 9. Gains of 100 to 10 have frequency bands that meet all audio and many other applications. The voltage follower has an impressive frequency band from 0 to 1 MHz when it is made with the operational amplifier whose open circuit half-power point is no more than 8 Hz. In this case we can see the value of negative feedback for expanding the frequency bandwidth of amplifiers.

SLEW RATE

When a voltage step is applied to the input of an operational amplifier or of an analog amplifier made with an operational amplifier, the voltage output moves quickly to a new level. The rate at which it responds is called the *slew rate*. The slew

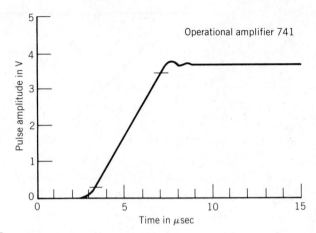

FIGURE 11-7

The Slew Rate for an Operational Amplifier

The rate of change in an output voltage with a step change in input voltage is the slew rate. The voltage change from 5 to 95%, divided by the time interval, yields the slew rate.

rate differs among operational amplifiers; a commonly used general-purpose operational amplifier has a slew rate of about 1 V/μsec, while particular operational amplifiers made to respond more quicky have slew rates of 50 to 500 V/μsec.

The slew rate is analogous to the propagation delay in digital gates. Figure 11-7 shows the response of an operational amplifier to a square wave input generated by digital element. The interval between the signal and the responses—the time delay—is not usually a problem in analog amplifier applications, but if operational amplifiers are used with digital gates, the relative time delays must be considered. In many applications that combine operational amplifiers and digital gates, however, the analog amplifier is used in the first or last stage. In speech synthesis, for example, the final digital-to-analog conversion may include an operational amplifier to drive the speaker. In this case the slew rate is not a limitation because there are no digital gates that require synchronization with other gates following the operational amplifier, and the time variations in speech are quite slow compared to the slew rate.

DIFFERENCE AMPLIFIER

An operational amplifier is a difference amplifier with auxiliary internal circuits to guarantee stable, linear, high amplification. The voltage difference between the inverting and the noninverting inputs is amplified with the open-circuit gain of the operational amplifier. The typical difference amplifier shown in Figure 11-8 has symmetrical inputs to both the inverting and the noninverting inputs.

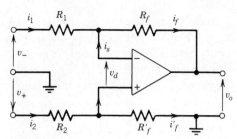

FIGURE 11-8
A Difference Amplifier
The operational amplifier may be used to give as an output the amplified difference between the two input signals.

The Kirchhoff loop and node equations for the difference amplifier are

$$v_+ - i_2 R_2 - i'_f R'_f = 0$$
$$v_- - i_1 R_1 - i_f R_f - v_o = 0$$
$$v_- - i_1 R_1 - v_d + i_2 R_2 = 0$$
$$i_2 - i'_f - i_s = 0$$
$$i_1 + i_s - i_f = 0 \qquad (11\text{-}16)$$

In these calculations we use our knowledge that i_s is negligible, so it is set equal to zero, and that $v_d = -v_o/A_{oc}$. The negative sign is used because of the polarity of v_d relative to the inverting input in Figure 11-8. If we eliminate the currents as the equations are solved simultaneously, we find that for the difference amplifier, when $R_f/R_1 = R'_f/R_2$,

$$v_o = (v_+ - v_-)\frac{R_f}{R_1 + \left(\dfrac{R_1 + R_f}{A_{oc}}\right)} = (v_+ - v_-)\frac{R_f}{R_1 + \delta} \qquad (11\text{-}17)$$

For an ideal operational amplifier, $A_{oc} = \infty$, $\delta = 0$; thus the difference voltage is amplified by the factor R_f/R_1:

$$v_o = (v_+ - v_-)\frac{R_f}{R_1} \qquad (11\text{-}18)$$

A real difference amplifier made with an operational amplifier may not be significantly poorer than the ideal difference amplifier. If the gain of an ideal difference amplifier is $R_f/R_1 = 100$, for example, and the open-circuit gain of the operational amplifier is $A_{oc} = 2 \times 10^5$, then the real difference amplifier amplifies the difference voltage by 99.95.

Both inputs have low input resistances because of the symmetric input arrangement. They are not equal, however, because the inverting and the noninverting inputs have different relations to ground, even when $R_f = R'_f$ and $R_1 = R_2$. The simple step of putting a voltage follower in each input line before R_1 and R_2, respectively, gives high input impedance to both signals.

COMMON-MODE REJECTION RATIO

Each of the voltage signals to the difference amplifier can be separated into two components. We assume that $v_+ = v'_+ + v_{cm}$ and that $v_- = v'_- + v_{cm}$ where v_{cm} is the common-mode voltage; that is, the voltage v_{cm} is a part of or common to each input signal. We see that the difference voltage, which is amplified, is

$$\begin{aligned} v_+ - v_- &= (v'_+ + v_{cm}) - (v'_- + v_{cm}) \\ &= v'_+ - v'_- \end{aligned} \qquad (11\text{-}19)$$

and that the common-mode voltage is not amplified. The common-mode voltage has the same amplitude and the identical phase on each input.

A common-mode voltage may arise from many sources. It may, for example, be the induced voltage on the leads to the inputs by the 60-Hz building power system. The ability of the difference amplifier to ignore the common-mode voltage, even though it may be orders of magnitude larger than the difference voltage, $v'_+ - v'_-$, and to amplify the difference voltage, is extremely important in many measuring tasks. In real difference amplifiers, however, the assumption made in Equation 11-19 that v_{cm} disappears entirely is not realistic. The quality of a difference amplifier in this respect is judged by the ratio of the amplified difference voltage to the amplified common-mode voltage. The common-mode rejection ratio (CMRR) was defined in Chapter 10, when we discussed a simple transistor difference amplifier. You may recall that

$$\text{CMRR} = \frac{A_v(\text{difference})}{A_v(\text{common-mode})}$$

In difference amplifiers made with high-quality operational amplifiers, the CMRRs may be quite good, perhaps 80 to 100 dB (10^4 to 10^5). The difference amplifier can be adjusted to maximize the CMRR by replacing the resistors R_2 and R'_f with a potentiometer whose adjustable center tap is connected to the noninverting input. The CMRR is maximized when a signal is connected to both input leads to create a common-mode input and when the output voltage is minimized by setting the potentiometer. The adjustment to improve the CMRR will alter slightly the gain provided by Equation 11-17.

ARITHMETIC USES OF OPERATIONAL AMPLIFIERS

Both the inverting and the noninverting amplifiers multiply the input voltage by a constant. The constant for the inverting amplifier is $-R_f/R_1$; the noninverting amplifier uses the factor $1 + R_f/R_1$. Other arithmetic operations are possible, such as addition, multiplication of functions, integration, differentiation, generation of logarithms and antilogarithms, and their use for generating powers and roots. In each case the output voltage of the operational amplifier circuit has the desired arithmetic relation to the input voltage signal.

Because the inverting input is at virtual ground and because the resistance through the operational amplifier inputs to ground is extremely high, all the input current to the inverting terminal passes through the feedback network. The key to understanding the arithmetic uses of the inverting amplifiers is the *current continuity*, $i_1 = i_f$. The amplifying circuit must satisfy this requirement; in doing so the output voltage becomes the desired function of the input voltage, $v_o = f(v_1)$.

From Figure 11-2, which illustrates the inverting amplifier, we extract the voltage-current relations shown in Figure 11-9. We assume that v_1 is positive at the moment of analysis; consequently the output voltage, v_o, is negative at the same instant because of the inverting property of the amplifier. The Kirchhoff loop and node equations are

$$v_1 - i_1 R_1 = 0$$
$$-v_o - i_f R_f = 0 \qquad (11\text{-}20)$$

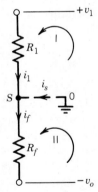

FIGURE 11-9

The Current-Voltage Relation in an Inverting Amplifier

The continuity of current in the resistor network of the inverting amplifier is the key to its use in performing arithmetic as the amplifier sinks the current.

and
$$i_1 - i_f + i_s = 0$$

We find with $i_s = 0$ that

$$\frac{v_o}{v_1} = -\frac{R_f}{R_1}$$

as before.

We can now make clear the dominance of the current through the input resistor and the feedback resistor in establishing the voltage relations in the inverting amplifier. The continuity of current, $i_1 - i_f = 0$, at the point S, which is the virtual ground, causes the input voltage and the output voltage to be related. The operational amplifier circuits that perform arithmetic functions have the identical property: all current reaching S through the input circuits must leave S through the feedback circuit that allows the functions to be generated.

We illustrate this point by examining the current-voltage relations in Figure 11-9. The voltage drop due to i_1 in R_1 reduces the input voltage to zero at S from the value v_1 (which we assume is positive at the moment of analysis). The continuity of the current, $i_1 = i_f$, provides an additional voltage drop in R_f to the negative value of v_o shown in the figure. The magnitudes of the voltage drops are proportional to the respective resistors, R_1 and R_f, and the output voltage must acquire the negative magnitude, which sinks the current. (The term "sink" means to provide a means by which the electrical charges can continue to move through the circuit.) The very high open-circuit gain of the operational amplifier ensures that the difference signal, v_s in Figure 11-2, which arises from a difference in the voltage drops between the input and the feedback circuits, is so minuscule that the gain $-R_f/R_1$ pertains.

SUMMING AMPLIFIER

In many applications of analog electronics an important function is the addition of signals from various sources. This is sometimes called *signal mixing*, and it is useful in audio and control circuits. In addition, the summing of digital signals is an important means of digital-to-analog conversion (DAC).

The *summing amplifier*, shown in Figure 11-10, has several input circuits to a single inverting amplifier. As we recall, the input resistance of the inverting amplifier, Equation 11-8, is effectively the input resistance, R_1, and the input current is $i_1 = v_1/R_1$. Because the node S in Figure 11-10 is at virtual ground, we can generalize to the condition of several inputs. The input current from each input signal through its input resistor, $i_n = v_n/R_n$, is independent of every other current because of the virtual ground; thus all the input currents combine and pass through the feedback resistor. S is called the *summing point* or *summing node* because $i_f = \sum i_n$. The output voltage generated by the operational amplifier acquires the

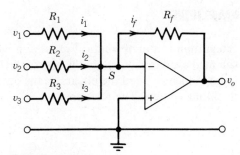

FIGURE 11-10

A Summing Amplifier

The current continuity through the summing point at virtual ground yields as an output the sum of the input voltages weighted by the input resistors.

magnitude, sign, and time variation that can sink the algebraic sum of the input currents through R_f.

The current relation in Figure 11-10 is

$$i_f = i_1 + i_2 + i_3$$

or

$$i_f = -\frac{v_o}{R_f} = \frac{v_1}{R_1} + \frac{v_2}{R_2} + \frac{v_3}{R_3} \quad (11\text{-}21)$$

Simply to add the input voltages, all the input resistors, R_n, are set equal to the feedback resistor, R_f. Equation 11-21 becomes

$$-v_o = v_1 + v_2 + v_3 \quad (11\text{-}22)$$

In all the operational ampifier circuits and equations, we have consistently used lowercase letters for voltages and currents because that practice designates time-variable values. This is not to say that the voltages cannot be constant (dc), but only that they are not restricted to dc. Only the frequency limits of the operational amplifier restrict the variability of the voltages and currents. Equation 11-22 indicates that the voltages of different time dependence and magnitude may be added by summing amplifier as long as the sum remains within the voltage range of the operational amplifier, about 80 percent of V_{cc}.

A weighted sum can be formed by using input resistors that have the desired ratio to the feedback resistor. One useful method of weighting converts numbers represented as binary bits into an output voltage proportional to the numerical value. For example, the values $R_f/R_3 = 4$, $R_f/R_2 = 2$, and $R_f/R_1 = 1$ with the LSB (least significant bit) on input 1 converts an octal number into its voltage analog. This application is discussed in the next chapter, which deals with digital-to-analog conversion.

INTEGRATING AMPLIFIER

Integration is a mathematical process in which a function, called the *integrand*, is summed over an interval. In the case of the *integrating amplifier* the integrand is the input voltage, v_1, and the integral is the output voltage, which is the sum in time of the input voltage. The relation is

$$v_o(t) = -\int_{t_0}^{t_f} v_1(t)\, dt \qquad (11\text{-}23)$$

where t_0 and t_f are the initial and final times, respectively, between which the integration is performed. As indicated in Equation 11-23, $v_1(t)$ is usually a function of time itself. Figure 11-11 shows an integrating amplifier.

The capacitor, C, in the feedback circuit is the element that relates the input voltage to its integral over time. The basic relation of voltage and charge in a capacitor (Equation 7-13) may be rewritten for time-varying terms to give

$$q(t) = Cv(t) \qquad (11\text{-}24)$$

where $q(t)$ is the charge on the capacitor in coulombs, C is the capacitance in farads, and $v(t)$ is the voltage across the capacitor. The term $q(t)$ is the time integral of the current into the capacitor,

$$q(t) = \int_{t_0}^{t_f} i(t)\, dt \qquad (11\text{-}25)$$

The charges moving as the independent-input current, which depends only upon $v_1(t)$ and R_1, pass through the summing node, S, and accumulate on the feedback capacitor. The output voltage, $v_o(t)$, must change in an appropriate time pattern to

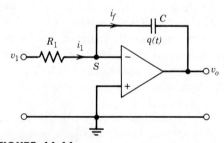

FIGURE 11-11

An Integrating Amplifier

In sinking the current, the output voltage varies as the integral of the input voltage.

sustain the flow of charges into the capacitor; that is, it must sink the current. The output voltage, $v_o(t)$, is given by

$$v_o(t) = -\frac{q(t)}{C} = -\frac{1}{C}\int_{t_0}^{t_f} i_f(t)\, dt = -\frac{1}{C}\int_{t_0}^{t_f} i_1(t)\, dt$$

$$= -\frac{1}{R_1 C}\int_{t_0}^{t_f} v_1(t)\, dt \tag{11-26}$$

The term $-1/R_1 C$ is a scaling factor in seconds^{-1}. For real-time integration $R_1 C = 1$ sec is used.

An easily understandable example of integration occurs when the input voltage is taken as a constant, $v_1(t) = V_1$. We find, for $v_o(t_0) = 0$ with $t_0 = 0$ and $t_f = t$,

$$v_o(t) = -\frac{1}{R_1 C}\int_0^t V_1\, dt = -\frac{V_1}{R_1 C}\int_0^t dt$$

$$= -\frac{V_1}{R_1 C} t \tag{11-27}$$

The output voltage grows negatively and uniformly with time as the input voltage remains constant at a positive value. The relation between $v_1(t) = V_1$ and $v_o(t)$ is shown in Figure 11-12.

The applications of integrating amplifiers is considered later.

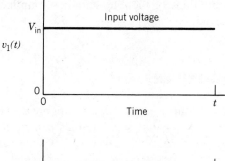

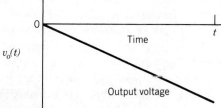

FIGURE 11-12

The Input and Output Voltages in an Integrating Amplifier

The integral of a constant v_1 is a linear change in the output voltage inverted by the amplifier.

An exchange of the positions of R_1 and C in the integrating amplifier circuit (Figure 11-11) produces a *differentiating amplifier*, whose output voltage is the time derivative of the input voltage. We use the definition of current

$$i_1 = \frac{dq_1}{dt} = \frac{dq_f}{dt} \tag{11-28}$$

and Equation 10-19 to write

$$v_o(t) = -R_f \frac{dq_f}{dt} = -R_f C \frac{dv_1(t)}{dt} \tag{11-29}$$

The output voltage, $v_o(t)$, is the time derivative of the input voltage, $v_1(t)$, times the scaling factor, $(-R_f C)$ seconds.

Differentiating amplifiers provide an output-voltage pulse when a discontinuity occurs in the input-voltage signal. Output-voltage pulses at each edge of a square wave voltage input signal can mark the occurrences over time of the transitions of the square wave, but the usefulness of a differentiating amplifier is limited by its response to noise on the input. Noise is any unwanted electrical transient, and it consists typically of high-frequency voltage excursions. The voltage amplitude of noise may be relatively low, but its rate of change, dv_{noise}/dt, is high. The output voltage is frequently overwhelmed by noise on the input, so that it is difficult to achieve the orderly time differentiation of a smoothly changing input voltage. The limitations of the differentiating amplifier suggest that we should not pursue it further.

LOGARITHMIC AMPLIFIER

Logarithmic amplifiers can compress the dynamic range of voltages over 4 to 7 decades, so that small signals can be observed with the same instrument that accommodates very large signals. They can generate logarithmic analogs of signals that may be added or subtracted to yield the logarithms of products or quotients, and they may be multiplied by constants to give powers and roots. The inverse process of generating antilogarithms, as exponential amplifiers do, restores the required function.

The generation of an output voltage, which is the logarithm of the input voltage, requires an element that has a logarithmic voltage-current relation. The logarithmic relation can be found in the current dependence on voltage in a *pn* junction. The junction can be a diode, but a bipolar junction transistor also has this characteristic; thus either may be the element. The current-voltage characteristic of the *pn* junction is

$$I = I_s \left[\exp\left(\frac{qV}{kT}\right) - 1 \right] \tag{11-30}$$

where I is the junction current and I_s is the saturation current, both in amperes, V is the junction voltage, $q = 1.6 \times 10^{-19}$ coulombs, $k = 1.38 \times 10^{-23}$ Joules/Kelvin is Boltzmann's constant, and T is the absolute temperature in Kelvin (Moll, 1958).

Equation 11-30 can be inverted to give

$$V = \left(\frac{kT}{q}\right) \ln\left(\frac{I}{I_s} + 1\right) \tag{11-31}$$

We denote the natural or Naperain logarithm by ln. The quantity $kT/q = 25.3$ mV at ambient temperature (20°C); it is clear that the voltage has a direct temperature dependence, about 86 μV/K. Because the saturation current, I_s, depends upon the density and lifetime of holes and electrons in the semiconductor, it is also temperature-dependent. I_s is typically a fraction of a microampere; thus the ratio I/I_s is usually much greater than unity, and the logarithm in Equation 11-31 is taken by dropping the 1. Additional circuitry provides the temperature stability in available integrated-circuit logarithmic amplifiers and ensures the linearity over many decades. In Figure 11-13 a logarithmic amplifier with a junction transistor relates the input voltage to the output voltage by

$$v_o = -\ln(v_1) \tag{11-32}$$

Two connections of the transistor base may be used. The base tied to the collector is equivalent to using a diode as the element, but the base-grounded configuration gives a better approximation to the logarithmic relation. A second operational amplifier (not shown) is required to establish the zero voltage level to ensure the desired mathematical relation. Integrated circuits with excellent logarithmic behavior are available.

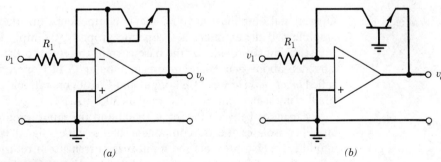

FIGURE 11-13
Logarithmic Amplifiers

(*a*) A bipolar-junction transistor as an equivalent diode. (*b*) A bipolar-junction transistor as a logarithmic element. A diode or a bipolar-junction transistor as the feedback element uses its exponential current relation to give as an output voltage the logarithm of the input voltage.

The antilogarithmic or exponential amplifier transfers the diode or transistor from its position in the logarithmic amplifier to the input circuit. Integrated circuits that have multiplying and dividing capabilities through the generation of logarithms also contain the exponential amplifiers necessary to restore the products or quotients.

INPUT-OFFSET VOLTAGE

An operational amplifier is a difference amplifier made with many components that are ideally matched but that actually differ slightly in performance. Because of the differences, the voltage between the input terminals of the operational amplifier is not truly zero but a finite value. Thus, when the inputs are tied together, a nonzero output voltage exists, as if a very low-voltage battery were lying between the inverting and the noninverting inputs. This small voltage is called the *input-offset voltage*.

This problem was anticipated in discussing the TTL difference amplifier; Figure 10-5 shows a small adjustable resistor R, which allowed the amplifier to be balanced. Many operational amplifiers as integrated circuits have terminals across which an adjustable resistor can be placed to null the input-offset voltage so that the output voltage is zero, or at least very small, when both inputs have the same signal. In a commonly used operational amplifier, the wiper of the variable resistor is attached to $-V_{cc}$. The ends of the resistor are attached to the offset null pins on the DIP, the inputs are tied together, and in this way the output voltage is nulled. The practice of nulling the input-offset voltage is important in many applications of operational amplifiers, and provides the best CMRR (common-mode rejection ratio).

INPUT-BIAS CURRENT

Operational amplifiers have no magic components, but they do have transistor amplifiers. If the amplifier is to operate properly, the input lead to the base of a transistor must provide for the base current to return to the emitter. These small currents, about one microampere in bipolar (TTL) operational amplifiers, are called *input-bias currents*. FET-input and CMOS operational amplifiers have much smaller input-bias currents, as small as 10^{-11} A.

A capacitor between the input signal and the input terminal of an operational amplifier isolates the dc component of the input signal from the operational amplifier. It also prevents the input-bias current from returning to the amplifier circuit. A resistor must be connected between the capacitor and the input terminal to ground. We noted this practice in capacity-coupled amplifiers; in Figure 10-1, for example, R_1 and R_2 provide paths for the base current to return to the emitter. If the path to ground is not present, the coupling capacitor will charge and shift the quiescent operating point of the amplifier. In vacuum-tube technology the resistor is called a *grid-leak resistor* and serves the same purpose.

In dc measurements with a very high resistance sensor, such as a PH probe, the input-bias current may overwhelm the voltage being measured. An FET-input or a CMOS operational amplifier is preferable to a TTL operational amplifier for this kind of measurement.

MATCHING OUTPUT CURRENT REQUIREMENTS

Operational amplifiers not only have low output impedances, but also limits on the current they can supply to the circuits they drive. Additional current capability is provided by buffer amplifiers, which are placed between the operational amplifier and the load. Figure 11-14a shows a noninverting amplifier with an emitter follower used to increase the output current as long as v_o is positive. The current to the load with this arrangement can be the transistor's β times the base current provided by the operational amplifier. Because β, defined in Equation 8-4, usually lies between 50 and 200, the current gain is significant. In this case the emitter follower is incorporated into the feedback network.

Figure 11-14b uses an emitter follower after the amplifier. The signal range of the emitter follower in this circuit is approximately equal to the positive and negative voltage range of the operational amplifier.

Function generators are integrated circuits that provide sine waves, square waves, and triangle waves outputs. Many function generators regularly require buffer amplifiers in their use. The buffer amplifier must be placed on the output

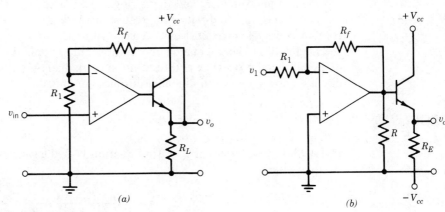

FIGURE 11-14

Amplifiers with Emitter-Follower Buffers

The output-current limitation of an operational amplifier can be offset by a simple transistor emitter-follower buffer amplifier; (a) the single-polarity buffer is inside the feedback network; (b) both positive and negative output signals can be buffered by using bipolar voltage sources.

because there is no way to incorporate it in the internal feedback networks of the integrated circuits unless terminals are provided for that purpose. In many applications an inverting or a noninverting amplifier or an operational amplifier used as a voltage follower is an effective buffer amplifier. A transistor emitter follower, as shown in Figure 11-14b, may be the simplest buffer.

The operational amplifier circuits that we have discussed here are the bases for many remarkable circuits with temperature and other compensations available as DIPs. These circuits, generally called *linear circuits*, perform many control and measurement functions. Operational amplifiers themselves can be used in many ways, both because of their unique properties and because it is easy to imagine a simple operational amplifier circuit as the solution to an otherwise difficult electronic problem. A wide range of parameters, such as, slew rate and input impedance, can be selected to meet diverse needs by choosing from the large variety of operational amplifiers available as integrated circuits.

SUMMARY

The integrated-circuit operational amplifier is a truly amazing electronic element that replaces many other amplifying arrangements for small signals with frequencies of less than 1 MHz. Control, measuring, and instrument needs are usually more easily met by constructing them with operational amplifiers to improve impedance matching, remove degradation from component aging, and give linear or other desired amplification than by employing discrete amplifiers and circuit elements.

Operational amplifiers and digital logic gates used together allow data to be processed and analyzed with the greatest advantages of both technologies. The combined use of these devices in voice synthesis, for example, illustrates their special abilities to accept analog information and to apply the greatest advantages of digital signal processing and restoration to an analog output. The next chapter discusses how analog and digital electronics are united.

PROBLEMS

11-1 Show that by setting $i_s = 0$ in Equation 11-3, the gain with feedback may be written

$$A_f = -\frac{R_f}{R_1} \frac{1}{1 + \frac{1}{A_{oc}}\left(1 + \frac{R_f}{R_1}\right)}$$

If the open-circuit gain $A_{oc} = 10^5$ and the ratio $R_f/R_1 = 10$, find the percentage difference between the gain of inverting the real operational amplifier and the ideal inverting operational amplifier. Express the gain difference in decibels.

11-2 Using the currents and voltages in Figure 11-4 and v_0, show that the input resistance is $R_{if} = $

11-3 Use the Kirchhoff equations to find the gai amplifier when the operational amplifier is not

11-4 In the difference amplifier (Fig. 11-8) the input r ing input and of the inverting input are different r relations among the resistances to make the inp that the gain R_f/R_1 is unchanged by the values given in terms of R_1 and R_f.

11-5 Find the region where the logarithmic amplifier is linear if $I_s = 10^{-7}$ A. Beyond 20 mA the transistor circuit is not linear because of its internal resistances. If $I_s(T) = 10^{-9}$ A/K, what will be difference in the logarithms between operation at 20°C and at 50°C? Describe the effect of temperature difference in terms of voltage error if the inverse process is taken: that is, $V_1 \to \ln V \to \log^{-1} v = v_o$.

11-6 Show that the T-filter as the feedback network in an inverting amplifier allows modest resistors to replace an equivalent high resistance.

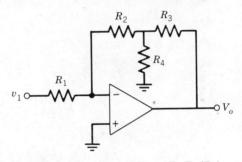

Find that R_f (equivalent) $= (R_2 + R_3 + R_2 R_3/R_4)$ so that $A_f = (R_2 + R_3 + R_2 R_3/R_4)/R_1$.

Hint: The current distribution dominates the circuit. Show that R_4 is a gain control. Find the range of R_4 for $10 \le A_f \le 1000$ with $R_1 = 1$ kΩ.

11-7 Show that multiple inputs (as in a summing amplifier) to the noninverting amplifier gives an average of the voltages weighted by the input resistors; that is, it becomes a voltage-averaging amplifier.

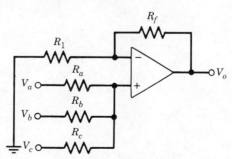

11-8 Show that the noninverting amplifier can be a summing amplifier if

$$\left(1 + \frac{R_f}{R_1}\right) = \text{number of inputs}$$

11-9 Show that the summing amplifier will give an average of the input volts where $R = R_1 = R_2 = R_3$, etc., and

$$\frac{R_f}{R} = \frac{1}{\text{number of inputs}}$$

REFERENCES

1958 Moll, J. L., "The Evolution of the Theory for the Voltage-Current Characteristics of P-N Junctions," *Proceedings of the IRE*, **46**, 1076 (1958).

1980 *The Linear Control Circuits Data Book*, Second Edition, Texas Instruments, Dallas, Texas. The frequency response in Figure 11-6 is taken from this book.

CHAPTER 12

ANALOG-DIGITAL CONVERSION

Many quantities have continuous values, including temperature, pressure, displacement, rotation, voltage, current, and intensity of light and sound. The digital representation of the values is important for digital processing and analysis, for regulation and control, and for recording and transmitting the data that quantize them. The task of quantizing the continuous values into a binary scale is called *analog-to-digital conversion* (ADC).

Digital-to-analog conversion (DAC) is the inverse process, in which data in discrete values are converted or restored to a continuously variable form. The clarity and fidelity of laser-read digital recordings of music and speech and the freedom of these recordings from background noise are recent advancements, but the human ear cannot convert binary data recorded on a disc to music or speech. The electronic conversion of binary data to analog signals is necessary.

Some ways of sampling analog signals cannot be considered analog-to-digital conversion because the amplitude of the signal, an important aspect of the information content, is not expressed in binary form. In the straightforward sampling of a sine wave, the amplitude is an essential part of the information content. Figure 12-1 shows the result of sampling a sine wave at discrete sampling intervals, but the pulses are not binary signals. Binary signals (bits) have no amplitude variation; they are off (0) or on (1). In contrast, the conversion from an analog signal to a digital signal results in a binary number at each sampling point.

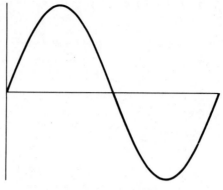

Sinusoidal analog signal

Sampling clock pulses

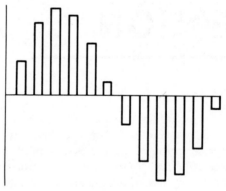

Sampled sine wave

FIGURE 12-1

Amplitude Sampling of Sine Wave

The amplitude is determined at regular intervals as a step in a analog-to-digital conversion.

If the sine wave amplitude at each sampling point is expressed as a binary number and the sampling time is similarly expressed as a binary number (perhaps as a sequential memory location), then analog-to-digital conversion has been made. Several other methods of sampling a sine wave are discussed below.

In our discussion of analog-to-digital conversion and digital-to-analog conversion we assume that the variable to be converted is voltage. The *transducer* is the name of the device that produces a voltage or a current proportional to the physical phenomenon to which it responds. A temperature transducer, for example, can generate a voltage related to the temperature. In some cases the voltage is derived from the sensitive element in an electrical circuit, as when a temperature-sensitive resistor and a constant resistance form a voltage divider whose voltage is related directly to the temperature. Some transducers generate voltages directly, such as photoelectric elements and piezoelectric devices. Many voltage sources require voltage amplification to facilitate conversion from their analog values to binary numbers.

ANALOG-TO-DIGITAL CONVERSION

Only two basic techniques exist for analog-to-digital conversion (ADC). One is to compare the analog voltage amplitude to a binary voltage scale in which the match yields the binary number that corresponds to the amplitude. The other technique is to integrate the analog signal and to use the measured time (a given number of clock pulses) for the amplitude of the integral to reach a value to establish an equivalent binary number. Each of the systems discussed below uses one or the other of these techniques.

Two important parameters must be considered in selecting a conversion technique or a variation of a conversion technique. One is the precision required in the analog-to-digital conversion; the other is the speed or the time interval allowed for the conversion. These two parameters are essentially incompatible because high-precision and high-speed conversions are difficult to achieve concurrently. High-speed or fast analog-to-digital conversion is a relative term, but in the context of digital computer conversion of binary data from analog to digital, "high-speed" sampling intervals are about ten computer clock cycles rather than tens of thousands. At the present, conversion in the interval of one microsecond is moderately fast but not limiting.

Because many analog-to-digital conversions need not be made quickly, a high degree of precision is possible. The complexity and cost of attaining high precision, however, may modify the goal. The application intended for the conversion of the analog variable may determine what degree of digital precision is required.

The precision of analog-to-digital conversion is established by the number of binary bits, which correspond to maximum or full-scale analog value. Four bits allow the quantization in 0 to 15 equal intervals. For the binary representation of the amplitude to change by one bit, an analog amplitude must change by 6.25 percent. A byte (eight bits) allows a precision of 0.4 percent; seven bits of binary code correspond to approximately 1 percent encoding accuracy. Table 12-1 lists the number of binary bits that encode the rotation of a shaft in degrees, minutes, and seconds.

The smallest change in an analog value that can be resolved by an n-bit converter is the value carried by the least significant bit (LSB). The LSB carries 2^{-n} of the full-scale value.

TABLE 12-1
Encoding Shaft Rotation

Quantum	Number of Binary Bits	Number of Intervals	Rotational Precision (θ/bit)
Degree of arc	9	512	0.70 degrees
Minute of arc	15	32,768	0.66 minutes
Second of arc	21	2,097,152	0.62 seconds

COMPARATOR LADDER

A *comparator* is an operational amplifier circuit in which a transition in the output voltage from one extreme of one voltage polarity to the other extreme of the opposite voltage polarity ($+80\%$ of V_{cc} to -80% of V_{cc}) occurs when the signals on the inverting and the noninverting inputs change their relative values. In a comparator ladder, as shown in Figure 12-2, several comparators are provided with a sequence of relative reference voltages from a series resistor voltage divider. Each operational amplifier changes its output state; that is, the output switches when the analog signal that is being converted to a binary equivalent exceeds the relative reference voltage on the input of the particular comparator. In this situation all the comparators whose reference voltages are below the voltage of the analog signal are in one state while those with reference voltages greater than the analog signal are in the opposite state. The process of analog-to-digital conversion is completed by a binary encoder. An octal-to-binary encoder is shown in Figure 3-12.

The conversion from analog to binary values occurs continuously. The instantaneous analog-to-digital conversion is the distinctive characteristic and the important advantage of the comparator ladder. Frequently, however, the output of the encoder is latched and supplied to the data bus. The output of the encoder may be updated by an enable signal, and the new value to be read may be latched at the clock frequency. Thus, in the applications where the binary values are processed by a computer, the necessity of reading at specific intervals introduces an effective finite conversion time.

Figure 12-2 shows a number of components whose precise values determine the accuracy of the analog-to-digital conversion. The three-bit analog-to-digital converter yields only 12.5 percent resolution and requires seven comparators. A resolution of about two percent can be obtained with 64 comparators and a six-bit encoder ($2^6 = 64$). It is clear that moderate precision involves a long comparator ladder and that precision resistors must be used in the voltage divider. The reference voltage source that drives the resistor voltage divider must meet the same high standards of constancy and magnitude.

Even if the desired precision is present, there is a quantization uncertainty. All analog values within a given range are represented by a single binary number. In

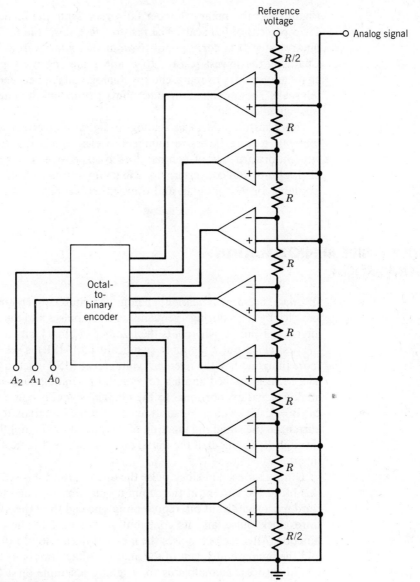

FIGURE 12-2
A Comparator Ladder
Eight operational amplifiers quantize an analog voltage into one of eight amplitude values encoded as an octal number.

Figure 12-2 the binary number 001_2 represents the analog voltage from 6.25 to 18.75 percent of full scale. The resistor, $R/2$, gives the 6.25 percent offset, and the binary value 111_2 corresponds to an analog voltage from 93.75 to 106.25 percent. Changing the lowest resistor to R and removing the top resistor in the resistor chain allows 001_2 to represent any analog voltage between 12.5 and 25 percent of full scale. The quantization uncertainty is one-half the interval represented by the least significant bit.

Comparator ladders as analog-to-digital converters with small quantization intervals require excessive numbers of elements and extreme linearity; thus the straightforward use of comparators is expensive. Even so, an eight-bit (0.2%) analog-to-digital converter with 256 comparators operating from dc to 40 MHz is available for video, radar, and transient signal processing.

SUCCESSIVE-APPROXIMATION CONVERSION

In *successive-approximation* analog-to-digital conversion the analog signal is sampled by steps during a sequence of clock pulses and the conversion is completed after $n + 1$ pulses, where n is the number of binary bits to be determined. During the first clock pulse the most significant bit (MSB) is set. If the analog voltage is more than one-half the reference voltage, the MSB is set to one; otherwise it is zero. The MSB is latched and fed to a digital-to-analog converter, which generates an analog signal proportional to the bit. The signal is one-half the voltage when the MSB = 1. This voltage is subtracted from the initial analog voltage by applying a current to the summing junction of the operational amplifier to reduce the current through R_1 to one-half the initial analog voltage. Figure 12-3 shows this arrangement.

During the second clock pulse the one-quarter bit is set, and the contribution of this bit to the current at the summing junction is added to that of the MSB. The third most significant bit, representing one-eighth of the full scale, is set during the third clock pulse, and its contribution is added to the current at the summing junction. After n clock pulses the n bits of the analog-to-digital conversion are set, and the binary equivalent of the analog voltage can be read on clock pulse $n + 1$.

This process is analogous to weighing a sample on a balance. The balance is tested starting with the largest standard mass. If the added standard mass exceeds the mass of the sample, it is removed (set to zero), and the next smaller fractional standard mass is tried. The comparison is continued until the smallest fractional mass is tried, and then the process is terminated. The mass of the sample is determined by adding together all the standard masses on the balance pan.

The number of bits depends upon the size of the counter or register used to hold the binary value. Twelve-bit analog-to-digital converters that use the successive-approximation process can complete the conversion in as little as 3 μsec; typical conversion times range from 15 to 25 μsec.

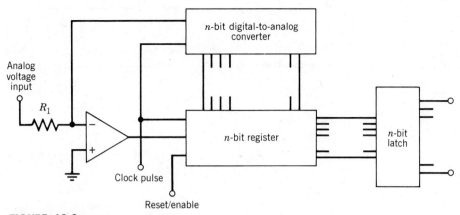

FIGURE 12-3

Successive-Approximation ADC

The quantized magnitude of the analog voltage is found in successive steps by subtracting the fraction of the magnitude established in earlier steps.

SUBRANGING

Subranging may be considered a variation of the successive-approximation analog-to-digital conversion. In this technique the binary bits are determined in only two steps. The most significant portion of the analog signal is converted at a fast rate by an *n*-bit comparator ladder. The *n*-bits are used to generate an analog voltage by a digital-to-analog converter, which is subtracted from the initial analog input voltage. Typically $n = 5$; thus the $1 - 2^{-5} = 0.96875$ of the full-scale value is established in the first step. The residual or difference analog signal is amplified; then a comparator ladder establishes the binary equivalent of the residual. The two sets of binary numbers are combined to yield the binary equivalent of the initial analog voltage.

The two-step analog-to-digital conversion combines speed and resolution. Subranging converters provide 12-bit accuracy at a word rate of 5 MHz.

SAMPLE-AND-HOLD CIRCUITS

Many analog-to-digital converters use a *sample-and-hold circuit* so that the voltage being converted will not change during the conversion interval. For fast-changing analog signals the sampling interval is as small as a microsecond or less.

A sample-and-hold circuit is shown in Figure 12-4. The operational amplifier charges or discharges the capacitor so that the voltage of the capacitor tracks the analog input voltage. The output of the second operational amplifier is returned to

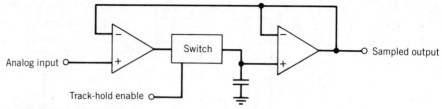

FIGURE 12-4

A Track-and-Hold Circuit

A changing analog signal is sampled at a fast rate, and the value is held during the analog-to-digital conversion interval.

the input of the first, making the combination a voltage follower. The low-impedance, high-current output of the operational amplifier optimizes the process of charging the capacitor, but the ability to track a rapidly changing analog-input voltage is also governed by the slew rate of the operational amplifier.

At the beginning of the analog-to-digital conversion process the track-hold-enable signal opens the switch between the first operational amplifier and the capacitor. The charge on the capacitor is captured to preserve the voltage that it has at the beginning of the hold interval. The analog-to-digital converter has a steady sampled output voltage to process. A voltage drift can result from the leakage of charge from the capacitor and from the possible charging or discharging of the capacitor through a less-than-ideal switch. The ratio, however, between the charging current in the tracking mode and the current in the hold mode can be as high as 10^7; thus, with high-quality components, the sampled voltage can be held quite steady for the short period required for the analog-to-digital conversion. Manufacturers specify that their integrated-circuit track-sample-hold elements have maximum voltage drift rates of no more than a few millivolts per second.

COMPRESSION AND EXPANSION

Because some analog signals have large dynamic ranges, linear analog-to-digital encoding of the signals will not provide the desired resolution for the small-amplitude components of the signals. In speech and music the large amplitudes represent loud portions; the details and quality of the sound are carried by the small-amplitude signals. The small-amplitude portions of the analog signals can be encoded more sensitively with nonlinear processes of analog-to-digital conversion.

Nonlinear processing has been used in the recording industry for a number of years. Before the master record is cut, the sound is amplified in nonlinear fashion to enhance the small amplitudes. Thus the small signals produce significantly greater variations in the grooves of the record than would have occurred otherwise. The low-amplitude variations of the unwanted but unavoidable noise from the cutting stylus are made relatively less important. The playback amplifies the sound with the inverse nonlinear amplitude relation; as a result, the noise is suppressed effectively and the quality of the sound is retained.

Two basic techniques exist for compression and expansion in analog-to-digital conversion to improve the encoding sensitivity to small-amplitude signals. One compresses the analog signal by nonlinear analog amplification and then encodes it with a linear analog-to-digital converter. The second uses a nonlinear analog-to-digital converter to increase the precision of encoding the small-signal portions of the analog signal. In each system the analog signal is restored by an inverse process; in the first case the encoded signal is restored by a linear digital-to-analog converter and then the analog voltage is amplified in nonlinear fashion. In the second case, a nonlinear digital-to-analog converter with the inverse nonlinearity yields the analog signal.

The increase in sensitivity to small-amplitude components of analog signals can be illustrated if we consider a poor-resolution four-bit analog-to-digital converter when the low-amplitude signals are expanded before or during the conversion process. Two different nonlinear expansions are used in the illustration. Both the logarithmic and the square root expansions of the analog signal emphasize the low-amplitude components relative to their linear values. Figure 12-5 shows the

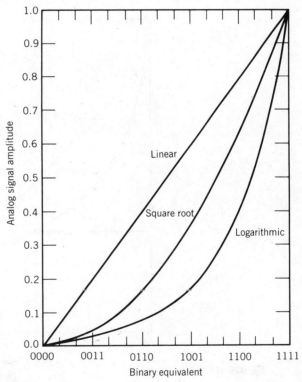

FIGURE 12-5

A Nonlinear-to-Digital Conversion

The logarithm and square root nonlinear analog-to-digital conversion provide greater resolution for low-amplitude signals than does linear conversion.

portion of the analog signal represented by the four-bit binary numbers for linear, square root, and logarithmic analog-to-digital conversion. The expansion of the low-amplitude portions with the nonlinear analog-to-digital conversion is significant. For the square root conversion the lower half of the binary numbers covers approximately 20 percent of the analog amplitude. When the nonlinear conversion uses the logarithm, approximately one-tenth the amplitude is expanded over the lower half of the binary numbers.

Figure 12-6 shows the increased resolution of the small-amplitude components of the analog signal by nonlinear expansion. In this figure the effective bit number, n, is plotted versus the four-bit binary code. At all magnitudes of the analog signal the four-bit linear conversion gives the same resolution; thus the horizontal line at $n = 4$ represents the case of no expansion, or linear conversion. The square root and the logarithmic expansion both give significantly higher resolution, $n > 4$, for small signals. The square root nonlinearity is less sensitive than the logarithmic compression for all small values. For large-amplitude analog signals, however, the

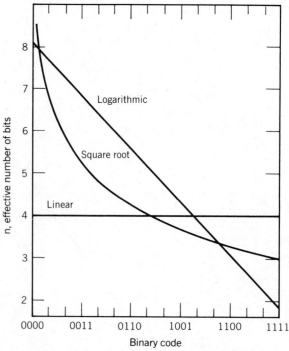

FIGURE 12-6

An Effective Resolution in Analog-to-Digital Conversion with Expansion

Nonlinear conversion provides eight-bit resolution for small-amplitude signals in a four-bit binary conversion.

square root process maintains substantially better resolution after the respective curves cross on the right-hand side of Figure 12-6.

The illustration uses a four-bit analog-to-digital converter, but many applications use more bits in the conversion. A useful arrangement uses an eight-bit converter with one bit to designate the sign of the analog signal. The remaining seven bits are used to encode an expanded analog signal. With nonlinear square root expansion, the seven bits can yield an effective 12-bit resolution for amplitudes that are encoded in the interval $000\,0001_2$ to $000\,0010_2$, and a six-bit resolution in the highest range from $111\,1110_2$ to $111\,1111_2$.

NONLINEAR ANALOG-TO-DIGITAL CONVERSION

The analog-to-digital converter itself can provide the nonlinear expansion in the process of establishing the binary equivalent to the analog signal. The technique described above as subranging can be modified to give expansion during the conversion. The first three most significant bits (A_6, A_5, A_4) establish $2^3 = 8$ segments of conversion sensitivity. (In this example we are considering the seven-bit conversion discussed in the previous paragraph.)

Figure 12-7 is a schematic of an analog-to-digital converter, which selects the analog amplitude segment to be encoded by the remaining bits of the n-bit converter. The octal value established by the three most significant bits is converted by a digital-to-analog converter to give the analog voltage, which is subtracted from the original analog voltage to make the residual analog voltage extend from

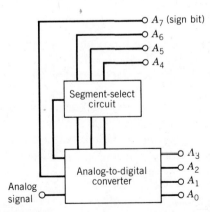

FIGURE 12-7

A Segmented Analog-to-Digital Converter

The three most significant bits select the linear scale to convert the residual portion of the analog signal.

328 Analog-Digital Conversion

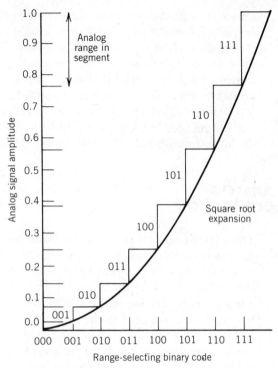

FIGURE 12-8

An Analog Range Selection in a Digital Expansion

The segment determined by the three most significant bits with a square-root nonlinear conversion establishes the range for converting the residual portion of the analog signal.

the base to the top of the appropriate segment. Figure 12-8 shows the segments identified with the octal values having a square root nonlinearity for expansion. Within the segment the analog voltage is encoded in linear fashion in the remaining bits.

In this process, unlike subranging, the residual voltage is treated as if the reference voltage of the converter extended only over the segment. In subranging the treatment of the residual analog voltage is uniform and independent of the amplitude of the analog voltage portion encoded in the first step.

TRACKING ANALOG-TO-DIGITAL CONVERTER

In many applications where digital values are desired the analog value changes slowly because the voltage of a battery or a power supply, for example, is being monitored or the temperature of an oven is being controlled at a set value. The

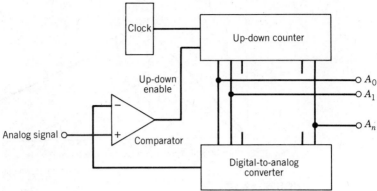

FIGURE 12-9
A Tracking Analog-to-Digital Converter
The up-down counter follows the change in the analog signal without starting over to reestablish the value at each sampling interval.

analog-to-digital converter needs to update the digital output as it changes rather than remeasure during each sampling interval.

The tracking or servo analog-to-digital converter uses a single comparator whose inputs are the analog voltage and the output of a digital-to-analog converter. The latter is a summing circuit whose input is the accumulated binary count in an up-down counter. A clock pulse to the up-down counter is added to or subtracted from the count value by the up-down enable signal from the comparator. Figure 12-9 shows this arrangement.

When the analog signal is larger than the feedback signal generated by the digital-to-analog converter, the up-down counter accumulates counts. When a balance is achieved, the up-down enable signal is reversed on alternate clock pulses, the LSB (least significant bit) flips, and the total count is the digital value that corresponds to the analog signal. A drift of the analog signal either up or down will be followed by the converter. A reference binary number ANDed with the up-down counter output can activate a warning or a control circuit.

DUAL-SLOPE ANALOG-TO-DIGITAL CONVERSION

Dual-slope analog-to-digital conversion is completely different from the techniques discussed above because it uses an integration rather than comparison to establish the digital equivalent of the analog amplitude.

The analog-to-digital converter integrates the average analog signal from zero to the value acquired over a preset interval of time. At the end of the first integration interval the accumulated voltage (or charge) is reduced by integrating "down" with a reference voltage as the signal source until zero is reached. During

330 Analog-Digital Conversion

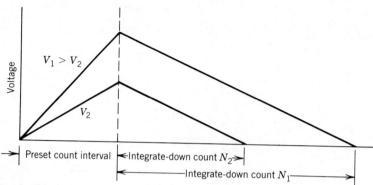

FIGURE 12-10

Voltage-Time Relations in a Dual-Slope Conversion

The count accumulated in counting down from the voltage as established in a fixed interval measures the analog voltage.

the integrating-down interval a counter or scaler counts a clock signal. The net count is proportional to the average of the analog signal. Figure 12-10 shows the voltage-versus-time relations in both the integrating-up interval, during which the analog signal is used as the input, and the integrating-down portion, when the reference voltage is the source. It may be noted that the integrating-down time is proportional to the amplitude of the analog voltage.

At the end of the sequence the count is latched and displayed, and the cycle is repeated. The rate of sampling—the reciprocal of the period of the cycle—is usually low, and may be 1 to 10 Hz. That is, a measurement of the analog signal may be updated once a second or once each one-tenth of a second.

Figure 12-11 shows the circuit arrangement for a dual-slope analog-to-digital converter that is free of the errors arising from drift in the components or from

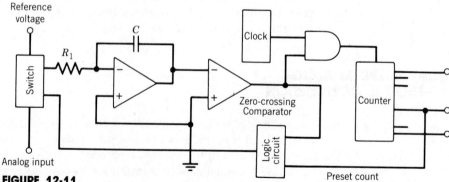

FIGURE 12-11

A Dual-Slope Analog-to-Digital Converter

The circuit components achieve dual-slope conversions.

change in the clock frequency because the elements are used in both the integrate-up and the integrate-down parts of the sequence. A change in the clock rate, for example, alters the preset time for integrating the analog signal in the same ratio as the integrate-down time. The logic circuit responds to the input-signal amplitude and polarity as well as governing the sequence.

The number of bits in the conversion can be quite large. A typical use of the dual-slope analog-to-digital converter is in a digital voltmeter. The bits may be counted by a BCD (binary-coded decimal) counter and be displayed as decimal digits. In many cases the displays are $3\frac{1}{2}$ or $4\frac{1}{2}$ LED (light-emitting diode) displays, and a cycle time of one-tenth of a second or longer is acceptable. The sign of the analog signal and the relative amplitude is established by the internal logic circuit, and the sign and decimal point in the display are set accordingly.

The $3\frac{1}{2}$-digit or $4\frac{1}{2}$-digit displays have three or four full digits; that is, digits with values 0 to 9. The $\frac{1}{2}$ digit is 1. To facilitate reading the instrument, for example, this arrangement allows a value of $+0.9998$ to change to $+1.0003$ without a shift in the decimal point as the analog voltage changes from one decade to the next. The 1 is not shown unless it is required. The maximum value displayed in the sample without shifting the decimal point is 1.9999 in a $4\frac{1}{2}$-digit display. The maximum value that can be displayed, of course, is 19999. A display of $4\frac{1}{2}$ digits requires 15 bits in binary code and 17 bits in BCD; these are equivalent to counting $20{,}000 - 1$ clock pulses in the integrate-down time for the maximum number of bits, regardless of the decimal position.

SINGLE-SLOPE ANALOG-TO-DIGITAL CONVERTER

Figure 12-12 shows the circuit of a single-slope analog-to-digital converter. The analog amplitude is measured by integrating it until its integral reaches a preset value; as an alternative, a reference voltage may be integrated until its integral reaches the analog amplitude. In either case, a comparator terminates the counting of a clock pulse, which began with the beginning of the integration interval. The integration of the reference voltage is preferable because the analog signal may be less constant.

A typical application of this converter is a pulse-height analyzer, which sorts out the decay spectra of radioactive isotopes. The nuclear detector produces a pulse whose amplitude is proportional to the energy of the nuclear event. A peak-voltage detector holds the pulse as it is used as the input to the single-slope analog-to-digital converter. When it is triggered by the nuclear event, the reference voltage is integrated from zero until it reaches the pulse-height voltage held at the input, and the condition is sensed by the comparator. A scaler accumulates clock pulses during the interval; a one, which represents the pulse, is added to the bits stored in the memory location that correspond to the number of clock pulses counted. The converter is reset until the next nuclear event is detected.

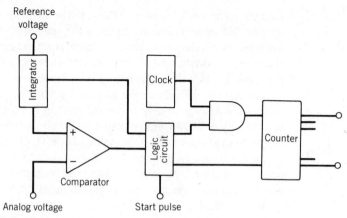

FIGURE 12-12
A Single-Slope Analog-to-Digital Converter
Reliable conversions are provided by the intergration of a reference voltage until the comparator terminates the counting sequence. The circuit is typical of pulse-height analyzers.

If a nuclear event occurs immediately after the event being converted but before the conversion is completed, it is ignored, and the corresponding nuclear datum is lost. In pulse-height analyzers a measurement of the "dead time"—the portion of the time during which events cannot be recognized—is available to correct the data for the missed events.

The single-slope analog-to-digital converter, unlike the dual-slope converter, does not have the advantage of freedom from many stability problems. The quality and stability of the components and the constancy of the clock pulse rate must be ensured for reliable conversion, but it may process the next pulse immediately rather than after a count-down interval, which contributes to "dead time."

STAIRCASE ANALOG-TO-DIGITAL CONVERTER

The staircase analog-to-digital converter uses clock pulses to drive both a counter and a digital-to-analog converter. The digital-to-analog converter increases its analog output with each clock pulse in a stairstep fashion. A comparator senses the instant at which the stairstep voltage exceeds the original analog voltage and stops the counter at that point. The content of the counter is the binary equivalent within the least significant bit of the original analog voltage. The binary value is latched, the counter and the digital-to-analog converter are reset to zero, and the sequence is repeated. This arrangement, useful for monitoring analog voltages (see Figure 12-13), is reset after each measurement in contrast to the tracking ADC.

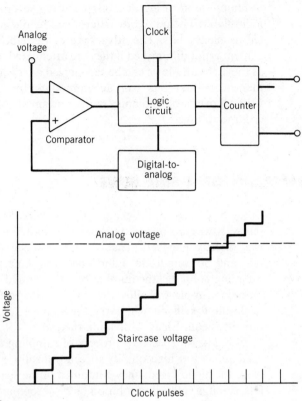

FIGURE 12-13
A Staircase Analog-to-Digital Converter
A voltage increment is added at each clock pulse until the comparator signal establishes the digital equivalent of the analog signal by latching the counts and restarting the sequence.

VOLTAGE-CONTROLLED OSCILLATORS

Voltage-controlled oscillators change their frequency of oscillation by a factor of about 10 in response to a control voltage change by a factor of two. These integrated circuits can be monitored by a frequency meter or other means to yield a measurement of the analog voltage or to track a change in the analog voltage. The voltage-controlled oscillators are sometimes called *function generators* if their output signals are available in different forms, such as square or sawtooth waves. The other forms include clocklike pulses, in which the period is divided unequally between the positive and the zero portions.

As a means of analog-to-digital conversion, voltage-controlled oscillators may be less sophisticated than other methods, but they are a simple and effective means of remote monitoring and of measuring several different analog voltages with a multiplexer. The pulse rate determined by the voltage being measured is transmitted as binary bits. The advantage of the clock pulses is that their information content is not diminished if they are attenuated or distorted in transmission as long as the highs and lows can be recognized as separate states. The digital detection of the transmitted bits can accommodate the attenuation that plagues analog voltages over long and noisy paths, and small voltages can be monitored as reliably as large voltages.

DIGITAL SAMPLING OF A SINE WAVE

In the beginning of this chapter it was noted that sampling a sine wave at discrete intervals would not constitute analog-to-digital conversion unless the amplitude was directly encoded. That is, a digital signal has only two values—on and off—and no amplitude significance. The binary code for the amplitude at each sampling point has the number of bits required to give the desired accuracy. Four bits, for example, will allow the amplitude to be divided into eight equal spaces (2^3), while the fourth bit will carry the sign ($+$ or $-$) of the voltage polarity. This process is equivalent to dividing the range, $-V$ to $+V$, into 16 segments.

In digitizing speech, a process of sampling and digitizing is called *pulse-code modulation*. For high-quality speech transmission the number of sampling times is twice the highest frequency contained in the speech; the data are represented by 8 bits, or an amplitude resolution of $2^8 = 256$ parts. Because speech has frequencies of up to 4 kHz, the sampling rate is 8 kHz. Each sampling point is represented by 8 bits; thus even for a zero amplitude sample, 8 bits (0000 0000) are generated and transmitted. The bit rate $8 \times 8 \times 10^3 = 64$ kbits/sec. The sampling frequency, 8 kHz, requires an analog-to-digital conversion of 125 μsec, a rate easily achieved. A person who communicates over a high-speed digital channel would hear natural speech, but would not be aware of the digital processing, transmission, and restoration by the digital-to-analog converter at the receiving end. (Commercial systems may use a frequency band from 300 to 3000 Hz to increase channel capacity.)

As suggested in the example above, the clock rate that is considered optimum is at least twice the highest frequency component in the wave. Twice the highest frequency is the minimum sampling rate proved by Nyquist for a sine wave to ensure its reconstruction. The sine wave in Figure 12-1, of course, is oversampled.

Because human hearing has a frequency range to about 20 kHz, the sampling rate for music reproduction or transmission must be 40 kHz; that is, once each 25 μsec. This rate is also achieved readily by available analog-to-digital converters. It is clear that waves associated with speech and music—that is, audio signals—can be converted by analog-to-digital converters rapidly enough to avoid loss of the information required to reconstruct the analog signal. The problems arise when the data are to be stored in digital memory; a few seconds of audio information fill a large memory array. One saving feature of speech recognition, however, is that low-quality speech can be understood even if it sounds mechanical or flat.

The bit storage requirements are so extreme that ways to compress the number of bits have been devised. Bit rate always competes with the quality or naturalness of reconstructed sound, but analog-to-digital audio signals can be encoded more efficiently (in terms of the number of bits) than by sampling the amplitude directly at a high clock rate.

To encode the sine wave in Figure 12-1 with far fewer bits, it is possible to sense the change in the amplitude between sampling points rather than to encode the amplitude at each sampling point. If we let zeros indicate a rising amplitude and ones a falling amplitude, only one bit needs to be determined and saved for each clock pulse. Twelve bits are found in the 12 clock pulses, and the digitized equivalent of the sine wave in Figure 12-1 becomes 000111111000. This method of encoding signals is called *delta modulation*. Figure 12-14 shows a delta-modulation system; the low-pass filter eliminates high-frequency noise in the input signal.

As seen in Figure 12-15, delta modulation is not ideal. The signal detail is lost in determining whether each value is larger or smaller than the one just evaluated. It is clear that the signal has higher-frequency components than the frequency of the clock. A faster clock will improve the analog-to-digital conversion because it is closer to the condition for sampling set forth by Nyquist and will increase the rate of bit production. This may not be desired even though the additional data can restore the original wave form more faithfully. In the original wave form of Figure 12-15, the components with frequencies higher than the clock frequency are not

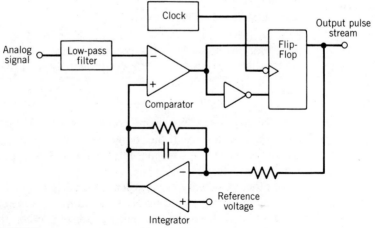

FIGURE 12-14

A Delta-Modulation System

The analog signal is encoded into an efficient binary code by comparing the change in signal amplitude rather than the amplitude itself.

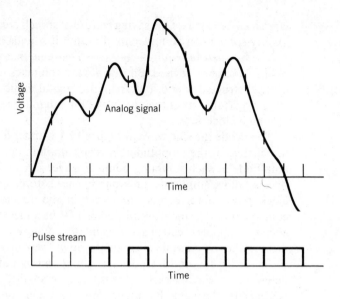

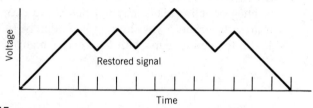

FIGURE 12-15

A Delta-Modulation Processing

At each sampling interval the change in the analog signal is recorded by a zero if the amplitude increases from the previous value and by a one if the amplitude decreases. The signal is restored by intergrating the bit pattern.

sampled adequately, but the reproduced wave form may be acceptable in exchange for fewer bits to store. If the sound is speech, it may be understandable, but it will not sound natural. A low-pass filter in the audio output will "round" the restored signal.

Another possibility exists when a signal is sampled and encoded at a rate less than the minimum stated by Nyquist. Sometimes the effect is to encode an analog signal different from the one being processed. When the analog signal is restored from the binary data, it is unlike the original; this effect is called *aliasing*. The restored signal has assumed a different identity, or an alias. Aliasing is illustrated in Figure 12-16, where it can be seen that if the clock rate were twice the highest frequency of the analog signal, aliasing would not be possible.

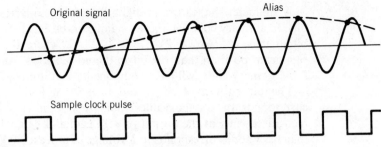

FIGURE 12-16
Aliasing
An inadequate sampling rate can result in encoding a signal different from the original. Sampling at the Nyquist rate prevents aliasing.

RESTORATION OF THE ANALOG SIGNAL

Analog signals are converted to binary code for many reasons. These include the need to give digital readout of measurements, to record on high-density laser-read discs, to transmit audio signals by fiber optics, and to process the signals to increase contrast or other properties of the signal, as is done in case of satellite-acquired data. In many cases the digital form is an intermediate state for the data, and the restoration of the analog signal is required. An inverse process of digital-to-analog conversion restores the analog signal to its initial form. If the restoration is to be faithful, the inverse process of digital-to-analog conversion must compress the output in a manner reciprocal to the encoding expansion (if it has been used). Even so, quantization uncertainty, unwanted nonlinearities, drifts in reference voltages, and other phenomena always contribute to degrade the analog data from its original form.

DIGITAL-TO-ANALOG CONVERSION

We now turn to digital-to-analog conversion (DAC), the process of generating an analog signal from a digital (binary) code. It has been shown that several of the systems that convert analog signals to digital signals use digital-to-analog conversion. In the successive approximation and subranging by analog-to-digitial converters, for example, the most significant bits are used to generate an analog voltage, which is subtracted from the original analog voltage to yield a residual voltage. The residual voltage is used to establish the remaining bits in the conversion.

A characteristic feature in digital-to-analog conversion is the relative value of bits. In a binary code each more significant bit has twice the value of the previous bit; that is, $2^{n+1} = 2 \times 2^n$, $n = 0, 1, 2, \ldots$, where the $n + 1$th bit is in a more significant position than the bit in the nth position. An analog voltage generated from a binary code will have components that are related by powers of 2.

The bit pattern of zeros and ones from logic gates or counters actually corresponds to voltage values that are not necessarily 0 or 5 V. Zero is the recognized state at the output of a TTL digital gate when its voltage is no more than 0.5 V; one corresponds to a voltage of at least 3.0 V but less than 5.0 V at the gate output. Because the logic gates have nonzero output voltages for the zero state and output voltages from 3.0 to approximately 5.0 V for the one state, practical digital-to-analog converters do not use the output voltages of logic gates directly as inputs. Instead, they use the output voltages to activate electronic switches that apply zero voltage to the inputs of the digital-to-analog converter for the zero state and 5.0 V or another reference voltage for the one state. Other means, such as feedback to the operational amplifiers in the digital-to-analog converters, ensure that the analog signal output is related properly to the digital input. In presenting the principles of digital-to-analog converters we assume that the voltage levels are zero or V_{cc} when applied to the converters.

BIT-WEIGHTED SUMMING AMPLIFIER

The summing amplifier shown in Figure 11-10 is an obvious choice for converting binary code into an analog signal. The input resistors in Figure 12-17 are scaled so that the current to the summing junction, s, of the operational amplifier is proportional to the magnitude of the binary code. In order to weight the bits the input resistances are found from the current relations, $i_0 = V_{cc}/R_0$, $i_1 = V_{cc}/R_1$, etc., where $i_1 = 2i_0$, $i_2 = 2i_1 = 4i_0$ and $i_3 = 8i_0$; thus $R_3 = R_0/8$, $R_2 = R_0/4$, and $R_1 = R_0/2$.

The current in the feedback resistor is the sum of the input currents through the summing junction. Thus we can write

$$i_f = -\frac{V_0}{R_f} = \frac{V_{cc}}{R_0} + \frac{V_{cc}}{R_1} + \frac{V_{cc}}{R_2} + \frac{V_{cc}}{R_3}$$

$$= \frac{V_{cc}}{R_0} + \frac{2V_{cc}}{R_0} + \frac{4V_{cc}}{R_0} + \frac{8V_{cc}}{R_0}$$

$$= (1 + 2 + 4 + 8)\frac{V_{cc}}{R_0}$$

$$= 15\frac{V_{cc}}{R_0} \tag{12-1}$$

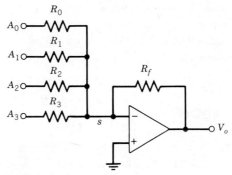

FIGURE 12-17

A Bit-Weighted Summing Amplifier

The output voltage is proportional to the value of the binary code.

as the maximum current when each binary bit has the value 1 in the four-bit number. The magnitude of the output voltage can be scaled relative to the supply voltage V_{cc} by the choice of R_f and R_0. For the greatest magnitude of V_0 we have

$$V_0 = -V_{cc} \frac{R_f}{R_0} \times 15 \tag{12-2}$$

A binary signal with n bits as inputs to a bit-weighted summing amplifier will have the largest magnitude of the analog voltage, V_0, when all bits are in the one state. Equation 12-2 becomes

$$V_0 = -V_{cc} \frac{R_f}{R_0} \times (2^{n+1} - 1) \tag{12-3}$$

where $n = 0, 1, 2, \ldots$ to correspond to 2^n as the positional value of the bit.

The bit-weighted summing amplifier is less practical than the method described next, the R-2R current ladder. The bit-weighted summing amplifier is less than ideal for two important reasons. One is the need for a number of input resistors with different but exactly related values. The fabrication of resistors within integrated circuits becomes very difficult when the values differ by so many powers of 2. A 10-bit digital-to-analog converter must have the resistance of the largest resistor, $2^{10} = 1024$ times that of the least resistor, and eight intermediate values must be set precisely. The second reason is the time required for several least significant bits to contribute fully to the analog signal. The largest resistor in a 10-bit converter has a resistance of 1MΩ or more, and the circuit to which it belongs has an inherent capacity of about 100pF. Together these yield a time constant of about 0.1 msec, which is far longer than the desired conversion time. The effect is described as causing a long "settling" time; that is, the time for establishing the final magnitude of the analog signal. These effects are not so important for four-bit converters, as discussed in the previous paragraph. Commercially available integrated circuits use only the two resistor values discussed below.

R-2R CURRENT LADDER

In Chapter 11 we emphasized the role of current to the summing junction in an operational amplifier as an inverting amplifier. The *R-2R current ladder* is an arrangement of resistors of only two values, R and $2R$, which divides the current from the binary input, V_{cc}, in the ratios of $1:2:4:8$, etc., to the summing junction. Figure 12-18 shows the R-$2R$ network as the source current to the summing junction. An alternate arrangement uses an R-$2R$ ladder as the feedback circuit from the output of the operational amplifier to the summing junction. In either arrangement the equal splitting of the current at each node is the key to converting the binary number into its analog equivalent. A typical value of R in the R-$2R$ network is $2.5\ k\Omega$. The resistors are thin films formed on an integrated circuit wafer and laser-trimmed to the precise values. With the modest resistance values the inherent time constant is small and the settling time is acceptable. Successive 10-bit conversions are achieved in $5\mu sec$ or less.

The binary input $A_3 A_2 A_1 A_0$ in Figure 12-18 actuates the switches $S_3 S_2 S_1 S_0$ and either grounds the current for a zero or directs it into the summing junction. In the R-$2R$ resistor network the current at each node—that is, the current junction

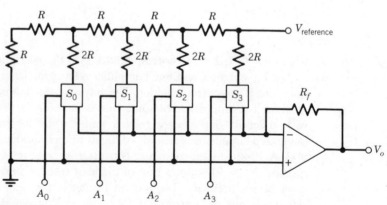

FIGURE 12-18

R-2R Ladder Digital-to-Analog Converter

The *R-2R* ladder weights the bit value by splitting the current to the summing junction. The *R-2R* ladder is used in commercially available digital-to-analog converters.

corresponding to each bit—is split into two equal parts. The node corresponding to the most significant bit carries one-half the current to the summing junction (if the bit is high) or to ground (if the bit is low), while the other half of the current goes to the MSB-1 node. There it is split into two parts; one-fourth of the current goes to the summing junction or to ground and the other fourth to the MSB-2 node. The equal splitting of the remaining 2^{-n}th portion of the current occurs at each node. R-$2R$ current ladders have 10 or more nodes; thus large binary numbers can be converted to analog signals. Commercially available elements can convert 16- or 18-bit numbers.

The stability of the digital-to-analog converters is maintained by using bit-activated switches on the line from each node. To maintain the current at all times the bit-activated switch places the current on the line to the summing junction when the bit is high and shunts it to ground when the bit is low. The switches are

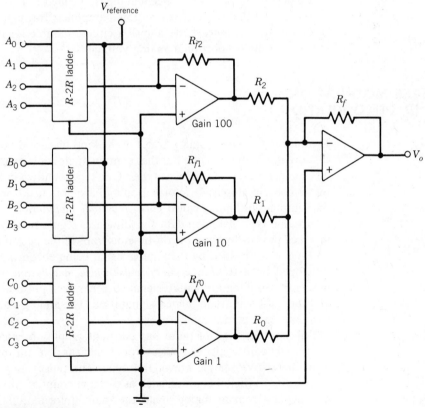

FIGURE 12-19

A Three-Decade BCD Digital-to-Analog Converter

The binary-coded decimal bits are weighted by the R-$2R$ ladder, and each decade is weighed by the inverting amplifiers to give an output voltage proportional to the numerical value of the binary code.

bistable arrangements of transistors in which the current is carried by one element or the other, depending upon the value of the bit on the switch.

A difficulty (called a "glitch") can occur when the binary number changes in a manner that involves the most significant bits. A binary input of 01 1111 1111 that changes to 10 0000 0000, for example, requires one switch to close and nine switches to open. If the one switch closes before the others open, a full-scale analog value will occur momentarily; if it closes slowly, a zero analog value is generated. In high-quality digital-to-analog converters the manufacturers provide a sample-and-hold circuit to avoid the problem. These elements are said to include "deglitchers."

Several bit-weighted summing amplifiers may be used in combination to convert a large binary code into an analog voltage. For example, one may convert three decimal digits encoded in binary-coded decimal (BCD). In this case one bit-weighted summing amplifier is used for each decimal digital, and the output voltages themselves are weighted by the factors 10^n, $n = 0, 1$, and 2 to correspond to the positional value of the decimal digit. The factors 1, 10, and 100 are obtained by the gain of the respective operational amplifiers through the choice of the ratio of the resistors. Figure 12-19 shows a circuit suitable for converting three BCD inputs into the proportional analog voltage.

PULSE MODULATION AND DEMODULATION

In addition to regular binary codes of the kind discussed above, pulses may occur that are not in binary code but that contain analog information. These pulses may be classified as several different types. One type is shown in Figure 12-1, where the pulses contain amplitude information and are obtained from the analog signal by the sampling technique called *pulse-amplitude modulation*.

Another type of analog-data encoding uses regularly occuring squarelike pulses in which the on-time and the off-time portions of the period vary. The width of the on time is determined by the analog signal being encoded. These pulse trains are developed by a technique called *pulse-width modulation*. Figure 12-20 shows a means of sampling an analog signal to generate a pulse-width modulated signal and the pulse-width modulated signal itself. The analog signal and the triangular signal are inputs to a comparator.

Still another kind of pulse has constant amplitude and width but occurs at random time intervals. The pulses from a nuclear detector that measures the decay of a radioactive sample are of this type. The pulses may be recorded with an asynchronous digital counter, such as a ripple counter. Although these pulses are not encoded from an analog signal, the analog information that is contained is the average rate of decay or the radiation level. This information may be displayed on an analog meter, as shown in Figure 12-21; however, a more practical rate meter is presented in laboratory Exercise 28.

The meter may respond to different rates of events by selecting the time constant $\tau = R_f C_f$ to give a reasonable meter current. Each pulse causes an equal quantity

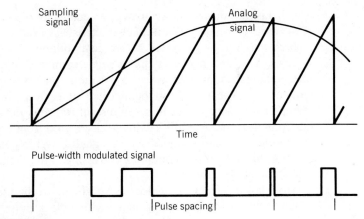

FIGURE 12-20

A Pulse-Width Modulation

A pulse-width distribution proportional to an analog amplitude can be achieved for an appropriate sampling signal.

of charge, q, to be placed on the feedback capacitor so that the average voltage across the feedback circuit is qpR and the average current is qp. Here p is the average pulse rate (events per second) into the meter. The mechanical inertia of the meter causes the position of the meter needle to remain relatively constant for a given pulse rate. The uncertainty or fractional standard deviation of an instantaneous reading is $1/\sqrt{2pR_f C_f}$, which indicates that a short time constant and a slow pulse rate give an uncertain instantaneous reading of the rate.

The conversion of analog signals into the pulse trains that carry the analog information is called *modulation*; the recovery of the analog signal is *demodulation*. It is useful to encode analog signals into pulses for several reasons.

The two most important reasons concern the transmission of the information.

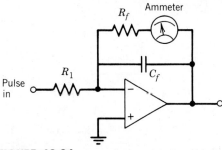

FIGURE 12-21

A Rate Meter for Asynchronous Pulses

A meter responds to the rate of equal-sized pulses that occur in a randomly timed or asynchronous manner.

The first is that the transmission of any signal gives rise to noise as an inherent part of the received signal. The on-off characteristics of pulses, however, allow the noise to be recognized and rejected before demodulation into the analog signal at the receiver. A second reason is that a single communication channel (such as wire, fiber optic, radio, or microwave) can be multiplexed to carry hundreds of signals sequentially. The noise can be rejected at each point at which the signal is reamplified to account for losses in distant transmissions. Even in signals that are buried in noise, such as the satellite transmission from deep space, it is possible to extract the information content if the signals travel in pulse form. The information is extracted by extensive computer processing of the signal, which uses a technique called autocorrelation.

Modulation, which is analogous to analog-to-digital conversion, is typically achieved by mixing the analog signal with a clock pulse. Figure 12-22 shows two pulse modulation arrangements. The arrangement in Figure 12-22a yields pulse-width modulation with constant pulse spacing, like that seen in Figure 12-20. Figure 12-22b is an astable (free-running) oscillator, whose period is also maintained by the clock pulse to give constant pulse spacing. In voltage-controlled oscillators or function generators, for example, the pulse-on time can be varied by an analog voltage while the pulse frequency is maintained by a short-triggering clock pulse. This mixing of the analog voltage with the clock pulse is a different mode of operation from the single use of an analog voltage. An analog voltage would change the frequency, as in the voltage-controlled oscillators discussed

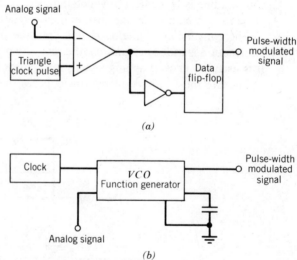

FIGURE 12-22

Pulse-Width Modulator Arrangements

(*a*) Comparator-based and (*b*) voltage-controlled oscillator-based. Two clock-driven systems provide pulses with regular time intervals. The pulse duration is determined by the analog signal amplitude.

earlier, but would not vary the distribution of the on-time and off-time portions of the pulse from a function generator.

The demodulation of the pulses is equivalent to the digital-to-analog conversion of the quantized (but not necessarily binary) pulses. Demodulation is essentially the integration of the on-time portion of the signal, and it may be performed by a low-pass filter. The filter capacitor is charged in a manner proportional to the width, amplitude, and sign of the pulse. The low-pass filter may be an operational amplifier integrator circuit with a discharge path across the capacitor (see Figure 12-21). A means of discharge must exist so that the analog signal can diminish in accordance with a reduction in the on-time of a pulse. The time constant, RC, of the filter must be related to the sampling period of the modulated signal in order to allow the voltage to drop rapidly enough to follow the original analog signal. Excessive ripple is unwanted, however; thus the smoothing of the final signal may either require another low-pass filter or may occur "naturally" because of the mechanical inertia of the needle in an analog meter. In the case of audio signals the ripple frequency may lie beyond the channel bandwidth (of the telephone) and may not be transmitted from the demodulator to the listener.

SUMMARY

The important tasks of analog-to-digital conversion and digital-to-analog conversion are analogous to modulation and demodulation. In each case the conversion can be achieved in only two independent ways.

Analog-to-digital conversion uses either a comparator or an integrator-controlled counter. All devices use one of these two techniques in ingenious ways to improve resolution or to accomplish other desired features in digitizing the analog signals.

Digital-to-analog converters use either a low-pass filter or a resistance ladder with a summing amplifier to generate the analog signal related to the data in pulse form. Various circuit arrangements are employed in digital-to-analog conversion to restore the original content of an analog signal.

Many advantages of transmission and signal processing are made evident when the data are in binary form, but the data storage spaces in semiconductor memory arrays are filled quickly. Data-sampling techniques can alleviate the storage requirements at the expense of some quality in the restored signals.

PROBLEMS

12-1 Find the number of bits needed to encode a rotation of a shaft to an accuracy of 0.01 percent of a degree. If the positioning is used to machine gear teeth, and if a tooth at a radius of 0.2 m is to be machined to within 100 μm, what is the required encoding?

12-2 Show the arrangement for setting and controlling a lower and an upper temperature on a baker's oven whose temperature is monitored by a tracking analog-to-digital converter.

12-3 In the bit-weighted summing amplifier (Figure 12-17), find the resistor values so that the output-voltage magnitude for the bit pattern 1000_2 is -6 V. Show that the maximum output voltage is less than twice this amount.

12-4 Starting from the right-hand side, show that at each node in the R-$2R$ ladder (Figure 12-18) the resistance is equal to $2R$ in both paths and that the current is split into two equal parts at each node so that the bit represented by the node is weighted properly.

12-5 Design a rate meter (Figure 12-21) in which the rate is presented as a logarithm so that a meter can display a wide range of pulse rates without a scale change. What is the standard fractional deviation versus rate in the logarithmic display?

12-6 Design a logic circuit (Figure 12-13) for a staircase analog-to-digital converter.

CHAPTER 13

"SMART" ELECTRONICS

In this chapter we discuss a sampling of options that are available in the use of analog and digital electronics for a wide variety of purposes. Our purpose is to illustrate that the potential of electronics is nearly unlimited and that the options in the use of electronics depend solely on the innovative talents of the user.

In many situations the interaction among the electronic elements and components of the system give rise to special problems. The suppression of such difficulties is one aspect of "smart" electronics. Each of the simple analog and digital topics in this chapter offers a means of meeting a particular problem or of simplifying a complex electronic system.

SUPPRESSION OF TRANSIENTS

Many sources of electrical transients produce multiple input states to logic gates. The most obvious source of transients is the closing of a switch or the depressing of a key on a terminal. When two metal conductors are brought together in this fashion, the phenomenon known as *contact bounce* is unavoidable. A number of random electrical pulses occur before the contact is firm and the electrical signal is steady. Contact bounce may exist for a millisecond, during which a digital counter will register many counts, perhaps 5 to 100, as it records the electrical spikes.

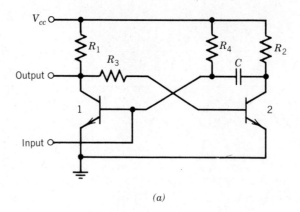

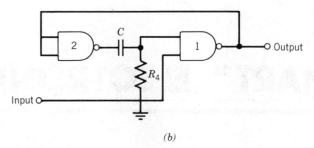

FIGURE 13-1

A Single-Shot Multivibrator

The single-shot or monostable multivibrator generates a single pulse when triggered. The circuits (a) and (b) are equivalent. Circuit (b) is limited in operation.

Two fundamentally different means exist for defeating the noise and recognizing the presence of the steady signal. The first is a flip-flop or a single-shot multivibrator, which responds to a pulse and will not accept another input for the short interval during which the contact becomes firm. The second, discussed below, uses software rather than hardware for the same purpose.

Figure 13-1a shows a monostable or one-shot multivibrator made with bipolar junction transistors. The stable state occurs when transistor 1 is conducting because the voltage on its base is transmitted by the resistor R_4. The voltage to the base of transistor 2 is the low voltage of the collector of transistor 1, transmitted by resistor R_3. Transistor 2 is not conducting. A negative signal or transient of sufficient amplitude on the base of transistor 1 causes a regenerative change that turns on transistor 2, as the collector voltage of transistor 1 rises in response to the negative signal. As transistor 2 starts to conduct, its dropping collector voltage, transmitted by the capacitor C, turns off transistor 1; this is a self-reinforcing process.

The state of high voltage at the collector of transistor 1 is temporary. The decay of the voltage across capacitor C permits the base of transistor 1 to rise until conduction starts in transistor 1. Capacitor C discharges as its charge flows through R_4 and through the parallel combination of R_2 and the conducting transistor 2. The duration of the single pulse is governed by the time constant CR_4. The return to the stable state is also regenerative. The output pulse, the high on the collector of transistor 1, lasts for approximately $0.7CR_4$ seconds.

An examination of Figure 13-1a reveals that the transistor amplifiers are essentially inverters interconnected to become a monostable circuit. Two NAND-gates interconnected in the same manner would appear to be an equivalent arrangement, but this is not the case because of the limitations of the TTL elements. The NAND-gates in Figure 13-1b are numbered to correspond to the transistors in Figure 13-1a, which have identical roles, but R_4 must be limited to a few hundred ohms if the combination of NAND-gates is to work.

Monostable or single-shot multivibrators are available as integrated circuits without the limitations of the interconnected gates. Many integrated circuits, such as keyboard decoders, include a single-shot multivibrator as a debounce element.

To verify the validity of data to be used by a computer or by a microprocessor-based control system, a subroutine can be used rather than electronic components. The data may be an interrupt request from a sensor, data from a keyboard, or a signal arising from a switch closing. The intelligent receptor responds to the indication of a signal by testing repeatedly for the presence of the signal. The software routine checks the signal and then generates a delay and makes another check. The sequence is repeated, perhaps 30 times, to determine that the signal is indeed present on the line. The delay is frequently generated by loading a register and decrementing its contents to consume a number of microseconds between checks. The reader will recognize that the software approach is applicable to any switch closing, while each switch requires a single-shot multivibrator to defeat contact bounce.

MISSING-PULSE DETECTION

Many digital circuits have a clock to generate pulses at a constant rate. A *missing-pulse detector* recognizes a departure from the expected rate and indicates the new condition.

One can imagine the damage that may occur if a machine tool directed by a microprocessor-based controller fails to receive the normal sequence of instructions either from outside or from a built-in position sensor. In the case of uncertain control, an alarm or a system disable must occur.

A sequence of pulses can trigger a monostable or one-shot multivibrator. When the period of the one-shot multivibrator is slightly longer than the interval between pulses, one of two things may happen. Either the one-shot multivibrator may ignore the subsequent pulse and complete its cycle, or the subsequent pulses may retrigger the device and restart its cycle so that it remains high (1) as long as the pulse train continues. In the latter situation an interruption in the pulse train will allow the cycle to finish, and the output will fall to a low (0). This condition clearly indicates a missing pulse or the interruption of the pulse train, and an alarm can be given.

Active security systems that use ultrasonic signals or pulsed light beams respond when the pulse train between the transmitter and the receiver is interrupted by an object or a person for a period longer than that of the monostable multivibrator. Active systems are difficult to defeat.

CIRCUITS WITH HYSTERESIS

A circuit with hysteresis is useful because it responds to a signal at one level and ends its response only when the signal is at a different level. (The term "hysteresis" applies to the situation in which a response lags the application of the cause. Hysteresis initially pertained to the delay in the magnetization of a piece of iron as it lags behind the magnetic field causing the magnetization.) Hysteresis in the responding circuit prevents multiple transitions between states because of noise on the signal or because a sensor at a threshold provides uncertain signals. A temperature controller, for example, should not turn a heat source rapidly on and off a number of times while it is seeking a steady state. Bimetallic thermostats have a temperature differential between "on" and "off," and this property is often desirable in circuits that respond to electrical signals.

A Schmitt trigger is a useful circuit because its response to an input signal includes hysteresis. Figure 13-2 shows a Schmitt trigger made with transistors and illustrates the relation between the input signal and the output signal. The hysteresis arises because transistor 1 has two levels of collector voltage. Transistor 1 is nonconducting until the "on" threshold is reached. When it conducts, the voltage divider R_3 and R_4 drops the base of transistor 2, which stops conducting. This state remains until the input signal drops to the point at which transistor 1 is cut off. When R_1 is greater than R_2, the emitter voltage is less when transistor 1 is conducting. Hysteresis is created because the input voltage must drop below V_{on} to V_{off} to cut off transistor 1. The time interval is determined by the wave form and by the amount of hysteresis. This device is unlike the monostable or one-shot multivibrator, in which the time interval is governed by the decay of an RC circuit.

Figure 13-3 shows an operational amplifier voltage comparator with hysteresis. The reference voltage from the voltage divider R_a and R_b for the comparator, V_{on} or

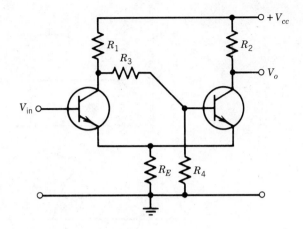

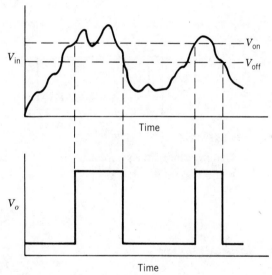

FIGURE 13-2

A Schmitt Trigger

The Schmitt trigger responds to a signal of sufficient amplitude and returns to its initial state only after the signal amplitude has dropped to a lower level.

V_{off}, depends upon V_0. The difference in the two states of V_0 provides the hysteresis.

Schmitt trigger inputs are available on inverters, logic gates, monostable multivibrators, and other integrated circuits to provide hysteresis in their response. The Schmitt trigger input is shown by a square, which represents a hysteresis loop within the symbol for the element. In TTL integrated circuits the hysteresis is typically 0.8 V.

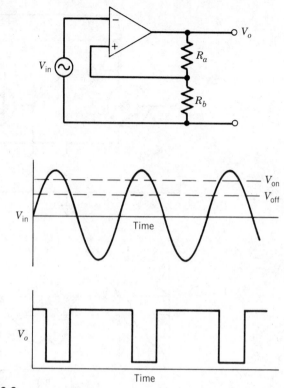

FIGURE 13-3
A Comparator with Hysteresis
The operational amplifier circuit has the same characteristics as the Schmitt trigger.

IMPEDANCE IMPROVEMENT

High input impedance and low output impedance are among the important properties of a noninverting amplifier made with operational amplifiers. As shown in Chapter 11, however, the inverting amplifier does not have a high input impedance; it was suggested in that chapter that a voltage follower, a noninverting amplifier with a gain of one, can be used as an input stage to provide an effective high input impedance. The voltage follower may be used with other analog circuits and elements to gain the same advantage.

If true voltages are to be determined, voltage measurement must not load (draw significant current) from digital circuits or shunt (bypass) current around high impedances. An inexpensive volt-ohm multimeter or a simple voltmeter made with a microammeter and a series resistor has low impedances and load circuits in voltage measurements. An operational amplifier–voltage follower placed on the

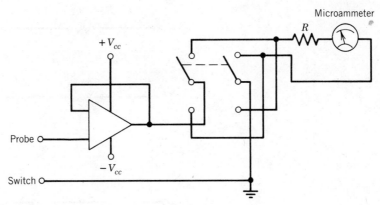

FIGURE 13-4

An Inexpensive Voltmeter

A voltmeter with high input impedance suitable for digital circuit measurements uses a voltage follower as an input element.

input probe makes either of these inexpensive circuits suitable for voltage measurements for digital circuits.

In the arrangement shown in Figure 13-4 the voltage measurements are restricted to about ± 80 percent of V_{cc}, the voltage of the two-sided power supply for the operational amplifier. When the probe is not in contact with the circuit whose voltage is being measured, the "floating" probe does not govern the output voltage. The feedback circuit causes the output to become approximately $-V_{cc}$, and a current exists through the microammeter. The switch is positioned to permit either a positive or negative voltage to be measured.

MOSFET SWITCHES

Enhancement metal-oxide-semiconductor field-effect transistors (MOSFETs) are excellent switches with a resistance typically less than $100\,\Omega$ when conducting and $10^8\,\Omega$ or more when cut off. These states may alternate at many megahertz when driven by a clock, and may change from off to on and vice versa in a few nanoseconds when a signal is applied to the gate. The MOSFETs (Figures 8-20 and 8-21) permit switching times and arrangements that were unimaginable only a few years ago.

These switches may be used in creative ways; they are particularly useful when combined with operational amplifiers and analog signals for sampling signals and for analog-to-digital and digital-to-analog conversion.

Figure 13-5 shows a circuit with MOSFET switches. An equivalent resistor is created by switching the input line to a capacitor and then switching the capacitor to the output line, as shown in Figure 13-5b. The circuit in Figure 13-5c consists of

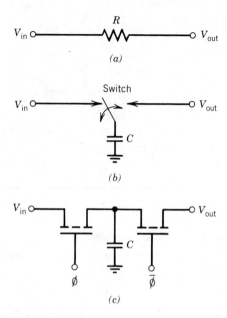

FIGURE 13-5

A Switched Capacitor Resistor

(a) Equivalent resistor. (b) Schematic of an equivalent resistor. (c) Circuit with MOSFET switches. MOSFET switches and a capacitor act as a resistance. The resistance value is inversely proportional to the frequency of switching and to the magnitude of the capacitance.

two enhancement MOSFETs whose gates are driven by a square wave (ϕ) and its complement ($\bar{\phi}$) so that the switches conduct alternately. The equivalent resistance is inversely proportional to the frequency and the capacitance. A 33-pF capacitor in a circuit driven at 50 kH is equivalent to about 10 kΩ, as measured in the voltage-dividing circuit in Figure 1-1. Laboratory Exercise 26 explores this phenomena.

It is predicted that this circuit will be used in large integrated circuits as an alternative to the difficult task of creating resistors within the circuits.

CONTROLLABLE BUFFER-INVERTER

An exclusive-OR-gate can be used as a bit-controlled element that passes an input signal (buffer) or its complement (inverter). The truth table for the XOR-gate in Table 13-1 reveals how the function (buffer or inverter) may be selected.

Either input A or B may be used as the control input. If B is the control, for example, we have

$$B \text{ low } (0) \quad C = A$$

TABLE 13-1
An XOR-gate Truth Table

A	B	C
0	0	0
0	1	1
1	0	1
1	1	0

and

$$B \text{ high } (1) \quad C = \bar{A}$$

Suppose that information is to be transmitted as encrypted binary data so that it can be read only by the intended receiver. A very long binary word, such as a prime number of a hundred or more decimal digits in binary form, may be shifted bit by bit as the control input to modify the transmitted message. The receiver with the key, the prime number, can decipher the message, which is unintelligible to others.

PARITY CHECK

In transmitting data, a number, or a letter from a keyboard, a check of the integrity of each datum is desirable. Noise or other sources of electrical interference may change a zero to a one in the code before it is received. One means of checking is to be certain that an even or odd number of ones is provided to the transmission system in a byte and that the same number of ones is received.

In transmitting a number, letter, symbol, or control character by American Standard Code for Information Interchange (ASCII), seven bits designate one of the 128 possible items. An extra bit (1) is added to the ASCII code if the addition will give the byte an even number of ones for even parity or an odd number of ones for odd parity. The bit, if it is added, does not affect the ASCII code, but if the byte does not satisfy the parity check, the character is not accepted and an error is indicated. Figure 13-6 shows a parity generator for a four-bit number.

One role of parity check is found in reading bar codes by scanners for automatic cash registers in supermarkets. One half of the bar code has even parity and the other half has odd parity, which allows the "smart" cash register to determine the order in which the bar code is to be interpreted. This arrangement eliminates the need for the clerk to position the package before passing it over the scanner.

Because parity checking consumes time and bit space, a slightly different and less precise procedure is used for applications such as writing binary data on discs.

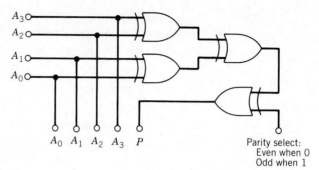

FIGURE 13-6

A Parity Generator

The number of ones in a transmitted binary signal is made even or odd. The parity check of the received signal confirms transmission integrity.

A *check sum* is made by adding all the ones in all the bytes in a block of data stored in a track. The lowest group of eight bits (or byte) is retained from the sum and recorded with the other data. When the data are read, the ones are summed and compared to the check sum. A discrepancy causes a reread or a fault signal.

ERROR CORRECTION

A parity check indicates that an error has occurred, but it will not allow the error to be corrected. If additional parity bits are generated from the datum, however, and transmitted with the datum, an error can be both detected and corrected. The datum bits and the parity bits form algebraic relations among themselves, which permit the position of an error bit to be designated. An inverter can return the bit to its true value. The Hamming code is the usual prescription for error correction.

The source of data includes a parity generator, which transmits the datum bits and the parity bits in a particular order, as shown in Figure 13-7. The parity bits are produced by exclusive-OR gates as follows:

$$P_1 = A_3 \oplus A_2 \oplus A_0$$
$$P_2 = A_3 \oplus A_1 \oplus A_0$$
$$P_3 = A_2 \oplus A_1 \oplus A_0$$

Table 13-2 shows the bit values for several hexadecimal numbers and for the parity bits associated with them. The column number, it can be seen, is nonzero for any number in which a bit error occurs. If no error exists in transmission, the zero

TABLE 13-2
Data and Parity Bits from the Hamming Code

Hexadecimal Number	Column							
	0	1 P_1	2 P_2	3 A_3	4 P_3	5 A_2	6 A_1	7 A_0
0		0	0	0	0	0	0	0
1		1	1	0	1	0	0	1
2		0	1	0	1	0	1	0
3		1	0	0	0	0	1	1
4		1	0	0	1	1	0	0
⋮								
C		0	1	1	1	1	0	0
⋮								

column will be identified. The binary code $(B_2 B_1 B_0)$ of the column number is found from

$$B_0 = P_1 \oplus A_3 \oplus A_2 \oplus A_0$$
$$B_1 = P_2 \oplus A_3 \oplus A_1 \oplus A_0$$
$$B_2 = P_3 \oplus A_3 \oplus A_1 \oplus A_0$$

The three XOR-gates at the top of Figure 13-6 will perform the logic of generating any one of the *B*s.

The reader should verify that the bits in Table 13-2 are correct by ascertaining that the zero column is designated by the *B*s, and that if a change takes place in any bit, datum or parity, in any number, its column will be identified. In Figure 13-7 the elements on the right (at the receiving end of the transmission lines) ensure accurate data.

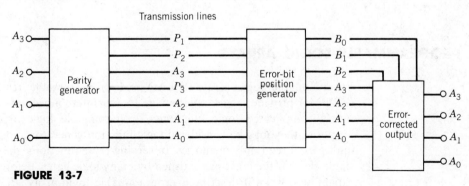

FIGURE 13-7

A Error-Correction Transmission System

Information for correcting errors in transmission is carried by several parity bits.

READ ONLY MEMORIES

Read only memories (ROM) and programmable read only memories (PROM) are used for nonvolatile storage of computer startup routines, monitor routines, interpreters such as BASIC, tabular data, and other information. ROMs and PROMs are also used to generate logic functions, to decode signals, and to perform similar tasks of interconnected logic gates. In this case the ROM is very much like a programmable logic array (PLA), discussed next, but it has a larger bit capacity.

One smart use of ROMs is as a storage array for a lookup table. There are many ways to generate a function like sin x, where x is the angle in radians. It may be found, for example, by calculating the power series

$$\sin x = x - \frac{x^3}{3!} + \frac{x^5}{5!} - \frac{x^7}{7!} + \cdots$$

Many products and sums are required to obtain an accurate value of the sine for values of x that cause the series to converge slowly. An alternate way, which uses much less computer time, is to look up values in a table and interpolate to find the precise value of the sin x.

The first step in using a table is to normalize the angle so that it lies between 0° and 90°. The normalization process determines the quadrant, 0°–90°, 90°–180°, 180°–270° or 270°–360°, to generate the appropriate angle to address the ROM. If the angle is 179°, for example, the appropriate angle to find the sin 179° is 1°. The angle in degrees as a binary number generates the address for the ROM. The values of the sine for discrete angles are stored in the ROM at the address. These values and the value of the sine for the next angle, which is stored in the adjacent ROM address, are read and used in interpolating the value of the sine at the desired angle. The two ROM addresses bracket the desired angle.

An alternative method is to store the sine values at equal magnitude intervals so that the interpolation is equally rapid and precise over all angles. The cosines can be found from the same table by using the relation cos x = sin $(90° - x)$; the other trigonometric functions can be generated with the sine and cosine.

PROGRAMMABLE LOGIC ARRAYS

Programmable logic arrays (PLAs) and programmable read only memories (PROMs) provide smart alternatives to the interconnections of logic gates to satisfy complicated logic functions. Programmable logic arrays serve several purposes; they may be considered logic-function generators, code converters, and multiplexed read only memories. In serving these functions, they compress into a single element the functions satisfied by many logic gates. Before illustrating these capabilities, we will describe a programmable logic array. PLAs may be made according to the pattern supplied by the user to the manufacturer, or field-programmable logic arrays (FPLA) may be configured to suit the need.

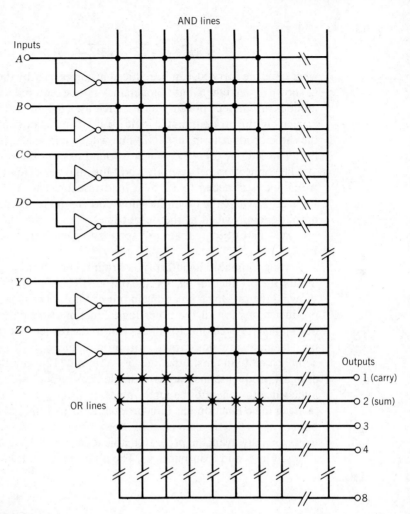

FIGURE 13-8

A Programmable Logic Array

A programmable logic array is an alternative to an array of individual logic gates. AND-logic and OR-logic in logic functions are made by corrections to lines.

Figure 13-8 shows that a programmable logic array is formed by crossing rows of input lines with columns of AND-lines. Associated with each input are its line and the line of its complement. The AND-lines also cross output lines; the connections to these lines generate OR-logic. The means of using a programmable logic array is illustrated with the familiar logic functions used in Chapter 3 to configure a full adder with basic logic gates. The functions

$$ABZ + \overline{A}BZ + A\overline{B}Z + AB\overline{Z} = C \text{ (carry)}$$

and
$$ABZ + \bar{A}\bar{B}Z + A\bar{B}\bar{Z} + \bar{A}B\bar{Z} = S \text{ (sum)}$$

where Z is the carry-in, can be reduced in complexity by the use of Karnaugh maps, as shown in Chapter 3, but this step may not be required with PLAs unless the total number of different product terms, such as ABZ, in all the functions exceeds the capacity of the programmable logic array. If more than one logic function contains the same product term, its AND-line will be connected to more than one output line. This is done for ABZ, which is found in both the sum and the carry.

The connection between an AND-line and each input line is a diode, and the AND-line is attached to V_{cc} by a load resistor. The AND logic is identical with Figure 3-2, where the input lines now correspond to inputs A and B and where the output C is provided at the output lines. An emitter follower on each AND-line provides the OR logic that corresponds to each state C on the output line of the PLA.

A programmable logic array is programmed from a truth table that lists the inputs, the product terms, and the outputs. Table 13-3 is a portion of a truth table showing the two logic functions in our example. The complete table would list the product terms for all the other logic functions and designate the corresponding output line for each one. It will be clear to the reader that the two logic functions in our example are not typical for programming a PLA because they are less complicated than most functions. Functions of 12 or more input variables, as well as many product terms, can use a single PLA rather than numerous interconnected logic gates to generate the logic values.

Each logic function has four product terms, a total of eight in our example, but one term, ABZ, is common to both; thus only seven AND-lines are used. Commercially available programmable logic arrays have 48 or 96 AND-lines with 14 input lines and 8 output lines. The x's in the truth table indicate no connection

TABLE 13-3
Truth Table for Programming a PLA

Terms	Inputs											Outputs							
	A	B	C	D	E	F	G	H	W	X	Y	Z	1	2	3	4	5	6	7	8
ABZ	1	1	x	x	x	x	x	x	x	x	x	1	1	1	x	x	x	x	x	x
$\bar{A}BZ$	0	1	x	x	x	x	x	x	x	x	x	1	1	x	x	x	x	x	x	x
$A\bar{B}Z$	1	0	x	x	x	x	x	x	x	x	x	1	1	x	x	x	x	x	x	x
$AB\bar{Z}$	1	1	x	x	x	x	x	x	x	x	x	0	1	x	x	x	x	x	x	x
$\bar{A}\bar{B}Z$	0	0	x	x	x	x	x	x	x	x	x	1	x	1	x	x	x	x	x	x
$\bar{A}B\bar{Z}$	0	1	x	x	x	x	x	x	x	x	x	0	x	1	x	x	x	x	x	x
$A\bar{B}\bar{Z}$	1	0	x	x	x	x	x	x	x	x	x	0	x	1	x	x	x	x	x	x

to the AND-lines. Output 1 is the carry; output 2 is the sum. The dots on the intersecting lines in Figure 13-8 are a map of the truth table for AND-logic; the crosses map the truth table for OR-logic.

As described above, the programmable logic array can be used as a function generator; now we will discuss another application. A code converter accepts an input word (bit pattern) in one form and produces an output in another form. One application of a PLA, which is fortunately disappearing from use, is to read a column on an IBM card and to produce an ASCII code. An IBM card has 12 rows in each column; each column is punched in one place to designate a number and in two or three places for a capital letter, a punctuation mark, or a symbol. The ASCII code uses seven bits to designate a letter, number, or symbol; the eighth bit is used to satisfy the parity of the byte. Because the card does not accommodate lowercase letters and because it can show only a few of the symbols, the pulses from the card reader can specify only a six-bit code.

The possible combinations of the 12 inputs, which are limited in number, are decoded by a programmable logic array to provide the six-bit code. A read only memory can do the same task equally well, but the ROM will accommodate $2^{12} = 4096$ words, while the card reader can identify less than 50 words from the 12 rows in the column. The PLA is, in effect, a compact ROM.

SUMMARY

Many special tasks and problems have emerged as the electronic revolution leads the technological revolution. Before the last decade the most reliable components in technology were mechanical. More recently, because of limitations in speed of response, the size of mechanical controls, the cost of fabricating mechanical components, and other factors such as wear, maintenance, and repair, it has become advantageous to turn to electronic systems as components, especially as integrated circuits became reliable, versatile, and less costly. Today, electronic controls are not installed on appliances, automobiles, and mechanical systems to impress the customer (although they may serve that purpose) but because electronic systems are cheaper in cost and labor and make the industry more profitable.

Electronic systems are smarter and more adaptable than mechanical systems. It is possible to use a microprocessor-based or other system that has the capacity to perform more than the task at hand; the cost may be no greater than for a single-purpose electronic system, and the smart system can be programmed to do just its job. The few techniques discussed in this chapter only serve to illustrate ways to approach special situations and are presented to stimulate the imagination.

PROBLEMS

13-1 Expand Table 13-2 to show all parity bits of the 16_{hex} numbers.

13-2 Assume an error in A_{hex}. For example, assume $A_{hex} = 1110$ rather than the correct value 1010_2 and use the Hamming code to identify the column to be corrected.

13-3 Design a parity checker to receive the data transmitted from the parity generator in Figure 13-6.

13-4 Program a programmable logic array like the one shown in Figure 13-8 to decode a four-bit binary signal to the hexadecimal numbers, 0 to F, for a seven-segment LED display. See Table 3-16 and Figure 3-14.

CHAPTER 14

POWER SUPPLIES

In all previous chapters, the voltages and currents required to operate the digital gates, the analog circuits, and the microprocessors were assumed to be available. In this chapter we examine how currents and voltages are provided and the means by which they are regulated; that is, how they are kept within the precise magnitudes that are desired or needed to avoid damage to integrated circuits and to ensure proper operation.

Many electronic devices, such as hand-held calculators and small radios, operate from batteries that provide steady (dc) voltages and currents. Here, however, we are interested in the circuits that provide dc voltages and currents from alternating power sources, usually 60 Hz and 120 V in the United States and Canada and 50 or 60 Hz and 220 V in the rest of the world. Our power supplies should be as close to batteries as we wish in terms of constant voltage and current, but they will not become exhausted after a period of use.

Power supplies may be characterized by the dc voltages and currents they produce and by the electric circuitry that they employ. They may be half-wave, full-wave, voltage doublers, radio-frequency, or photovoltaic. Whatever type they may be, their fundamental characteristics are the voltages and the maximum currents they provide. The product of the voltage and the maximum current is the power capability of the unit, $P = VI_{max}$ in watts.

Integrated circuits are attractive because they operate at a low power, but most have strict limits on the voltage that they require. They are powered as the voltage obtained from an ac building power outlet is transformed in magnitude and rectified to become dc voltages and currents. A typical power supply for a small computer will provide unregulated dc voltages and currents rated at ± 18 V at 2 A and at $+8$ V at 20 A. The voltages and currents represent the unregulated dc power supplied to the circuit boards of the computer. Each board has an individual regulator to ensure that the voltages to its electronic components are held closely to ± 12 V and $+5$ V.

Power sources that include regulators are called *regulated power supplies*. The difference between the basic unregulated power source and a regulated power supply is the presence of an electronic element with feedback to hold the output voltage at the desired value and to minimize any variations in the voltage as the current varies between the minimum and the maximum rated values. The regulator circuit also reduces the short-time variations in voltage that are present in the output of the unregulated power supply. The unregulated power supply must produce a higher voltage than the regulated output because the regulator absorbs a portion of the power in the form of the voltage drop across it times the current through it. Thus the regulator is a source of power loss and an unwanted heat source.

In this chapter we examine the electrical principles involved in transforming ac power to high-quality dc power and discuss several techniques of transformation. We consider the voltages typically used by integrated circuits. Higher-voltage power supplies do not differ in electrical principles, but the demands upon the elements, such as the inverse voltage that a rectifier must sustain, are much greater than in low-voltage supplies. Finally we examine a relatively recent development in power-supply technology called a switching regulator, which successfully combines analog and digital concepts with efficient energy conversion.

ROOT-MEAN-SQUARE VOLTAGE

To consider the process of generating dc voltages and currents from ac power lines, we need the relation between the ac wave form and the ac power.

In Chapter 7 we noted that the ac and dc power in a resistor R is the same if

$$P_{dc} = \frac{E_{dc}^2}{R} = \frac{E_{ac}^2}{R} = P_{ac} \text{ W} \qquad (14\text{-}1)$$

and that the ac voltage is described by

$$e(t) = E \sin 2\pi ft \text{ V} \qquad (14\text{-}2)$$

where E is the amplitude, f is the frequency, and t is the time.

The time average of the square of $e(t)$ is given by

$$\langle e(t)^2 \rangle = \frac{1}{\tau} \int_0^\tau e(t)^2 \, dt = \frac{E^2}{2} \tag{14-3}$$

where τ is the period of time for one full cycle. We conclude that

$$\frac{E^2}{2} = E_{ac}^2 = E_{dc}^2 \tag{14-4}$$

so $E_{ac} = E/\sqrt{2}$ and E_{ac} is called the root-mean-square voltage, 70.7 percent of the ac voltage amplitude E.

The 120-V magnitude of E_{ac} requires that the sine wave describing the ac voltage have a magnitude

$$E = \sqrt{2} E_{ac} = \sqrt{2} \times 120 = 169.7 \text{ V}$$

and the ac voltage reaches an equal negative value. Thus, the range of the sinusoidal voltage is 339.4 V.

The quantities involved are shown in Figure 14-1, where the crosshatched area under $e(t)^2$ is the integral in Equation 14-3. The values $e(t)$ and $e(t)^2$ are plotted

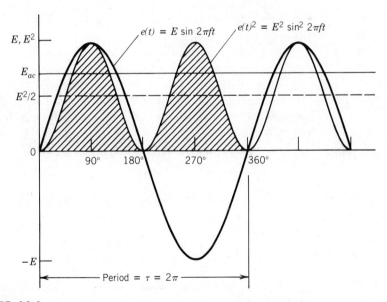

FIGURE 14-1

Voltage Relations in Finding RMS Voltage

The values $e(t)$ and $e(t)^2$ are plotted with normalized amplitudes for E and E^2.

with equal height to emphasize their time relationship. One can see that the portion of the crosshatched area about $E^2/2$ will fill in the plain area below $E^2/2$ to give the average value of $E^2/2 = E_{ac}^2 = E_{dc}^2$, as required by Equation 14-1 and given in Equation 14-4.

VOLTAGE AMPLITUDE TRANSFORMATION

Low-voltage power supplies, such as an 8-V unregulated voltage supply for integrated circuits, rectify ac voltages, which are reduced from the 120-V building power circuit. *Transformers* are used to transform the voltage amplitudes. A transformer has two separate windings, in which the numbers of turns are in the ratio of the primary voltage to the secondary voltage. For a voltage step-down the primary circuit with a larger number of turns, N_P, is connected to the ac power circuit, and the secondary with a smaller number of turns, N_S, is connected to the rectifying element. The transformer voltage ratio $V_P/V_s = N_P/N_S$ is a fixed value established by the manufacturer. V_P is the 120 V (rms) of the power circuit.

The primary and the secondary windings are insulated from each other electrically so that the reference voltage (ground) of the dc power supply is independent of the ground of the building power circuit. An improper connection or a fault between the primary and the secondary windings may tie the secondary circuit to the reference voltage of the primary winding. This situation is to be avoided.

The efficiency of voltage transformation is ensured by the low-reluctance iron path for the magnetic flux that passes through both the primary and secondary windings. In Figure 14-2 the iron core is symbolized by the parallel lines drawn between the symbols for the coils. If the lines are omitted, it signifies that no iron is present for the magnetic path; in that case the transformer is called an *air-core transformer*.

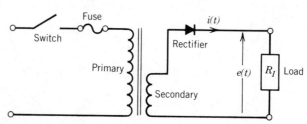

FIGURE 14-2

A Half-Wave Rectifying Circuit

The circuit produces the pulsing secondary voltage and current with one polarity.

RECTIFICATION

The process known as *rectification* of electric voltage and current is the selection of a single polarity of voltage and direction of current. This selection is made by a rectifier, an electrical device with the properties of a diode, which conducts with one voltage polarity and blocks conduction when the opposite voltage polarity exists.

A half-wave rectifying circuit connected to an ac source is shown in Figure 14-2. The switch enables the device to be activated, and the fuse protects the input (primary) side of the circuit from excessive current or power demands, which occur if the transformer has a fault or if the secondary terminals of the transformer are shorted. The fuse is chosen for the current magnitude, which in proper use corresponds to about 150 percent of the maximum operating current. At the instant when the switch is closed, the current may be several times higher than at all other times; thus it is common practice to use a fuse that requires a fraction of a second of overload before it interrupts the current by melting. "Slow-blow" fuses are made with added heat capacity, so that they take longer to reach the melting temperature during an overload. With a slow-blow fuse the current drops from its high initial value to a normal value before the fuse melts.

Figure 14-3 shows the normal voltage and current to the load resistor in the half-wave rectifying circuit of Figure 14-2. Even though a single polarity of voltage and current is maintained, the variations of amplitude with time indicate that the half-wave rectifier is not even approximately equivalent to a battery.

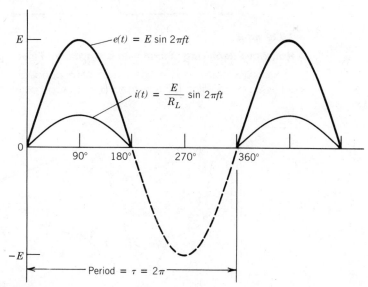

FIGURE 14-3

Voltage and Curent in a Half-Wave Rectifying Circuit

The alternating voltage e(t) and current i(t) with positive polarity are passed in alternate half-cycles.

CAPACITANCE FILTER

A large capacitor that follows the rectifying diode in parallel with the load, as shown in Figure 14-4, makes the voltage and the current in the circuit much closer to those of a battery. The capacitor is charged by current through the diode in excess of the load current. During the portion of the cycle when the rectified voltage is lower than the voltage reached by the charged capacitor, $v_C(t) = q(t)/C$, the charge stored in the capacitor can leave the capacitor and can flow through the load. The diode prevents the charge from the capacitor from flowing back through the transformer.

Figure 14-5 shows the new current-voltage relations with the capacitor in place. This figure contains a great deal of information about the properties of power supplies. The voltage across the load v_L (we will use v instead of e to correspond to the usual voltage designation on integrated circuits) has a higher average value $\langle v_L \rangle$, and is a smoother $v_L(t)$ than without the capacitor, although still rough. The

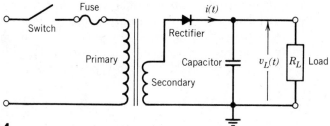

FIGURE 14-4

A Half-Wave Rectifying Circuit with a Capacitor Filter

The capacitor stores charge when the rectifier conducts, and it supplies current to the load when conduction is interrupted. The average current is increased.

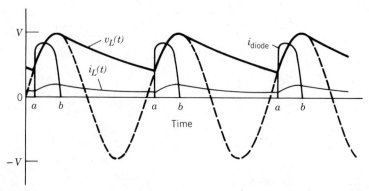

FIGURE 14-5

Voltage and Current in a Half-Wave Rectifying Circuit with a Capacitor Filter

The average voltage and the current to the load are sustained by current pulses.

average load current $\langle i_L \rangle$ has also risen, and the current $i_L(t)$ is steadier. The current through the diode, however, has become a large pulse with a peak current amplitude several times greater than without the capacitor. Since the cathode of the diode is at the voltage of the capacitor, we note that during the other part of the sine wave cycle the peak inverse voltage across the diode reaches approximately twice the peak of the ac secondary voltage. The diode must carry the high current pulse and withstand the inverse voltage.

This arrangement is closer to the equivalent of a battery, but the variations in $v_L(t)$ as it fluctuates above and below its average value $\langle v_L \rangle$ are undesirable for most applications. It is possible to separate $v_L(t)$ into voltage components by using Fourier analysis in the same way that gave the components of the square wave in Figure 7-17. The load voltage, $v_L(t)$, is found to have a dc part, which is the average voltage $\langle v_L \rangle$, and a series of sine waves with different amplitudes that, when added to $\langle v_L \rangle$, give $v_L(t)$. The largest of the sine wave components has a frequency of 60 Hz, the same as the supply voltage. The frequencies of all the other sine waves are integer multiples of 60 Hz. The variations, which are called the *ripple voltage*, indicate how unsteady the load voltage is. As the lowest-frequency component is reduced, all the higher-frequency components become even less significant, and the voltage $v_L(t)$ approaches the dc voltage of an equivalent battery. Another element, a *voltage regulator*, may be added to the circuit to reduce the ripple voltage. The operation of the voltage regulator is discussed below.

We will now consider how this circuit responds if the load current is changed —increased or decreased—or if the capacitor is replaced with a larger or smaller capacitor. If the load current increases, the average voltage $\langle v_L \rangle$ will drop and the ripple will increase because the voltage $v_L(t)$ drops more between the intervals in which it is being restored. Conversely, a reduction in the load current will allow the average voltage to increase, and in this case the ripple will be less. If the capacitor is made larger, the average voltage will increase and the charging time for the capacitor will diminish.

To describe the behavior of the circuit in quantitative terms, let us examine the voltage and currents in the two intervals shown in Figure 14-5. The interval *a* to *b* is the portion of the time in each cycle during which the diode conducts and the voltage of the capacitor is restored by the charge that flows into it. In the interval *b* to *a*, the capacitor supplies current to the load and its voltage drops.

First, let us consider the interval *b* to *a*, during which the capacitor discharges through the load resistor and its voltage falls. Because the load voltage $v_L(t)$ is also the capacitor voltage, we will examine it in the interval *b* to *a*. Recalling the equations in Chapter 7 for *RC* circuits, we write

$$v_L(t) = \frac{q(t)}{C} = v_C(t)$$

Ohm's law, however, states that $v_L(t) = i(t)R_L$ in the load resistance, and Kirchhoff's loop law requires that

$$v_C(t) + v_L(t) = 0$$

We find

$$\frac{q(t)}{C} = -i(t)R_L \tag{14-5}$$

Differentiating Equation 14-5 with respect to time gives

$$\frac{1}{C}\frac{dq(t)}{dt} = \frac{1}{C}i(t) = -R_L\frac{di(t)}{dt} \tag{14-6}$$

which is solved to give

$$v_L(t) = v_L(b)\,\exp\!\left(\frac{-t}{R_L C}\right) \tag{14-7}$$

where $v_L(b)$ is the voltage at the time b in the cycle when the conduction in the diode ends. The load voltage falls off exponentially in the interval when the diode is not conducting.

The time constant $\tau_L = R_L C$ governs the rate of falloff from b to a after the capacitor has been charged during the interval a to b. We see that the product of the load resistance, R_L, and the capacitance, C, characterizes the behavior of the half-wave rectifier with a capacitor filter. For a given load resistance, R_L, a larger capacitor makes the falloff less severe. Conversely, for a given capacitor, C, an increase in the load resistance—that is, a smaller load current—reduces the change in $v_L(t)$ between b and a. We conclude that either situation, which increases the time constant, $R_L C$, increases the average load voltage, $\langle v_L(t) \rangle$, and the average load current, $\langle i_L(t) \rangle = \langle v_L(t) \rangle / R_L$, and reduces the ripple. The increase in time constant, however, occurs at a price: the size of the current pulse through the diode in the interval $t(a)$ to $t(b)$. The interval diminishes as the time constant becomes larger and the average current increases; both of these increases make more demands on the ability of the diode to conduct.

The relation between the diode current and the average current is simple, as seen in Figure 14-4: all the current to the load must come through the diode. The magnitude of the diode current can be estimated from the average current by the ratio of the period of the rectifier to the interval of the diode conduction. We find

$$i_{\text{diode}} = \langle i_L(t) \rangle \frac{(1/60)}{t(b) - t(a)} \tag{14-8}$$

The period of the rectifier is 1/60 sec, but the interval $t(b) - t(a)$ becomes smaller as the time constant, $R_L C = \tau_L$, becomes larger. If the average current is substantial and if the time constant is large, the ability of the transformer and the diode to provide the peak current in the current pulse through the diode will be exceeded. At some point, it becomes unreasonable to add capacitance to increase the average load voltage, $\langle v_L(t) \rangle$, and the average load current, $\langle i_L(t) \rangle$, and to reduce the ripple. In the next section we show the limits on the time constant when a better rectifying circuit is presented.

FULL-WAVE RECTIFIER

One remedy for the limitations of the half-wave rectifying circuit is to use both portions of the supply-voltage sine wave to provide current to the load. Figure 14-6 shows a full-wave power supply. The secondary winding of the transformer is tapped at its center and the number of turns is doubled. This procedure is equivalent to adding another secondary winding back to back with the one shown in Figure 14-4 and requiring the common point—the center tap—to be the voltage reference or ground. The diodes alternate in conduction as the polarity of the voltage alternates.

Figure 14-7 shows the wave forms in a full-wave power supply with a capacitor filter. The average load voltage, $\langle v_L(t) \rangle$, and the average current, $\langle i_L(t) \rangle$, are

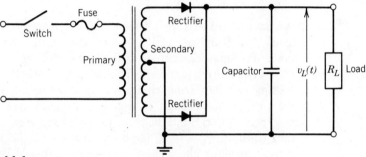

FIGURE 14-6

A Full-Wave Rectifying Circuit with a Capacitor Filter

The capacitor stores charge during conduction by the rectifier at twice the rate of the half-wave rectifier.

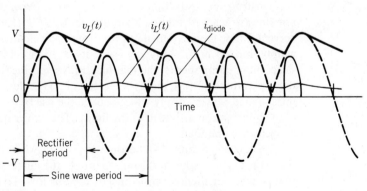

FIGURE 14-7

Voltage and Current in a Full-Wave Rectifying Circuit with a Capacitor Filter

The current pulses occur at twice the rate of those in the half-wave rectifier; thus the average voltage and current are relatively higher.

greater for the same $\tau_L = R_L C$, and the ripple is smaller. Note that the lowest frequency in the ripple is twice the frequency of the ac power line; that is, $f_{\text{ripple}} = 120$ Hz. The average current is achieved with twice as many diode current pulses per second as in the half-wave rectifier, even though the duration of each pulse is somewhat shorter for the same average current. The full-wave rectifier demands less from each diode and from the transformer. The transformer, as seen from the terminals of the secondary windings, is a Thevenin voltage source with an internal resistance; thus a large current pulse diminishes the secondary voltage by $i_{\text{peak}} r_{\text{Thevenin}}$. The maximum performance of a rectifying circuit depends upon the current-conducting ability and resistance of the rectifying element and the quality of the transformer.

RECTIFIER-CIRCUIT RESPONSE

We have found that the average load voltage $\langle v_L(t) \rangle$, the average current to the load, $\langle i_L(t) \rangle$, and the variation from the average (the ripple voltage), all depend upon the time constant, $\tau_L = R_L C$, and the type of rectifying circuit. We use $R_L = \langle v(t) \rangle / \langle i(t) \rangle$ and the capacitance C of the filter capacitor to find τ_L. The root-mean-square (rms) ripple voltage is commonly expressed as a percentage of the average load voltage. To quantize the performance of power supplies, $\langle v_L(t) \rangle$, $i_{\text{diode}}(\text{peak}) / \langle i_L(t) \rangle$, and the percent ripple have been calculated as a function of the time constant, $\tau_L = R_L C$, for both a half-wave and a full-wave rectifying circuit with a capacitor filter. These values are plotted in Figure 14-8, which shows the advantage of a full-wave rectifier over a half-wave rectifier for all time constants. For very long time constants that correspond to a very small load current, however, the advantages of full-wave rectification over half-wave rectification are not significant.

The model used to calculate the curves in Figure 14-8 does not include the voltage drop across the diode (which is current-dependent, as shown in Figure 8-9), nor does it account for the absolute limit in the magnitude of the peak rectifier current. These factors depend upon the specific elements used in making the power supply, but do not modify the contents of Figure 14-8. To guide the user in the selection of components, the manufacturer's specification sheet states the limits in the operation of the rectifying element. The design example of a full-wave unregulated power supply will illustrate the use of Figure 14-8 after we note some of the information provided in the figure. The values in Figure 14-8 are those at the capacitor or the load.

The data on Figure 14-8 become important when the load current is high. The time constant does not indicate separately the current level in the rectifier circuit; that is, whether the load resistance is high or low. To evaluate its significance, the information from Figure 14-8 must be related to the load current in a specific circuit. For a low current, $\langle i_L(t) \rangle = 0.01$ A, for example, a very large time constant that gives a factor of 25 in the ratio of diode current to average current does not imply undue demand on a high current-rectifying element. For an average load

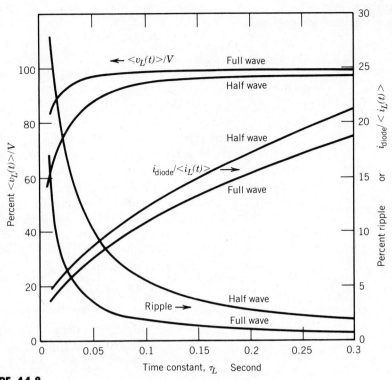

FIGURE 14-8

A Rectifier-Circuit Response versus a Time Constant

The time constant, $\tau_L = R_L C$, governs the behavior of both half-wave and full-wave rectifiers with filter capacitors. The parameters shown are used in power-supply design.

current of 10 or 20 A, however, one must be careful in choosing the time constant that will govern the operation of the circuit.

To illustrate the contents of Figure 14-8, we use $\tau_L = 0.1$ sec. The ratio of the time constant to the (1/60 sec) period of the half-wave rectifying circuit is 6; for a full-wave rectifier it is 12. Figure 14-8 shows that for $\tau_L = 0.1$ the ratio of the peak diode current to the average load current is more than 12 for the half-wave rectifier and 11 for the full-wave rectifier. The greatest advantages of the full-wave rectifier are the lower ripple voltage and the higher average load voltage.

Figure 14-8 shows that short time constants are disadvantageous except in the ratio of diode current to average load current. It also suggests that time constants greater than 0.3 sec are unreasonable for significant load currents because the magnitude of the current pulses through the diode is 20 or more times the average current. Nevertheless, if the time constant is suitable for the maximum rated average current to the load, it will be suitable for smaller currents to the load. If the diode current limit is involved, however, or if the voltage drop across the rectifier is significant, then the values from Figure 14-8 will overestimate the quality of the rectifying circuit.

EXAMPLE OF A POWER-SUPPLY DESIGN

Our objective is to design a 120-V, 60-Hz, relatively low-power unregulated power supply to give 8 V dc with a maximum current of 1 A dc. This circuit will be suitable for the addition of a voltage regulator, which will give +5 V, 1 A regulated output. The diode whose specifications are listed in Table 14-1 is selected as the rectifying element.

Although many approaches are possible in designing a power supply, we will use Figure 14-8 as the basis for our analysis. The starting point is the selection of a time constant, $\tau_L = R_L C$, to make an initial estimate of the diode current and the average voltage, $\langle v_L(t) \rangle$. A full-wave rectifier circuit is selected. An examination of the rectifier specifications in Table 14-1 alerts us to the need to keep the diode current within reasonable bounds; thus a time constant toward the left in Figure 14-8 is suggested. As a first attempt we use $\tau_L = 0.04$ sec.

We find $\langle v_L \rangle / V = 95$ percent, $i_{\text{diode}} / \langle i_L(t) \rangle = 7.8$ and 4 percent ripple. The following quantities may now be calculated.

$$V = \langle v_L \rangle / 0.95 = 8/0.95 = 8.5 \text{ V}$$

$$i_{\text{diode}} = 7.8 \times \langle i_L(t) \rangle = 7.8 \times 1.0 = 7.8 \text{ A}$$

$$V_{\text{diode}} = 0.8 \text{ V (slightly less than 0.85 V because 7.8 A}$$
$$\text{is less than 10 A; see Figure 8.9)}$$

$$V_{\text{secondary}} = 8.5 + 0.8 = 9.3 \text{ V}$$

$$V_{\text{secondary}}(\text{rms}) = 9.3/\sqrt{2} = 6.6 \text{ V}$$

$$R_L = 8.0/1.0 = 8.0 \text{ }\Omega$$

$$C = \tau_L / R_L = 0.04/8.0 = 5000 \text{ }\mu\text{F}$$

$$\langle i_{\text{primary}} \rangle = (N_S/N_P) \langle i_L \rangle = (6.6/120) \times 1.0 = 0.06 \text{ A}$$

$$\text{Transformer power} = i_P \times V_P = 0.06 \times 120 = 6.6 \text{ W}$$

$$\text{Fuse size} = 1.5 \times i_P \approx 0.10 \text{ A}$$

$$\text{Peak inverse voltage} = 2 \times 9.3 = 18.6 \text{ V}$$

The calculated values should be compared with the specifications of the power supply and the rectifying diode to see whether any unreasonable values appear,

TABLE 14-1
Rectifier Specifications

Average current	1 A
Maximum peak inverse voltage	50 V
Half-cycle peak surge current at 40°C	25 A
Peak surge current at 60 Hz for 1 sec	11 A
Forward voltage drop at 1 A	0.6 V
Forward voltage drop at 10 A	0.85 V

especially in the diode current and the inverse voltage. If they do, a new time constant can be selected and the calculations can be made again.

A transformer whose specifications match exactly the calculated values is not likely to be available, but a standard transformer listed by a supplier as

$$120 \text{ V}, 6.3 \text{ V} - CT - 6.3 \text{ V}$$

meaning that the primary voltage is 120 V and the secondary gives 6.3 V on each side of the center tap, is close enough to the calculated values to be recommended. A standard electrolytic capacitor with $C = 5000 \, \mu\text{F}$, 25 dc working voltage, is selected. The capacitor should have an operating voltage about 100 percent greater than the load voltage. Only electrolytic capacitors have such large capacitance at a reasonable size and cost, but they must be used within the voltage rating, and the proper voltage polarity must be observed. The half-cycle peak surge current can now be recognized as the current demand to charge the capacitor when the power supply is turned on. This is the current that the fuse must withstand momentarily.

Because the 5000-μF capacitor is quite large, a smaller value, perhaps 2000 μF, might be considered only if the substitution is clearly acceptable. The time constant becomes about half the initial trial value; that is, $\tau_L = 0.016$ sec. Figure 14-8 shows that the average voltage $\langle V_L \rangle = 0.88$ V, and the ripple is 8.7 percent. These values suggest that the smaller capacitor is not acceptable; the voltage regulator may not have adequate voltage margin to operate. This situation is discussed below in the example of a power-supply performance. Regulating elements are discussed shortly.

BRIDGE RECTIFIER

Another form of full-wave rectifier is a *bridge rectifier*, as shown in Figure 14-9. A bridge rectifier uses a single secondary winding rather than a center-tapped secondary, but the bridge contains four rectifying elements rather than the two

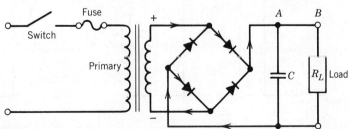

FIGURE 14-9

A Bridge Rectifier with a Capacitor Filter

The bridge rectifier provides full-wave rectification and an option in the voltage polarity of the filtered output.

shown in Figure 14-6. A bridge rectifier as a single integrated circuit includes all four rectifying elements.

The load current passes serially through two rectifying elements. The two elements that conduct in one half-cycle of the ac supply voltage alternate with the other two elements in the other half-cycle. In Figure 14-9 the arrows indicate the path of the current in the half-cycle when the positive polarity is shown for the secondary winding. Small bridge rectifiers are used in instruments to convert ac signals for dc metering.

An advantage of the bridge rectifier is that the voltage reference point (ground) can be at either end of the capacitor-bridge common point. That is, the circuit can be either a positive or a negative supply, depending on the grounding point. The ripple and its frequency, 120 Hz, are identical to those of the full-wave rectifier presented earlier, but the voltage drop across the rectifying element occurs twice in series with the load. The peak inverse voltage, however, is the same as a half-wave rectifier and half as much as a full-wave rectifier.

ADDITIONAL FILTERING

Filtering is the process of reducing the time variations in the voltage. As we have seen, the addition of a capacitor to the rectifier circuit greatly improves the average value of the load voltage and reduces the variations called "ripple." The insertion of additional passive elements—resistors, capacitors, and inductors—can provide further filtering; they are in common use in low-current (100 mA), high-voltage (100 V or more) power supplies. In high-current, low-voltage supplies, there is no advantage in using these additional components, for reasons stated below.

Figure 14-10 shows two different filter additions to the circuit discussed earlier. Now the load voltage V'_L is across the load and the added capacitor, C_1, and V'_L is a

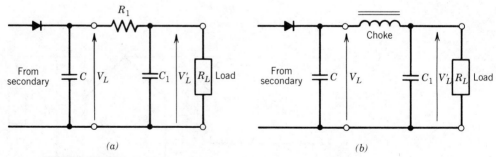

FIGURE 14-10

(a) Resistance-Capacitance and (b) Inductance-Capacitance Filter Sections

An additional filter section reduces ripple voltage and acts as a voltage divider for the dc component.

fraction of the load voltage V_L that appears on the initial capacitor, C; the dependence of V_L on operating conditions is shown in Figure 14-8. The added filter elements constitute a voltage-dividing circuit; thus we may write

$$V'_L = V_L \frac{Z_L}{Z_1 + Z_L} \qquad (14\text{-}9)$$

Z_L is the parallel combination of the load resistance, R_L, and the capacitative reactance of C_1, $X_{C_1} = 1/(2\pi f C_1)$. The frequency, f, is the supply-voltage frequency (60 Hz) for a half-wave rectifier and twice the value (120 Hz) for a full-wave rectifier. In Figure 14-10a the resistance $R_1 = Z_1$; this filter arrangement is called an RC filter section. C_1 and load resistor, R_L, yield the parallel impedance

$$Z_L = \frac{-jR_L X_{C_1}}{R_L - jX_{C_1}} \qquad (14\text{-}10)$$

Some algebraic calculation shows that

$$V'_L = V_L \frac{R_L + \dfrac{j2\pi f C_1 R_1 R_L^2}{R_L + R_1}}{R_L + R_1 + \dfrac{(2\pi f)^2 C_1^2 R_1^2 R_L^2}{R_L + R_1}} \qquad (14\text{-}11)$$

The dc portion of V'_L is found from Equation 14-11 by setting $f = 0$ to give

$$V'_L(\text{dc}) = V_L(\text{dc}) \frac{R_L}{R_L + R_1} \qquad (14\text{-}12)$$

which indicates that the dc voltage at the load is reduced by the fraction $R_L/(R_L + R_1)$ because of the added RC filter section. The ac or ripple fraction is given by Equation 14-11. In a full-wave rectifier, for example, $f = 120$ Hz, the lowest frequency component in the ripple, and if $R_1 = R_L = 8.0\,\Omega$, $C_1 = C = 2000\,\mu\text{F}$,

$$V'_L(\text{dc}) = V_L(\text{dc}) \times \tfrac{1}{2} \quad \text{and} \quad V'_L(\text{ripple}) = V_L(\text{ripple}) \times 0.077$$

All the other ripple components are reduced even more sharply.

The LC filter section shown in Figure 14-10b is more effective than the RC filter section in reducing the ripple voltage. It would be still more effective if the resistance in the windings of the inductance were small in relation to the load resistance, but this is not the case for high-current, low-voltage systems. By using the impedance of the inductance $Z_1 = R_1 + j2\pi f L$, where R_1 is the intrinsic resistance of its windings, Equation 14-9 can be rewritten to give

$$V'_L = V_L \frac{(X_{C_1})^2 [R_L(R_L + R_1)] - R_L^2 X_{C_1} X_L - jR_L X_{C_1}(R_L R_1 + X_{C_1} X_L)}{(R_L R_1 + X_{C_1} X_L)^2 + [R_L X_L - X_{C_1}(R_L + R_1)]^2} \qquad (14\text{-}13)$$

$X_{C_1} = 1/(2\pi f C_1)$, and $X_L = 2\pi f L$ pertains to the inductance.

The inductor has a high value of inductance, L, by virtue of the iron core. An inductor of this kind is commonly called a *choke*. A choke is analogous to a transformer on which only a primary winding is made to create high inductance. One disadvantage of the LC filter is the voltage drop, which is due to current times the resistance of the choke. The $f = 0$ behavior of Equation 14-13 is found first by dividing both the numerator and the denominator by X_C^2 and then by setting $f = 0$. We obtain

$$V'_L(\text{dc}) = V_L(\text{dc}) \frac{R_L}{R_L + R_1} \tag{14-12}$$

If the same values used in the example for the RC filter section are used for the LC filter, and $L = 0.1$ H, we obtain

$$V'_L(\text{dc}) = V_L(\text{dc}) \times \tfrac{1}{2} \quad \text{and} \quad V'_L(\text{ripple}) = V_L(\text{ripple}) \times 0.009$$

Chokes that have low resistance winding and high inductance are large, heavy, and expensive. In addition to the cost and the voltage drop (because of the resistance of the wire of the choke), another factor is involved in the effectiveness of a choke. The current through the choke is nearly all dc, and the large unidirectional current saturates the iron core and sharply reduces the inductance. The LC filter section is ineffective when the current is high, so it is not used in low-voltage power supplies for integrated circuits.

VOLTAGE DOUBLERS

The *voltage doubler* is a simple way to obtain twice as much dc voltage as a half-wave rectifier will yield from the secondary voltage of a transformer. The circuit of the voltage doubler shown in Figure 14-11 allows one capacitor to charge as a half-wave rectifier during one half-cycle and the second capacitor to charge in the other half-cycle. The sum of the voltages is double the voltage on each capacitor.

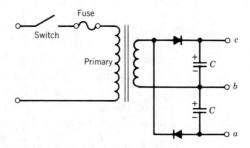

FIGURE 14-11

A Voltage Doubler with a Grounding-Point Option

Twice the voltage of a single half-wave rectifier can be acquired with either polarity, or bipolar voltages can be achieved, depending upon the grounding point.

In the voltage doubler three choices exist for the selection of the reference voltage or ground. (These are labeled *a*, *b*, and *c* in Figure 14-11.) The result of each choice is as follows:

Grounding Point	Voltage
a	Positive voltage doubler
b	Positive-negative supply
c	Negative voltage doubler

Options *a* and *c* are used to give relatively high-voltage amplitudes, perhaps 1000 V or more, for low-current applications, such as detectors in nuclear experiments. In the selection of components for high-voltage applications, one must consider the peak inverse voltage across the diodes and the voltage to be withstood by the insulation between the secondary and primary windings of the transformer.

Option *b* is used frequently to provide the symmetric $\pm V_{cc}$ required for many operational amplifiers. Positive and negative voltage regulators are added at points *c* and *a*, respectively, to ensure that the voltages are constant at the loads.

VOLTAGE REGULATORS

Steady, batterylike voltages, which are not quite achieved by rectifiers with a capacitor filter, can be provided by adding a voltage regulator between the capacitor and the load. The principle of voltage regulators was presented in Chapter 10; two simple configurations of voltage regulators are shown in Figure 10-3. The voltage regulator has three elementary components: a voltage reference, a feedback circuit with amplification, and a regulating element between the voltage source and the load.

In low-voltage applications, the most widely used voltage regulators are integrated circuits, which include all the elementary components in one package. These three-terminal devices are placed between the voltage source and the load, as shown in Figure 14-12. The voltage at the load will remain constant as long as the rated limits of current through the regulator are not violated and as long as the supply voltage is large enough to equal the sum of the load voltage and the voltage drop of the regulator.

The output voltage is well regulated because the regulator includes an active feedback circuit and a voltage reference. The active feedback circuit amplifies the error voltage, that is, the difference between the output voltage and the reference voltage. The greatly amplified error voltage is the input signal to the regulating element, which erases nearly all the voltage variations of the supply by increasing or decreasing the voltage drop across itself. The result is a steady load voltage, nearly independent of the current drawn and the ripple. (A small error must exist

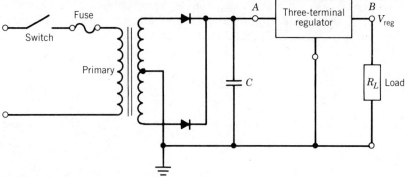

FIGURE 14-12

A Full-Wave Regulated Power Supply

The addition of a three-terminal regulator at the filtered output maintains a constant voltage independent of variations in load current.

because the amplification is not infinite.) Integrated-circuit voltage regulators typically ensure that the ripple is 0.1 percent or less of the rated voltage if the average unregulated supply voltage—the dc component—exceeds the regulated load voltage to a sufficient degree.

Because the regulator is interposed between the rectified voltage source and the load, the load current passes through it. The regulator and the load share not only the voltage from the source, but also the power. The power to the regulator is $P_{reg} = I_L V_{reg}$ in watts; the power to the load is $P_L = I_L V_L$ in watts. The proportion of the total power to the regulator and to the load is the same as the proportion of the supply voltage that they share. For a 5-V regulated load voltage the supply voltage must be about 8 V or higher; thus the power proportions are about 3/8 and 5/8 of the total power. The small integrated-circuit regulator package must withstand about 40 percent of the power and dissipate it as heat.

Integrated-circuit regulators frequently have an internal sensor, which disables the regulator if it overheats to prevent permanent damage. This action is to be avoided because it interrupts the power to the load. A heat sink is used to share the heat generated in the regulator; this device is a metal plate with a much greater area than that of the regulator, to which the regulator is attached thermally for heat transfer. Either natural or forced-air convection dissipates the heat from the combination.

Not only does the regulator lose power; so do the transformer and the rectifying elements. Typically, the efficiency of a regulated power supply is less than 50 percent. A power supply for a desk-sized computer may have 100 W to dissipate; an equal amount may be shared by a number of circuit boards. Efficient means of heat removal are used to avoid overheating the circuit boards and their electronic components. It is common practice to place individual voltage regulators with heat sinks on each circuit board to ensure that the voltage will be regulated. More efficient power supplies will be discussed below when we examine switching regulators.

EXAMPLE OF A POWER-SUPPLY PERFORMANCE

The regulated power supply in this example illustrates poor performance because of inadequate components. The example was chosen to show the problem and to suggest remedial measures for low-power regulated power supplies suitable for integrated-circuit systems.

The transformer, the bridge rectifier, and a capacitor filter were connected as in Figure 14-9, and a three-terminal regulator was placed between points A and B. (Figure 14-12 shows a regulator in place.) The nameplate values of the components are given in Table 14-2.

Figure 14-13 shows voltages in the circuit for different load currents. Several points may be made about this regulated power supply. First the transformer is of poor quality, as shown by the decline in the secondary ac voltage, V_{rms}, as the load current increases. The secondary is wound with wire of too small a gauge, and the small size of the transformer suggests that the small iron core for the magnetic field path gives only marginal performance.

Second the voltage at the bridge, measured with a dc voltmeter at point A in Figure 14-9, drops rapidly with increasing load current. Because the dc voltmeter averages the ripple, there is some uncertainty about the magnitude. It is clear, however, that the increasing load current rapidly exhausts the charge in the capacitor filter between charging pulses. A larger filter capacitor will offset this problem if the rectifier can carry still heavier current pulses.

TABLE 14-2
Nameplate Specifications of Power-Supply Components

Transformer	115 V primary
	6.3 V secondary
	60 Hz
	1 A
Rectifier	bridge
	1 A
Regulator	+5 V
	Type 7805 C
	1 A
Capacitor	2000 μF
	35 V

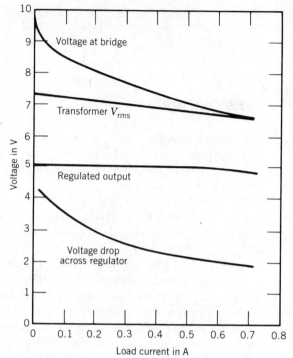

FIGURE 14-13
Performance Characteristics
The limitations of a small power source are illustrated by the measured values versus current in a regulated power supply.

The third point is that the voltage regulator reduces the voltage drop across itself as the unregulated voltage to it diminishes. At 0.6-A load current the regulator has minimized its voltage drop to about 2 V, about as small as it can become. The regulated voltage amplitude drops below the specified value of +5 V for load currents greater than 0.6 A, and the ripple voltage grows.

One can make an improved regulated power supply of about the same rating by changing components to remove the limitations mentioned above. A larger, more adequate transformer may be used, and a full-wave rectifier arrangement rather than a bridge rectifier. The bridge rectifier in Figure 14-9 has two diodes in series, which make the voltage drop across the bridge rectifier twice as great as a single-diode drop in the full-wave rectifier of Figure 14-12. A larger filter capacitor and heavier current diodes may be placed in the full-wave rectifier circuit.

A transformer with a much higher secondary voltage might be used to offset the limiting unregulated voltage available to the three-terminal regulator. If this is done, the regulator will be required to dissipate a great deal of additional power and will probably overheat.

REGULATION BY OPERATIONAL AMPLIFIERS

Operational amplifiers can serve in active feedback circuits to make excellent voltage regulators. The operational amplifier, however, is only one component of the regulator, which must include a voltage reference and a regulating element, usually a transistor. Both series and parallel voltage regulators with close tolerance in the voltage can be made, and the high gain of operational amplifier circuits can drive the regulating element. For most applications there are practical limits on the quality of regulation that should be achieved. In the thin wires in cables, for example, and in the printed lines on circuit boards in modern electronic devices, there occurs a voltage drop of about one percent or more of the regulated voltage that supplies the circuits.

Figure 14-14 shows a series voltage regulator with an operational amplifier circuit as the active feedback element in the regulator. The NPN power transistor is used as an emitter follower to provide the load current; the operational amplifier is a voltage follower. The voltage of the Zener diode is the reference voltage, which the output of the operational amplifier follows. The unity gain of the emitter follower ensures that the load voltage will stay equal to the reference voltage within the current-carrying limitations of the series transistor. The circuit is similar to that of three-terminal regulators.

Small variations in the regulated voltage, as provided by the circuit in Figure 14-14, arise as an error in the reference voltage because the unregulated voltage supplies current to the Zener diode. To the extent that the resistors and the operational amplifier are temperature-sensitive, temperature changes may shift slightly the regulated voltage. Except for special experimental and control applications, the truly small variations in the regulated voltage are unimportant. In the

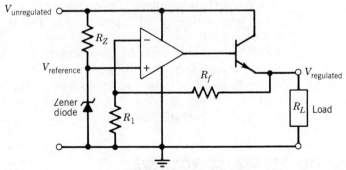

FIGURE 14-14

A Series Voltage Regulator with an Operational Amplifier

The operational amplifier controls the conduction of the transistor to regulate the output voltage. The arrangement is similar to the circuit of three-terminal regulators.

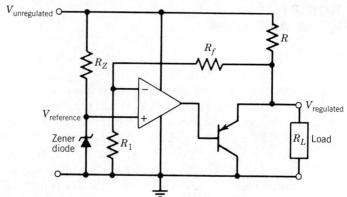

FIGURE 14-15
A Shunt Voltage Regulator with an Operational Amplifier
The power transistor shares the current with the load so as to maintain a constant voltage to the load.

integrated circuit-voltage regulators, each element experiences the same temperature change and there may be internal temperature compensation to ensure that the regulated voltage is insensitive to temperature variations.

The operational amplifier as a voltage follower with the voltage reference as the input is itself a voltage regulator for applications in which load currents are less than the maximum output current of the operational amplifier. In Figure 14-14 the series transistor is a current amplifier for the operational amplifier.

Figure 14-15 shows a shunt regulator made with an operational amplifier. The shunt regulator controls the voltage to the load by drawing additional current parallel to the load current, so that the voltage drop in the resistance R brings the voltage across the load to the desired value. The current parallel to the load current is called the *shunt current*. This current varies so as to nullify the ripple and the variations in the unregulated voltage that occur because of voltage variations in the ac line, temperature, or other phenomena. Note that the power transistor in this circuit is a *PNP* type.

The efficiency of a shunt regulator is poor if the load current becomes small. In this case, there is large power loss due to the high shunt current. On the other hand, the reference voltage is not coupled to the load voltage of the shunt regulator. As a result, the voltage regulator is more stable for some applications. The circuit has been called a "perfect" Zener diode.

SHIFTING THE REGULATED VOLTAGE

Integrated-circuit voltage regulators can be used to hold a voltage at a value different from their specified value. Figure 14-16 shows an arrangement of this kind. The regulated voltage output is set by the voltage difference between the

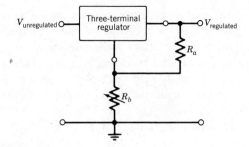

FIGURE 14-16

An Adjustable Voltage Regulator

A three-terminal regulator maintains an output voltage at a value governed by the resistor-voltage divider.

ground terminal of the regulator and the actual ground established by the current in R_b. The current in the path through R_a and R_b comes from the regulated output and is steady. Only a small fraction of the output current is required, and the regulated voltage may be set as desired by using a variable resistor for R_b.

CURRENT REGULATION

The simplest small-current regulator is an inverting operational amplifier in which the load is used as the feedback resistor. The constant current amplitude is established by the current from the input voltage through the input resistor to the virtual ground. The well-known feature of operational amplifiers, which ensures that the feedback current in an inverting amplifier is equal to the input current, sustains the current through the load independently of the load resistance within the current limits of the operational amplifier. (Figure 11-2 shows an inverting amplifier.)

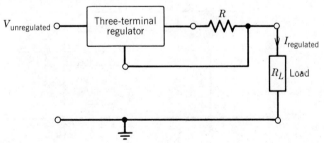

FIGURE 14-17

A Current Regulator

A three-terminal regulator maintains a constant current with the reference voltage drop across R.

Several other arrangements can be used to regulate currents with amplitudes greater than the capability of operational amplifiers. A convenient arrangement for modest currents uses a three-terminal voltage regulator for this purpose. (Figure 14-17 shows the arrangement.) The voltage drop across the resistance R in series with the load is the control voltage for the three-terminal regulator. If the load current changes, the regulator will respond by increasing or decreasing the voltage drop across itself to restore the current to its normal value. The current magnitude is limited to the rated current of the three-terminal integrated-circuit regulator.

SWITCHING POWER SUPPLIES

Switching power supply is the name given to an unregulated power supply equipped with a switching regulator. The switching regulator combines analog and digital technology to regulate the voltage in a most energy-efficient manner. The switching regulator uses a relatively high frequency, 5 to 50 kHz, which makes the filtering easier. A more important innovation is the use of an air-core inductor as an energy-storage device to sustain the current.

A switching regulator has four elements: a voltage reference element similar to a three-terminal voltage regulator, a pulse-width-modulation system, a switch, and an inductance-capacitance-diode output stage. The voltage regulator provides the reference voltage from which both the error signal and the stable voltage for the pulse-width-modulation system can be generated (Figure 14-18).

The pulse-width-modulation system consists of three elements: a differential or difference amplifier, a voltage comparator, and an oscillator. The differential amplifier, similar to that in Figure 11-8, compares the average output voltage with

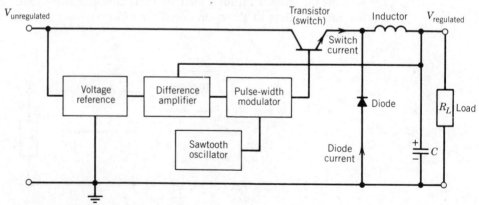

FIGURE 14-18

Schematic of Switching Regulator

An energy-efficient regulated power source is formed by combining digital and analog electronic elements.

the reference voltage to generate an amplified error signal. The error signal is one input to the voltage comparator. The other input comes from the oscillator, whose output is a stable triangular wave usually called a *sawtooth* wave. A typical oscillator frequency is 35 kHz.

The voltage comparator may be regarded as an operational amplifier with the amplified error signal attached to the inverting input and the sawtooth signal connected to the noninverting input, as in Figure 12-22a. As the sawtooth voltage increases, it equals and then exceeds the error voltage. This change in the relative values of the two inputs to the operational amplifier causes the amplifier to switch from the negative or zero-output voltage to full on or the maximum positive voltage. The output reverses the instant the decreasing triangular wave no longer exceeds the error voltage. The comparison of the two voltages generates a square wave. The duration of the on and off portions of the square wave depends on the error signal. Figure 14-19 illustrates the voltage comparison; it may be seen that the pulse remains on for a longer time when the output voltage drops relative to the reference voltage when the load current grows. The pulse-on interval corresponds to the conduction through the switch.

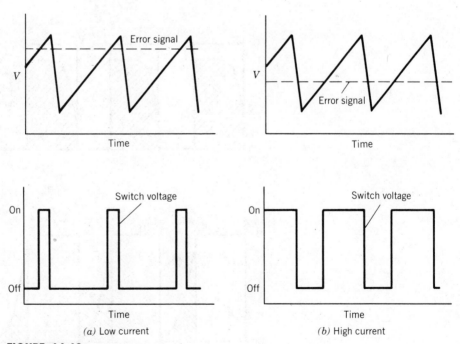

FIGURE 14-19

Pulse-Width Modulation

The "on" time of the transistor switch in Figuref 14-18 is determined by the pulse width derived from the error signal.

During the pulse-on interval the switch, a power transistor, connects the full unregulated voltage through an inductance to the capacitor and the load, which are in parallel. The load and the charging capacitor share the current. (Figure 14-18 shows the path for this current.)

When the square wave pulse is off, the switch can no longer conduct, but charges continue to flow to the load and the capacitor because the current through an inductance cannot change instantaneously. An induced electromotive force ensures that this occurs. Thus, when the switch ceases to conduct, the current comes through the diode; the capacitor transfers charge from its negative side to its positive side and to the load. (We note that this fundamental property of an inductance prevents the current both from turning on and from turning off instantaneously.) Figure 14-20 shows the currents during the pulse-on and the pulse-off intervals. The high frequency of the square wave requires an air-core inductance and makes the high-frequency ripple small because the time constant, $\tau_L = R_L C_1$, is larger than $1/f$.

The load current can vary widely, from a very narrow pulse-on and a wide pulse-off for a small-load current to almost constant conduction for high-load current

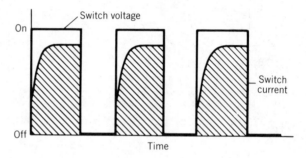

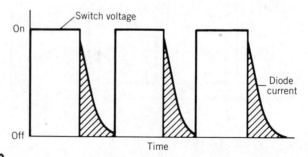

FIGURE 14-20

Current Relations in a Switching Regulator

The current after the "on" pulse is sustained through the diode by the inductance.

demands. In either case, efficiency is high because there is no major diversion of power into or across an element that shares the current and the voltage drop with the load. Only a small amount of power is lost from the integrated circuit containing the reference voltage element, the sawtooth oscillator, the voltage comparator, and the transistor switch. Switching power supplies that provide many amperes at regulated voltages have efficiencies as high as 85 to 90 percent, including the transformer and rectifier losses. When the small, lightweight switching power supplies are used, the removal of waste heat is a less serious matter. A suitable unregulated voltage source may be used, and parallel switching regulators may provide several different regulated voltages.

SUMMARY

Power supplies are essential parts of all electronic systems. Because the voltages of integrated circuits have narrow limits, reliable and well-controlled power supplies are necessary. The 7400-series TTL gates, for example, must be supplied with at least 4.75 and not more than 5.25 V.

Straightforward rectification and integrated-circuit elements for regulation facilitate the fabrication of power supplies. For systems that require many amperes of regulated power, standard performance-tested designs should be purchased or duplicated. Lesser power needs can be met in the manner suggested by the examples of power-supply design and performance. Underrated components should be avoided, since they frequently lead to disappointing results.

Designing and building regulated power supplies may be subject to the advice that is given about computer programming. Everyone should write a computer program once to learn what it is about and to get it out of his or her system, and then let the experts do all the other programming. Except for low-current, low-voltage supplies, this is also good advice.

PROBLEMS

14-1 Use Figure 14-8 to design an unregulated power supply to provide 2.0 A at +8.0 V. Specify the components and list the maximum performance requirements such as the peak current and peak inverse voltage.

14-2 A voltage-doubler power supply provides 1.0 mA at +1400 V for a nuclear detector. Design this power supply, find the maximum ratings, estimate the ripple voltage, and specify the components.

14-3 Design a power supply that uses a three-terminal regulator and the circuit of Figure 14-16 to give 0.5 A with the regulated voltage adjustable from +6 to +15 V. Consider the heat dissipation in the three-terminal regulator, and suggest how it might be minimized.

14-4 You have a power module (an encased transformer and power plug) left from the small hand-held calculator you lost. Its ratings are: input 120 Vac, 50–60 Hz, 3 W; output 5.6 Vac, 200 mA. You want to use it to make a power supply for a circuit containing three DIPs (25 mA per DIP) and a LED display (20 mA). Show the design using a Zener diode for regulating the voltage to about 5 V.

APPENDIX A

COMPUTER ARITHMETIC

Number systems and simple binary arithmetic were introduced in Chapter 3. Several other binary codes are given here, along with the means of converting among them. Integrated circuits convert from one code to another, and microprocessors have numerous operational codes devoted to the use of different number systems. Older microprocessors have simple operational codes such as rotate right, jump on carry, add, subtract, and other codes with which arithmetic operations can be programmed. Recent microprocessors have many additional operational codes (such as divide, decimal adjust for subtraction, ASCII adjust for addition, and integer multiply) and allow the arithmetic to be carried out in several number systems without conversion to binary numbers.

This appendix shows the means of conversion and some of the steps required to obtain the correct results when arithmetic is performed without conversion.

The Wallace algorithms are worth mentioning here because they illustrate alternate ways of doing arithmetic and because they point to ways in which the arithmetic in computers can be performed with great speed and accuracy.

BINARY CODES

The binary equivalents of decimal numbers are shown in Chapter 3. The bits are ordered to have positional values based upon the powers of 2; the least significant bit associated with $2^0 = 1_{10}$ is placed immediately to the left of the radix point. We now show variations in the binary codes that were created for two purposes. One reason for modifying the code is to avoid a binary number in which all the bits are zero, so that the absence of at least one high bit signals an error. An excess-three code with this property is made by adding $3 = 011_2$ to the BCD code.

A second reason for modification is to have a code in which sequential decimal-equivalent binary numbers are represented by a change in only one bit. This is the property of a Gray code. A rotating shaft, for instance, can be encoded with a Gray code so that the transition from $360°$ to $0°$ involves only a one-bit change in the binary code. It was pointed out in Chapter 12 that in digital-to-analog converters, the transitions from 0111 to 1000 as the midvalue is passed can give instantaneous transients in the conversion, depending upon whether the 1 in 1000 was produced before or after the ones in 0111 were extinguished. The former would yield 1111; the latter, 0000.

The decoding of any code for decimal display is equally easy with integrated circuits, programmable logic arrays, or ROMs.

Decimal	BCD	Excess-Three	Gray	Modified Gray
0	0000	0011	0000	0010
1	0001	0100	0001	0110
2	0010	0101	0011	0111
3	0011	0110	0010	0101
4	0100	0111	0110	0100
5	0101	1000	0111	1100
6	0110	1001	0101	1101
7	0111	1010	0100	1111
8	1000	1011	1100	1110
9	1001	1100	1101	1010
10 (A)	1010		1111	
11 (B)	1011		1110	
12 (C)	1100		1010	
13 (D)	1101		1011	
14 (E)	1110		1001	
15 (F)	1111		1000	

CONVERSION OF NUMBERS OF ONE BASE TO ANOTHER BASE

All positional-valued numbers have an integer part and a fractional part. The radix point, the decimal point for numbers of the base 10, separates the two parts. In the integer part the positional values given by the base are raised to a positive power;

the fractional part is related to the positional value of negative powers of the base. The zero power is immediately to the left of the radix point, as seen in the following lists of number systems.

Decimal

0 to 9	10^2	10^1	10^0 .	10^{-1}	10^{-2}	10^{-3}	
	100	10	1	0.1	0.01	0.001	Value in base 10

Binary

0 and 1	2^2	2^1	2^0 .	2^{-1}	2^{-2}	2^{-3}	
	4	2	1	0.5	0.25	0.125	Value in base 10

Octal

0 to 7	8^2	8^1	8^0 .	8^{-1}	8^{-2}	8^{-3}	
	64	8	1	0.125	0.0156	0.00195	Value in base 10

To convert integer numbers of one base to numbers of a smaller base, we divide the number by the smaller base and record the number and the remainder. The remainder yields the new number. The fractional part is converted by multiplying by the smaller base (which is the same as dividing by the base to the negative first power) and recording the overflow. The two portions are put together to yield the equivalent number in the new base.

Conversion from decimal to binary illustrates the algorithm. For this example we use 100.32_{10} and separate the integer part to find its equivalent.

$$100/2 = 50 + 0 \text{ LSB}$$
$$50/2 = 25 + 0$$
$$25/2 = 12 + 1$$
$$12/2 = 6 + 0$$
$$6/2 = 3 + 0$$
$$3/2 = 1 + 1$$
$$1/2 = 0 + 1 \text{ MSB}$$

The remainders listed from the bottom give the integer equivalent

$$100_{10} = 110\ 0100_2$$

The fractional part is multiplied by the base, and the fractional part of the product is used in the subsequent steps.

	Overflow	
$0.32 \times 2 = 0.64$	0	MSB
$0.64 \times 2 = 1.28$	1	
$0.28 \times 2 = 0.56$	0	
$0.56 \times 2 = 1.12$	1	
$0.12 \times 2 = 0.24$	0	
$0.24 \times 2 = 0.48$	0	
$0.48 \times 2 = 0.96$	0	
$0.96 \times 2 = 1.92$	1	
$0.92 \times 2 = 1.84$	1	
$0.84 \times 2 = 1.68$	1	
$0.68 \times 2 = 1.36$	1	
$0.36 \times 2 = 0.72$	0	LSB

The overflow is listed from the top to give the fractional equivalent.

$$0.32_{10} \approx 0101\ 0001\ 1110_2 = 0.319824219_{10}$$

The binary equivalent to $100.32_{10} = 110\ 0100 . 0101\ 0001\ 1110_2$. The calculation of the fractional part points out an important feature of converting one number system to another: in general, fractional values cannot be equated exactly with a limited number of bits; thus many bits may be required to reach the initial fractional value with an acceptable error.

Conversion from binary to decimal is achieved by summing the decimal equivalent of each bit according to its positional value.

BINARY-TO-BCD CONVERSION

A simple algorithm will convert binary numbers to binary-coded decimal (BCD) numbers, which then are decoded for display on seven-segment LEDs. Subroutines for computers and integrated circuits in DIPs, ROMs, and PLAs can make the conversion. We will illustrate the algorithm by converting $1101\ 0101_2 = 213_{10}$.

The first step is to examine the three most significant bits (MSB). If their value is 5_{10} or more, add the binary 011_2 (3_{10}) and shift left one bit. If their value is less than 5_{10}, merely shift left one bit. The next bit is added to the value from the first step,

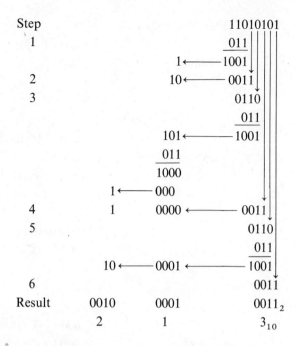

and the proccess is repeated. The same process is applied to the columns of bits derived by shifting to the left. When the least significant bit (LSB) is encountered, it is brought down, and the process is stopped.

BCD-TO-BINARY CONVERSION

The algorithm to convert BCD numbers to binary numbers requires right-shifting one bit at a time to generate the binary number, starting with the least significant bit. The bits shifted out form the binary number.

After a one-bit shift each new decade (BCD group) is examined. When a group is greater than $7 = 0111_2$, $3 = 0011_2$ is subtracted from it. Then the right shift gives the next bit. The new groups are inspected and 3 is subtracted again if necessary. The process continues until all high-BCD groups are zero and the rightmost group is less than $8 = 1000_2$. To give the binary number, the last group is copied with the bits generated earlier.

In the example below, subtraction of $3 = 0011$ is carried out by the two's-complement method.

Integrated circuits to convert BCD numbers to binary numbers are cascaded to handle BCD numbers of any size. The algorithm may be programmed for microprocessor-based systems, but hardware (integrated-circuit converters) is faster.

Computer Arithmetic

```
Step            Number      3      1     2₁₀    =    100111000₂
                           0011   0001  0010 BCD
 1 Shift                   0001   1000  1001                    → 0
 2 Subtract 3                     1100  1100
                                  0100  0101
                                     1     1
                           0001   0101  0110
 3 Shift                   0000   1010  1011                    → 00
 4 Subtract 3                     1100  1100
                                  0110  0111
                                     1     1
                           0000   0111  1000
 5 Shift                   0000   0011  1100                    → 000
 6 Subtract 3                           1100
                                        1000
                                           1
                           0000   0011  1001
 7 Shift                   0000   0011  1100                    → 1000
 8 Subtract 3                           1100
                                        1000
                                           1
                           0000   0001  1001
 9 Shift                   0000   0000  1100                    → 11000
10 Subtract 3                           1100
                                        1000
                                           1
                           0000   0000  1001
11 Shift                   0000   0000  0100                    → 111000
12 Copy last group                                              → 0100111000₂
```

SIGNED NUMBERS

Positive and negative numbers are represented by two parts, a sign and a magnitude. According to the convention, zero is used for a positive sign bit and one for the negative sign bit. The magnitude is $n - 1$ bits in length in signed numbers because the bit in the most significant position is the sign bit.

The three forms of signed binary numbers are signed-magnitude, signed-one's complement, and signed-two's complement.

The signed-magnitude binary numbers differ only in the sign bit. The positive zero is 0000 0000 and the negative zero in this case is 1000 0000. For example, 0000 1001 and 1000 1001 represent $+9$ and -9, respectively.

In a one's-complement number, a byte (eight bits) with the bit in the most significant position as the sign bit can represent 0_{10} through 127 positive values (0000 0000_2 to 0111 1111, or 00_{hex} to $7F_{hex}$) and 128 negative values (-0_{10} to -128, 1111 1111_2 to 1000 0000, or FF_{hex} to 80_{hex}). The negative numbers are the one's complements of the positive numbers. There are two zeros, one positive (0000 0000_2) and one negative (1111 1111_2). The one's complement is found by complementing each bit.

In Chapter 6, Table 6-4, program step EE gives the value $F5_{hex} = -11_{10}$ as the negative number which, when added to the program counter, produces an earlier address to which the program should jump in the iteration. A similar one's complement number is used in Table 6-5, step 01 EB, for the same purpose.

We showed in Chapter 3 that subtraction of a binary number is carried out readily by forming its two's complement, but we only illustrated the subtraction of a smaller positive number (subtrahend) from a larger positive number (minuend) to yield a positive difference. We now wish to consider the other possible combinations.

The two's complement form has only one zero, which is positive, $+0_{10} =$ 0000 0000_2. The two's complement of a binary number is formed by adding one in the least significant position to the one's complement; that is, the one is added to the number whose ones have been changed to zeros, and vice versa. The two's complement of zero is

$$\begin{array}{r} \overline{0000\ 0000} = 1111\ 1111 \\ 0000\ 0001 \\ \hline 1\ 0000\ 0000 \end{array}$$

The carry is discarded, making the two's complement of zero the same zero.

We will use two numbers, $A = 107_{10}$ and $B = 28_{10}$, with the two's complement to find

$$A + B = S$$

and

$$A - B = D$$

The two's complement is used in the examples that follow to express negative values of A and B where each may be positive or negative.

A and $B > 0$

A	0110 1011	107
B	0001 1100	28
S	1000 0111	135

The carry makes the sum appear to be negative, but it is recognized as an overflow, and the sum is produced as $S = 0\ 1000\ 0111$; the leftmost bit carries the sign.

A and $B < 0$

A	1001 0101	-107
B	1110 0100	-28
S	1 0111 1001	-135

Two possibilities exist in adding negative numbers. If the sign bit is 1, it indicates that the answer is correct. If the sign bit is 0, it indicates an overflow, as in this case. The signed magnitude of the sum is obtained by taking the two's complement of the number while retaining the sign bit.

	1 1000 0110	
	0000 0001	
S	1 1000 0111	-135

The addition of two small negative numbers will illustrate the case in which the sign bit is one and there is no overflow.

A	1111 1101	-3
B	1111 1000	-8
S	1 1111 0101	-11_{10}

The one in the leftmost position comes from the sign bits; it is discarded because the correct sign bit as been generated. The signed magnitude of the sum is

1000 1010	
0000 0001	
1000 1011	-11_{10}

$A > 0$ and $B < 0$

A	0110 1011	$+107$
B	1110 0100	-28
D	1 0100 1111	$+79$

The carry is generated if the difference is positive; it is discarded.

$A < 0$ and $B > 0$

A	1001 0101	-107
B	0001 1100	$+28$
D	1011 0001	-79

The sign bit is correct; no carry is generated. The signed magnitude of D is

	1100 1110	
	0000 0001	
D	1100 1111	-79

Table 6-7 shows that an arithmetic logic unit (ALU) does not generate a carry bit when $A < B$ in comparing magnitudes.

DECIMAL ARITHMETIC

In arithmetic operations, computers may use decimal numbers rather than converting them to binary numbers. The decimal numbers are used in binary-coded decimal (BCD) form, so that after an operation such as addition, the sum must be examined and corrected if it does not lie in the range 0000 to 1001 of BCD numbers. A number outside this range is corrected by subtracting $1010_2 = 10_{10}$ and adding that value to the next higher BCD group. Subtraction may be carried out by adding the two's complement of 1010_2, which is $0110_2 = 6_{10}$.

For example, we add the two numbers.

$$
\begin{array}{rcccc}
365 & 0011 & 0110 & 0101 \\
578 & 0101 & 0111 & 1000 \\
\hline
    & 1000 & 1101 & 1101 \\
    &   &   & 0110 \\
    &   & 1   & \overline{0011} \\
    &   & \overline{1110} & \\
    &   & 0110 & \\
    & 1   & \overline{0100} & \\
\hline
943 & 1001 & 0100 & 0011 \\
\end{array}
$$

Subtraction of a BCD number uses the nine's complement of the number. The nine's complement is obtained by inverting all bits, adding $1010_2 = 10_{10}$, and discarding the carry. For example,

$$
\begin{array}{rl}
\overline{3} = \overline{0011} = & 1100 \\
                                 & 1010 \\
\hline
                                 & 1\ 0110 \\
\end{array}
$$

The nine's complement of the BCD numbers are simply the values in the reverse order.

$$\overline{0000} = 1001$$
$$\overline{0001} = 1000$$
$$\overline{0010} = 0111$$
$$\overline{0011} = 0110, \text{ and so forth}$$

The decimal adjustment is required in subtraction as it is in addition, but a borrow is required. Microprocessors have decimal-adjust machine-language commands, which are used when decimal addition and subtraction take place. The decimal numbers are held with two BCD values in each byte. The format is called *packed BCD*.

DOUBLE-PRECISION ARITHMETIC

Some calculations cannot be made satisfactorily with 8 or 10 decimal digits for each number. Matrix inversions, which require numerous multiplications and divisions, and scientific computations, which entail finding the difference between two large numbers, become meaningless if an insufficient number of digits is used. Double-precision arithmetic is made possible by programming the arithmetic operations to use twice the number of digits for each number. In many computers a simple command will invoke double- or even triple-precision arithmetic.

A double-precision number is used by separating the upper half (the most significant part) and the lower half. In a 20-digit number the upper 10 digits are the most significant because they represent their value multiplied by 10^{10}. Let A and B be the double-precision numbers, $A = A_1 \times 10^{10} + A_2$ and $B = B_1 \times 10^{10} + B_2$. The product AB can be written

$$AB = A_1 B_1 \times 10^{20} + A_1 B_2 \times 10^{10} + A_2 B_1 \times 10^{10} + A_2 B_2$$

Because A_1, A_2, B_1, and B_2 are each 10-digit numbers, the product AB will be a 40-digit number, of which 20 digits will be retained for the next arithmetic operation. The product $A_2 B_2$ is not calculated because it makes a contribution only in rounding the least significant digit in the 20-digit product.

Double-precision division uses the algorithm

$$\frac{A}{B} = \frac{A_1 \times 10^{10} + A_2}{B_1 \times 10^{10}} \cdot \frac{1}{1 + \dfrac{B_2}{B_1 \times 10^{10}}}$$

$$\approx \frac{A_1 \times 10^{10} + A_2}{B_1 \times 10^{10}} \left[1 - \frac{B_2}{B_1 \times 10^{10}} \right]$$

The 20-digit quotient will be accurate to within the least significant digit.

The subtraction of double-precision numbers

$$A - B = (A_1 \times 10^{10} + A_2) - (B_1 \times 10^{10} + B_2)$$
$$= (A_1 - B_1) \times 10^{10} + (A_2 - B_2)$$

becomes more complicated if $(A_1 - B_1)$ and $(A_2 - B_2)$ have different signs. In that case $(A_2 - B_2)$ is complemented and added or subtracted, depending upon whether $(A_1 - B_1)$ is negative or positive, to form the difference.

Because some computers allow words (numbers) of any length, double-precision algorithms are not required for them. The arithmetic coprocessors extend the arithmetic ability of microprocessors by allowing word lengths of 64 or even 80 bits to be used without extensive programming. Division with 18-decimal digits can be carried out in about 40 μsec with a coprocessor, in contrast to about 3200 μsec with double-precision division algorithms programmed in software for the microprocessor.

WALLACE DIVISION ALGORITHM

The quotient, $Q = D/d$, can be found by subtracting the divisor, d, from the dividend, D, while counting the number of times, Q, until an underflow occurs. The divisor is added back to give the remainder. This method is unacceptable for high-speed computers because of the large number of steps required for large dividends.

Wallace's algorithm (Wallace, 1964) substitutes multiplication for division by finding the reciprocal of the divisor in a very rapidly converging process. The quotient is the product of the dividend and the reciprocal of the divisor.

The first step normalizes the divisor; that is, both the dividend and the divisor are shifted until $0.5 \leq d \leq 1.0$. For binary numbers this step places the most significant bit of the divisor just to the right of the radix point. An estimate of the reciprocal, $p \approx 1/d$, is obtained from a lookup table or from a logic-gate array. (To try the algorithm, assume $p = 1/\langle d \rangle$, where $\langle d \rangle$ is 0.75; thus $p = 1.333$. $\langle d \rangle$ is the mean value of the normalized divisor.) The Wallace algorithm can be written

$$
\begin{aligned}
b_0 &= p \\
a_0 &= b_0 d = pd \\
b_1 &= b_0(2 - a_0) \\
a_1 &= a_0(2 - a_0) \\
&\vdots \\
b_{i+1} &= b_i(2 - a_i) \\
a_{i+1} &= a_i(2 - a_i)
\end{aligned}
$$

It is found that b_i converges quadratically to $1/d$, and a_i converges quadratically to 1.0. By finding p from lookup tables or logic arrays so that $|1 - pd| < \frac{1}{32}$, a 40-bit reciprocal can be obtained in three iterations.

WALLACE SQUARE ROOT ALGORITHM

The square root of a positive number may be found by the method shown in Chapter 6. There we programmed an iteration by using an algorithm based upon the algebraic relation

$$N = (x + a)^2$$

to find $x \approx N^{1/2}$. The square root may also be found by a variant of the Wallace division algorithm. For a positive number N whose root is to be found, an estimate of $1/N^{1/2}$ is made. $N/2$ is used in the first iteration. The algorithm is

$$\begin{aligned} p &\approx 1/N^{1/2} \\ b_1 &= p \\ a_1 &= p^2 N/2 \\ &\vdots \\ b_{i+1} &= b_i(1.5 - a_i) \\ a_{i+1} &= a_i(1.5 - a_i)^2 \end{aligned}$$

Both b_i and a_i converge quadratically to $1/N^{1/2}$ and 0.50, respectively. A lookup table for estimating p can shorten the calculation to a few iterations. The Wallace algorithms become especially attractive when the terms b_i and a_i can be calculated in parallel processors.

REFERENCE

1964 Wallace, C. S., "A Suggestion for a Fast Multiplier," *IEEE Transactions on Electronic Computers, EC13*, **14** (1964).

APPENDIX B

LABORATORY EXERCISES

CONTENTS OF APPENDIX B

Exercise Number | **Exercise Title**

Exercise Number	Exercise Title
1	Basic Logic Gates
2	Logic-Gate Synthesis
3	Logic Functions
4	The Half Adder
5	The Full Adder
6	*RS* Flip-Flops
7	The Master-Slave *RS* Flip-Flops
8	The *JK* Flip-Flop
9	Digital Counters
9 Supplement	A Seven-Segment LED Display
10	Synchronous Counter and Word Timer
11	Shift Registers
12	A Serial Adder
13	An Elementary Memory Cell
14	Random Access Memory
15	Diode Characteristics
16	Common-Emitter Amplifiers
17	Multiplexing
18	Operational Amplifiers
19	Operational Amplifier Characteristics
20	Voltage Comparators
21	Summing Amplifiers
22	Peak-Voltage Detection
23	An Integrator
24	The Schmitt Trigger
25	Active Filters
26	MOSFET Electronic Switches
27	Signal Sampling
28	A Rate Meter
29	Interfacing
30	Voltage and Current Regulation

Student's Guide to Exercises

Data:
Seven-Segment Displays
Resistor Color Code

LABORATORY EXERCISE 1
Basic Logic Gates

All digital circuits involve the binary logic provided by AND-gates, OR-gates, and inverters. This exercise reveals and verifies the basic logic functions satisfied by these elements. The gates in dual in-line packages (DIPs) can be used on a solderless breadboard with electrical power, and temporary connections may be made with short lengths of wire.

AND-gates, OR-gates, and Inverters

The properties of logic gates are summarized in a tabulation of the output state for each combination of input states. The tabulations are called truth tables. Table B-1 is the truth table for all two-input AND-gates and all two-input NAND-gates. The difference between an AND-gate and a NAND-gate is that the output of the NAND-gate is the complement (the inverse) of the output of the AND-gate.

The truth table will be verified during the exercise. The DIPs used in this exercise and shown in the circuits are NAND-gates and NOR-gates. Boolean algebra can be satisfied equally with AND-gates or NAND-gates and with OR-gates or NOR-gates. That is, a gate or its complement may be used to solve binary logic equations.

Gate Logic Verification

- First we verify the operation of a two-input NAND-gate by measuring the output state for each combination of input states. Make a list of the input voltages and the output voltages; then list the input states and the output states as 0s and 1s. Both lists should be similar to Table B-1. Note that the voltage corresponding to the 0 state may not be 0 V and that the voltage corresponding to the 1 state may not be $+5.0$ V. Is it clear from the voltages you find that the states (0) and (1) are unambiguous?
- Connect an inverter to the output of the NAND-gate to make it into an AND-gate. Repeat the measurements to make the two lists, one with voltages and the other with 0s and 1s. Compare the results with Table B-1.

TABLE B-1
Truth Table for Two-Input AND-gates and NAND-gates

Input A	Input B	AND Output	NAND Output
0	0	0	1
0	1	0	1
1	0	0	1
1	1	1	0

- Replace the NAND-gate with a NOR-gate and repeat the steps to find the truth table for the NOR-gate. Then use an inverter to make an OR-gate and find its truth table. Note that the input and output pins on the NOR-gate DIP are not the same as on the NAND-gate DIP.

- Place a resistor between the output line of either the NAND-gate or the NOR-gate to ground on the breadboard and measure the voltage at the output. Use resistor values of approximately 1.0, 0.5, and 0.1 kΩ. (In most cases exact values of resistance in electronic circuits are not required to ensure proper operation.) Plot output voltage versus resistance. Does a value of resistance exist for which the current that passes through it loads the output so that a high state at the output may be ambiguous? The Resistance Color Code, in the Data section following the exercises, enables you to identify the values of resistors.

Use the data in your truth tables to find the truth table for an inverter.

Experimental Details

Power connections to the DIPs are not usually shown in the wiring diagrams for logic-gate circuits in order to reduce the complexity of the diagrams and allow the logic signals to be followed easily. The logic gates will NOT work, however, without the power connections, so the connections must be made as you wire the circuit. The signals and power-supply voltages used with the DIPs must not exceed +5.0 V. Higher voltages may destroy the integrated circuits in the DIPs.

The solderless breadboard provides the voltages to power the DIPs and for the logic signals. The voltmeter measures the input and output states when a wire is brought from it to the pin of the DIP. The DIPs are to be inserted carefully into the breadboard and short lengths of thin wire are to be used to make connections from the ground and voltage sources for power and input signals. As you add and connect elements, check their operation. Some DIPs may have been damaged and will not operate properly, or you may have made an error in your wire connections.

In Figure LE-1, the small circles adjacent to the element mean that the complement (inverse) of the logic-gate output will be provided. The symbol for the NAND-gate has a circle on the output line at the symbol, which means that the complement of the output of the AND-gate will occur. The small circles on the ends of the lines merely indicate connection points and do not have any other significance.

In logic-gate diagrams, wires that cross are NOT connected unless there is a dot on the intersection.

Points at Issue

Examine the truth tables to find how two-input NAND-gates and two-input NOR-gates may be made into inverters. Are there different ways of doing this?

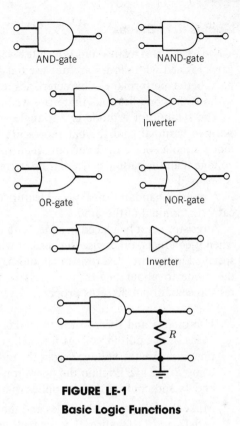

FIGURE LE-1
Basic Logic Functions

Components

7400 Quad two-input NAND-gate
7402 Quad two-input NOR-gate
7404 Hex inverter
Assorted $\frac{1}{4}$ W resistors

LABORATORY EXERCISE 2
Logic-Gate Synthesis

In this exercise we use the simplest electronic elements, diodes, to synthesize logic gates. The possibilities and the restrictions in the synthesis are illustrated. We also construct circuits with DIPs made with open-collector gates. The open-collector gates are integrated circuits that can have their output lines connected together, a practice not allowed with other kinds of gates.

AND-gates and OR-gates from Diodes

The earliest AND-gates and OR-gates were made with resistors and diodes. The gates formed with diodes are not used at present because integrated-circuit gates have better performance characteristics and may be placed in logic-gate circuits without the complications that arise when diode-made gates are interconnected.

The symbol for a diode is a triangle and a bar. The triangle is the anode, or positive terminal, and the bar represents the cathode, or negative terminal. The diode conducts electrical current when the anode is at a more positive (higher) voltage than the cathode. There is no current through the diode when the polarity is reversed.

A painted band or other special feature (a short lead or a flat area) indicates the cathode terminal of the diode.

A resistor must be placed in series with the diode to limit the current through it when it conducts; otherwise the current will destroy the diode. Unless a resistor is specified in the circuit, a resistor should be selected that limits the current through the diode to about one-third or less of the rated current. In 5.0-V systems a resistance of about 500 Ω is proper.

- Use diodes and the resistor to make the AND-gate shown in Figure LE-2. Measure the output voltage for each combination of input states. At the same time measure the voltage across the diodes and record them in the truth table, using $R = 2$ kΩ. Explain the operation of the gate. Make a truth table with 0s and 1s and confirm that it applies to an AND-gate.
- Make the OR-gate with diodes and the resistor and prepare a truth table for it with 0s and 1s. Confirm that the truth table applies to an OR-gate. Explain the operation of the circuit. Note that the polarity of the diodes is reversed from that in the AND-gate circuit.

Open-Collector Gates

The term "open-collector" simply means that the connection to the collector (a part of the transistor in the output circuit of the integrated circuit) has not been completed as usual in the integrated circuit within the DIP. An external resistor must be connected between the supply voltage and the output pin on the DIP to allow current to pass through the transistor.

Many different integrated circuits are available with open-collector outputs. They are made this way to allow greater currents in the output circuits for illuminating digital displays or to permit the interconnection of the output stages of logic gates. Chapter 3 discusses the problem of interconnected output stages.

Wired OR-gate with Open-Collector Inverters

The interconnection of the inverters shown in the circuit diagram (Figure LE-2) is allowed because they are open-collector (OC) inverters. It must be noted that there

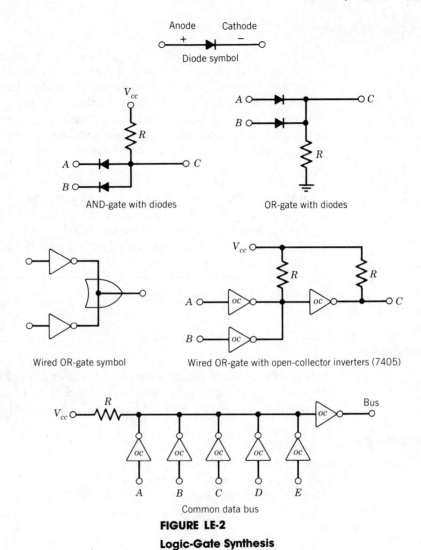

FIGURE LE-2
Logic-Gate Synthesis

is a difference between the wired OR-gate and the OR-gate made with diodes because the electronic properties of the integrated circuits make them superior to diodes. The symbol for the wired OR-gate shows the leads entering the sides of an OR-gate. No OR-gate is present because the symbol stands for the circuit made only with resistors and OC inverters.

- Make the circuit shown in Figure LE-2, using about 500-Ω resistors. Verify that the circuit obeys OR-gate logic by constructing a truth table.

Common Data Bus

Binary data may arise from many different sources in a computer or control system, and they must be transmitted. A single bus line is used in order to avoid wiring complexities. The common data bus in the circuit diagram allows bits (0 or 1) to be transferred to the bus from the single source, A, B, C, D, or E, that is enabled to provide data. All inputs except the data source must be held off; their inputs must be 0 or else the source of data will be uncertain. Data-source priority and sequencing must be part of the system for using a common data bus.

- Construct the common data-bus circuit and verify that the output is high (1) when each input line is raised to 1 and when the others are kept at 0. The resistor R is the resistor required for the OC inverter with the input, but the resistor for the OC inverter at the bus is not shown in the diagram. Compare the signal on the bus without a resistor to the signal with the required resistor in place. Use $R = 470\,\Omega$ for each resistor.

Points at Issue

Is it possible that the common data-bus circuit is a wired OR-gate? Answer this question by making a truth table for combinations of inputs. Does this table suggest that a wired OR-gate can have more than two inputs?

Components

1N941 or similar diodes
7405 Open-collector Hex inverter
Assorted $\tfrac{1}{4}$ W resistors

LABORATORY EXERCISE 3
Logic Functions

Logic gates may be formed in circuits that satisfy the algebraic equations of digital logic. The equations that satisfy the axioms of Boolean algebra are called logic functions. The gate arrangements duplicate exactly the digital logic required for arithmetic, control, and other operations. This exercise provides experience in connecting gates and verifying their logic.

Logic Function $ABC + \bar{A}BC + A\bar{B}C + AB\bar{C} = X$

The logic function X is very useful because it pertains to binary addition; in fact, it is the carry-out bit. A and B are the two bits being added, and C is the carry-in from the addition of the next least significant bits, that is, the bits just to the right of A and B. In a later exercise, an adder circuit will illustrate how the carry-out bit X is generated.

- Construct the circuit for X. Make a truth table for the circuit and verify that the logic function is satisfied exactly. In making the truth table, show the output state of each gate throughout the circuit (Figure LE-3). The independent variables are A, B, and C, so that there will be eight lines in the truth table.

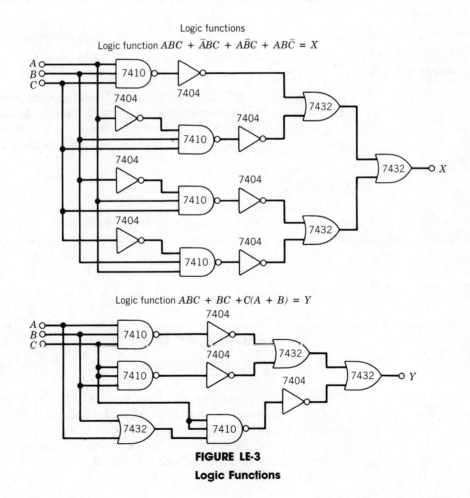

FIGURE LE-3
Logic Functions

Logic Function $ABC + BC + C(A + B) = Y$

The logic function Y is not related to binary arithmetic. Before you construct the circuit for Y, draw the logic gate arrangement that you would use to satisfy the logic function Y.

- Make your circuit or the one shown in Figure LE-3. Find the truth table for Y.

Experimental Details

As the circuits become more complicated it is necessary that the wire connections to the DIPs on the solderless breadboard be made in an orderly and systematic manner. Plan the layout and simplify the connections that are common to many gates, such as power lines. Your wiring must be transparent for analysis and troubleshooting.

Points at Issue

Both logic functions X and Y can be satisfied by fewer logic gates. Find the reduced logic functions, that is, simpler logic equations that satisfy the logic functions X and Y identically. Make a Karnaugh map for each function and use the technique shown in Chapter 2 to find the reduced logic functions for X and Y.

Components

7404 Hex inverter
7410 Triple three-input NAND-gate
7432 Quad two-input OR-gate (or 7402 with 7404)

LABORATORY EXERCISE 4
The Half Adder

Arithmetic in computers requires very rapid addition of binary bits. Successive addition is equivalent to multiplication, and successive subtraction is a means of dividing one number by another. A half adder is a circuit that adds two bits and gives the sum bit and a carry-out bit, but it will not accept a carry-in bit. A logic element called an exclusive OR-gate (XOR-gate) is used in the circuits for adding binary numbers.

Exclusive OR-gates

Several configurations of basic logic gates can yield the Boolean algebra for the exclusive OR-gate. Two different XOR-gates will be made to show that there may be several ways of satisfying a logic function identically. The first implementation of the XOR-gate can be made with a single DIP containing four two-input NAND-gates.

- Make this XOR circuit. Measure and record the state on each intermediate gate as you find the truth table.

The second implementation of the XOR-gate uses different basic gates. In later exercises a single DIP with four XOR-gates will be used rather than the circuits shown in Figure LE-4. Use the logic equation for each of the gates and the Boolean algebra for the second configuration to show that XOR logic is formed.

- Make the second XOR-gate.

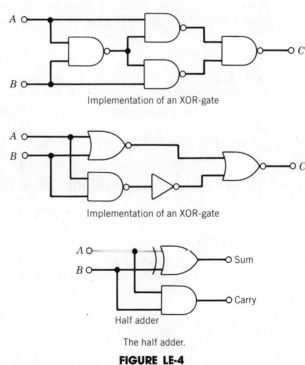

Implementation of an XOR-gate

Implementation of an XOR-gate

Half adder

The half adder.

FIGURE LE-4

The Half Adder

An XOR-gate satisfies the logic equation

$$C = A\bar{B} + \bar{A}B$$

which is also written

$$C = A \oplus B$$

The XOR-gate has a high (1) output if one input or the other is high (1), but a low (0) output if both inputs are low (0) or high (1).

Equivalence Gate

An inverter placed on the output of an XOR-gate complements the output, and the circuit becomes an equivalence gate or XNOR-gate. The XNOR-gate provides a high (1) when both inputs are the same or equivalent. That is, the output is one if both inputs are zeros or both are ones.

The XNOR-gate satisfies the logic equation

$$C = AB + \overline{AB}$$

which is also written

$$C = A \odot B$$

There are many applications for an equivalence gate or a bit comparator in digital logic. Can you suggest what some of these applications might be?

Half Adder

Table B-2 is the truth table for a half adder. A half adder adds the two least significant bits (LSB) to yield a sum bit and a carry-out bit. The LSBs are the rightmost bits in a binary number. In the truth table C is the carry-out and S is the sum.

An examination of Table B-2 shows that

$$S = A \oplus B$$

and that

$$C = AB$$

TABLE B-2
Truth Table for a Half Adder

A	B	C	S
0	0	0	0
0	1	0	1
1	0	0	1
1	1	1	0

TABLE B-3
Truth Table for a Half Subtractor

A	B	⟨B⟩	D
0	0	0	0
0	1	1	1
1	0	0	1
1	1	0	0

Half Subtractor

Table B-3 is the truth table for a half subtractor. In subtracting one bit from another, it may be necessary to borrow from the next higher bit pair, as when the bit being subtracted is larger (1) than the bit (0) from which it is being subtracted. In the truth table, B is being subtracted from A. $\langle B \rangle$ denotes the borrow and D is the difference, $D = A - B$.

An examination of Table B-3 shows that

$$D = A \oplus B$$

and that

$$\langle B \rangle = \bar{A}B$$

The sum and difference are satisfied by the same gate logic but the carry and borrow require different gate logic.

- Make the half adder and the half subtractor and confirm the arithmetic accuracy of their operation. Use either one of the XOR-gates that you made in the earlier part of the exericise.

Points at Issue

You will notice that the borrow $\langle B \rangle$ is generated when $A < B$. Use several four-bit numbers, such as 0110, 1010, 1001, and others, to find the sums and the differences. Can you draw a general conclusion about $\langle B \rangle$ and the relative magnitudes of A and B?

Components

7400 Quad two-input NAND-gate
7402 Quad two-input NOR-gate
7404 Hex inverter

LABORATORY EXERCISE 5
The Full Adder

A full adder adds two bits and the carry-in bit to produce a sum bit and a carry-out bit. It must be used for adding bit pairs to the left of the least significant bits. (There are three input bits and two output bits.) A full adder may be made with two half adders. Several bit pairs may be added by cascading full adders. In this exercise an adder for two two-bit numbers will be made by combining a half adder with a full adder.

Boolean Algebra of a Full Adder

The logic function or logic equation can be found from a truth table in which the variables and their complements, rather than zeros and ones, are written. In Table B-4 the variables A and B are the bits being added, and Z is the carry-in bit. The sum bit S and the carry-out bit C result from the addition.

The Boolean algebra equations for C and S are found by equating the terms in each row that give them. For example, rows 1, 2, 4, and 7 each gives S. In row 1 $\bar{A}\bar{B}Z = S$, in row 2 $\bar{A}B\bar{Z} = S$, and so forth. Using + for the Boolean algebra equivalent of OR, we find that S is given by

$$S = \bar{A}\bar{B}Z + \bar{A}B\bar{Z} + A\bar{B}\bar{Z} + ABZ$$

Similarly, since rows, 3, 5, 6, and 7 each yields C, we may write

$$C = \bar{A}BZ + A\bar{B}Z + AB\bar{Z} + ABZ$$

The equation for C was used in an earlier exercise. It can be reduced to

$$C = AB + AZ + BZ$$

TABLE B-4
Truth Table for a Full Adder

Row	A	B	Z	C	S
0	$\bar{A}$	$\bar{B}$	$\bar{Z}$	$\bar{C}$	$\bar{S}$
1	$\bar{A}$	$\bar{B}$	Z	$\bar{C}$	S
2	$\bar{A}$	B	$\bar{Z}$	$\bar{C}$	S
3	$\bar{A}$	B	Z	C	$\bar{S}$
4	A	$\bar{B}$	$\bar{Z}$	$\bar{C}$	S
5	A	$\bar{B}$	Z	C	$\bar{S}$
6	A	B	$\bar{Z}$	C	$\bar{S}$
7	A	B	Z	C	S

An economy of logic gates and time is attained by forming the reduced functions rather than the full expressions.

In the Karnaugh map for S there are no adjacent terms that can be combined to eliminate a common variable. Although Karnaugh map simplification cannot give a reduced equation for S, there is an equivalent equation for S, which is

$$S = Z \oplus (A \oplus B)$$

Chapter 3 shows how the reduced equation for S is found.

The equation reads S equals Z, XORed with the term formed by A XORed with B. This equation suggests that a full adder may be made with two half adders, each of which involves an XOR-gate. (See Figure LE-5.)

Parallel Adder for Two-Bit Pairs

Numbers in binary form consist of strings of bits. Two binary numbers are added by finding the sum and carry-out for each successive column of bits. Each bit pair except the LSB (least significant bit) requires a full adder because a carry-in may be generated in the previous step of addition. In this exercise, however, we use a half adder as the first stage because we assume that the first bits are LSBs.

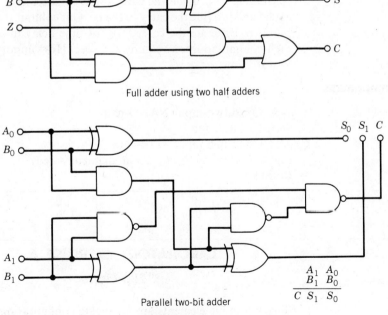

Full adder using two half adders

Parallel two-bit adder

FIGURE LE-5
The Full Adder

- Make a full adder by combining two half adders. Prepare a truth table for it by measuring C and S for each combination of A, B, and Z.
- Make the adder to find the sum and carry-out for two two-bit numbers, and write the truth table for it. Note that the carry generated in summing the LSBs is not involved externally. Can you find it? Show some examples of adding two two-bit numbers and compare them with the sum and carry from the adder you made.

Experimental Details

You may wish to use LEDs (light-emitting diodes) to show the state of each input and output. LEDs protected with a current-limiting resistor in series may be placed between the DIP pins and ground. They will be lighted when the state at the pin is high. A resistor value of about 500 Ω is suitable. LEDs have a short lead or a flat area to indicate the cathode lead to be placed at ground.

Points at Issue

Draw the logic-gate circuit that will allow two three-bit numbers to be added. Can your arrangement be extended to numbers of any bit size?

In this circuit the first carry had to be generated to be used in finding the sum and carry for the second pair. When each number has many bits, it may be time-consuming to provide the carry for each successive bit column so that it can be added and provide the carry to the next bit column. This is called the carry-forward problem. Estimate the carry-forward time delay in adding two 20-bit numbers if each column addition requires 0.1 μsec. How important do you think the carry-forward time delay is?

Components

7400 Quad two-input NAND-gate
7404 Hex inverter
7486 Quad two-input XOR-gate
Light-emitting diodes (0.125 in. diameter—red)
Resistors

LABORATORY EXERCISE 6

RS Flip-Flops

Binary bistable elements are logic-gate configurations that may change from one state to another when they receive an input. The name "flip-flop" describes the behavior of these circuits.

The gate arrangements in the earlier exercises are combinational circuits whose output is determined only by the input states and the logic equations they satisfy. Now we prepare to examine gate arrangements that are sequential logic systems in which the output state is determined by the previous state and the new input. An example of a sequential logic system is a scaler or counter whose content depends upon the sequence of earlier events.

Because a flip-flop placed in a desired state, high (1) or low (0), remains in that state until changed, the bistable elements are binary storage devices or digital memories.

An *RS* Flip-Flop

The name "*RS* flip-flop" comes from the property of the circuit to be reset into the low state (0) or set into the high state (1). One of the output terminals of a flip-flop is Q. The state of the flip-flop is the value of Q, namely, (0) or (1). The output of the other terminal is the complement of Q, called Q-bar, $\bar{Q}$. Both Q and $\bar{Q}$ are provided at pins on DIPs that contain flip-flops (Figure LE-6).

Basic *RS* flip-flops can be made with NAND-gates or with NOR-gates. This exercise shows that a fundamental difference exists between the NAND-gate flip-flops and the NOR-gate flip-flops because the truth tables for NAND-gates and NOR-gates are different. The difference requires that the input states be appropriate for the kind of flip-flop used. In commercially available *RS* flip-flops this is not a problem because of additional logic gates in the circuit. Later exercises will show how the additional gates solve the problem.

Unallowed input states exist in the basic *RS* flip-flops. Simultaneous set and reset signals, for example, require that the output state Q be both high (1) and low (0). It is uncertain which state will prevail when the signals are removed.

- Use the truth table for the NOR-gate and assume states for Q in the first basic flip-flop circuit; then use the possible values for R and S to determine the new value for Q (it may be the same or different). Now make the circuit and confirm its operation.
- Repeat the instructions in the previous paragraph for the NAND-gate basic flip-flop.

Clocked *RS* Flip-FLop

The usefulness of *RS* flip-flops is enhanced by adding logic gates on the R and S lines. The added gates allow an enable signal to be involved in the change of state of the flip-flop. The enable signal may be called a clock pulse (*CP*), even if it is not a reoccurring signal.

An advantage of the clocked flip-flop is that the input state on R and S may change, but unless there is also an enable pulse (*CP*), the flip-flop does not respond. The state Q is latched or preserved; the clocked flip-flop is an embryo digital memory element.

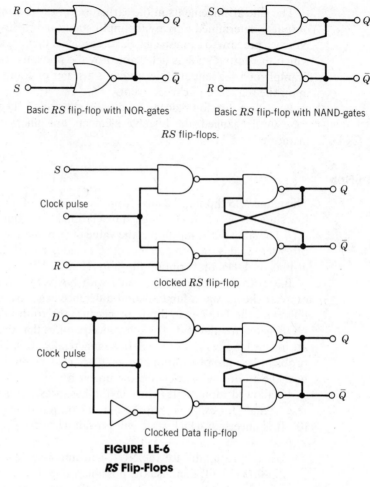

Basic RS flip-flop with NOR-gates

Basic RS flip-flop with NAND-gates

RS flip-flops.

clocked RS flip-flop

Clocked Data flip-flop

FIGURE LE-6
RS Flip-Flops

The basic flip-flops will be recognized as combinational circuits because the output state occurs in accordance with the gate logic and the input states. In the clocked flip-flop we find an initial state and a new state following the clock pulse. The new state, even if it is the same as the initial state, represents a sequential state. We recognize this new aspect of sequential circuits by designating the initial state by Q, as before, and the sequential state by $Q(t + 1)$.

- Make the clocked RS flip-flop and find the truth table for it. Because the clock or enable pulse is not an input state, it is not listed in the truth table. The sequential nature of the flip-flop is indicated, however, by the designation of the final state as $Q(t + 1)$. The truth table should have the following headings:

$$Q \quad S \quad R \quad Q(t + 1)$$

- Make a Karnaugh map from the truth table and use it to find the characteristic function for the clocked *RS* flip-flop.
- Add an inverter as shown in the circuit to change the clocked *RS* flip-flop into a data flip-flop. Find the characteristic function for the data flip-flop. Verify that the state Q is latched and does not follow D until there is a clock pulse. This circuit is sometimes called a data latch.

Points at Issue

Do the logic gates added to the basic *RS* flip-flop to make it a clocked *RS* flip-flop remove the possibility of unallowed inputs or the resultant indeterminate state?

Components

7400 Quad two-input NAND-gates
7402 Quad two-input NOR-gates

LABORATORY EXERCISE 7
The Master-Slave *RS* Flip-Flop

An important modification of the clocked *RS* flip-flop is the master-slave *RS* flip-flop. Master-slave flip-flops are used almost exclusively in digital circuits. DIPs contain master-slave flip-flops because they allow synchronization among all the components and are equally useful for all other purposes (Figure LE-7).

A Master-Slave *RS* Flip-Flop

The master-slave *RS* flip-flop is the first example of an element that involves a two-step sequential transfer of input states to the output Q. The configuration requires a fully completed clock pulse to advance the datum through the two-stage flip-flop.

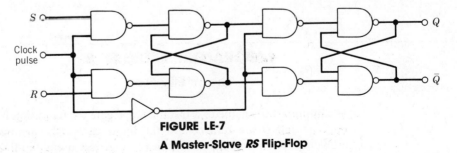

FIGURE LE-7
A Master-Slave *RS* Flip-Flop

The first stage accepts the input when the clock pulse goes high and places the state at the inputs of the second stage. The second stage cannot respond to its inputs until the clock pulse goes low; at this time the inverter in the clock line to the second stage causes the clock pulse to the second stage to go high. This change enables the second stage to respond to its inputs from the first stage.

- Make the master-slave *RS* flip-flop, and confirm that it requires a full clock pulse to transfer the input state to the output. Note that the output is the state $Q(t + 1)$ and that the intermediate states and the clock pulse are not listed in a truth table.

From the truth table find the characteristic function for the master-slave *RS* flip-flop. Is it the same as the one found in the previous exercise for the clocked *RS* flip-flop?

- Add an inverter on the input line to convert the master-slave *RS* flip-flop to a master-slave data flip-flop. Verify the sequential transfer of the datum from the first to the second stage. Find the characteristic equation for the *D* flip-flop. Is it the same as you found in the previous exercise?

Points at Issue

Does the master-slave arrangement for the *RS* flip-flop remove the problem of unallowed inputs?

The master-slave data flip-flop is also called a gated data latch. Suggest some uses for it.

Components

7400 Quad two-input NAND-gates
7404 Hex inverter

LABORATORY EXERCISE 8

The *JK* Flip-FLop

JK flip-flop is the name of a bistable logic gate configuration in which the output states Q and $\bar{Q}$ are returned to the input gates. The arrangement defeats the problem of unallowed inputs and results in a very useful flip-flop. *JK* flip-flops are widely used in digital circuits as integrated circuits in DIPs. Here we make them with elementary logic gates to find their properties and to learn the advantages of the JK flip-flop master-slave configuration.

The JK Flip-Flop

The basic JK flip-flop is made by returning the outputs of a master-slave RS flip-flop to the input (Figure LE-8). The result is a very useful flip-flop; the unallowed inputs are eliminated because the output signals of the second stage alter the input gate configuration. That is, the output signals of the second or "slave" stage are returned to the first or "master" stage, making the inputs depend upon Q and $\bar{Q}$.

- Make the master-slave JK flip-flop and find the truth table for it. Use the truth table and a Karnaugh map to find the characteristic function of the JK flip-flop. There is an unnecessary term in the characteristic function. Explain the reasoning that allows the unnecessary term to be dropped.

The truth table shows values for J, K, and Q that cause $Q(t + 1)$ to acquire a high or a low after a clock pulse. However, on all flip-flop DIPs there are pins that allow the state of the flip-flop to be placed initially either high or low. The pins are marked preset (the same as set) and clear (reset). Preset makes $Q = 1$, and clear causes $Q = 0$ independently of a clock pulse or other input; that is, they override any other signal. The symbols of the JK flip-flop and other flip-flops show the Pre and Cl pins and indicate that the action will occur if the pin is set low. The fact that a low (0) is required to set or clear is shown by a small circle on the symbol at the set or clear line.

The preset and clear pins must be tied to the supply voltage ($+5$ V) to ensure reliable operation of the JK flip-flop. One or the other is made low (0) for an instant to place the flip-flop into the desired initial state. The arrangement for allowing preset and clear is shown in Figure 4-9 in Chapter 4.

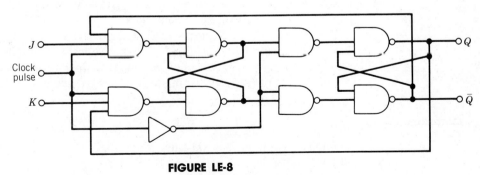

FIGURE LE-8
A Master-Slave JK Flip-Flop

Another feature of flip-flops, including integrated-circuit JK flip-flops, is that the change in state occurs when the clock pulse changes. This feature is achieved with additional gates in the flip-flop, which are not shown in Figure LE-8. The flip-flops and other elements with this property are called dynamic elements. The dynamic property is indicated at the symbol for the element by a > on the clock line. A small circle at that point indicates that the dynamic change occurs when the clock pulse returns to low (0). Synchronous operation and freedom from change of state during the clock pulse are achieved by dynamic transitions.

The Toggle Flip-Flop

The truth table for the JK flip-flop shows that when both J and K are high (1) the output $Q(t + 1)$ changes for each clock pulse. The JK flip-flop toggles from one state to another when a clock pulse is completed. That is, Q changes once for each clock pulse. Thus it takes two complete clock pulses for Q to complete a cycle of change. The output pulse rate is one-half of the clock rate. The division of the clock rate by two is the basis of binary counters or scalers.

- Make the J and K inputs high and confirm that the output toggles with each clock pulse. The truth table suggests the way that a JK flip-flop may be made to toggle ($J = K = 1$, as stated) and to stop toggling. What inputs would you use as control signals to start and stop toggling?

Points at Issue

The JK flip-flop circuits here use NAND-gates. Show the arrangement if NOR-gates are used in making the JK flip-flop.

The JK flip-flop made in this exercise is not dynamic, although it is a master-slave arrangement. Does this flip-flop change states if the inputs change while the clock pulse is high? Use the circuit you made to answer this question. Can one be sure of the state $Q(t + 1)$ if the inputs are changed while the clock pulse is high?

Components

7400 Quad two-input NAND-gate
7404 Hex inverter
7410 Triple three-input NAND-gate

LABORATORY EXERCISE 9
Digital Counters

Digital scalers or counters are important components of instruments and computers. As they record individual pulses or events, they generate a binary number that corresponds to the number of input pulses. Digital counters are made with flip-flops that are interconnected to follow a preset counting pattern. The pattern may be a simple binary scale, a binary-coded decimal (BCD) scale, or a count of another modulus. A modulus is the number that causes the counting cycle to repeat. A real time clock, for example, uses a modulus of 12 for hours and 60 for minutes and seconds. The latter is frequently made with a modulus 10 counter followed by a modulus 6 counter.

Digital counters may be asynchronous; that is, they may accept counts with random time intervals between them. Pulses from a radioactive source are asynchronous. Other counters may be synchronous and may record events only at times established by a generator of a continuous chain of clock pulses. Computers use synchronous counters.

Useful counters must be able to be cleared before starting to count and they must be able to be stopped without losing the count number they contain.

Integrated-Circuit *JK* Flip-Flops

Integrated-circuit *JK* flip-flops are available in different forms in DIPs. The one used in this exercise has AND-gates on the *J* and *K* input lines. The AND-gates can be used to establish logic conditions for the input pulse to reach the flip-flop. When the logic can be satisfied by the AND-gates on the input lines, other logic gates are not required in the circuit.

We selected these gates to illustrate an important property of transistor-transistor logic (TTL) gates: an unconnected input may be expected to act as if it were connected to the supply voltage. This is not always realized, however, and the DIP with floating leads may not work reliably or may not work at all.

This same possibility applies to preset and clear lines on the flip-flop. These two lines must be tied to the supply voltage. Table B-5 shows the conditions necessary for the *JK* flip-flop to operate. The *X*s indicate that the action is independent of the value of *J*, *K*, or the clock pulse (*CP*).

Digital Counters

The two counters shown in Figure LE-9 illustrate how the *JK* flip-flops are interconnected to give a specific counting sequence. The signals on the input lines *J* and *K* establish the sequence that is governed by the gate logic. At other times, different flip-flops and interconnections may be used for the same sequence; that is, there are no unique arrangements of gates and flip-flops for each counting sequence.

TABLE B-5
Function Table for a JK Flip-Flop

Preset	Clear	Clock	J	K	$Q(t+1)$
0	1	X	X	X	1
1	0	X	X	X	0
1	1	CP	0	0	No change
1	1	CP	0	1	0
1	1	CP	1	0	1
1	1	CP	1	1	Toggle

In this exercise both counters are asynchronous. They accept pulses that occur at random time intervals. The ripple counter records the events in a binary code. The BCD counter also generates a binary code, but its sequence is different because it resets to zero after it reaches nine. Both counters may be cascaded to increase their counting range.

Counters similar to the ones we make here are available in DIPs, and many integrated circuit counters have pin connections on the DIPs that allow the user to

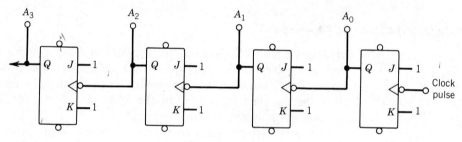

Four-bit ripple counter

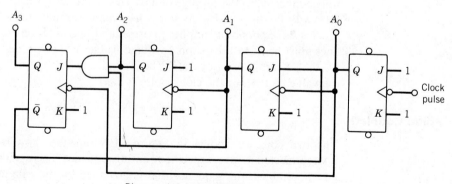

Binary-coded decimal (BCD) counter

FIGURE LE-9
Digital Counters

choose a counting sequence. A decade counter may allow a count of five and a count of two as well as a count of 10. It is usually more efficient in elements, reliability, time, and cost to use DIPs with the kind of counter required for a specific application than to make them, as we do here.

- Make the ripple counter and confirm that it scales in the binary code. Use LEDs on the output lines and observe the sequence when the pulses are provided by a bounceless switch. Then use a wire touched to the supply voltage to give pulses to the counter. The phenomenon of contact bounce should be evident. Is the number of counts with the wire reproducible?
- Rearrange the connections between the JK flip-flops to make the BCD counter, and confirm its counting sequence. Add the necessary connections to the clear lines to allow the counter to be reset to zero, and see that it works.

Points at Issue

A useful component of a counter is a means of starting and stopping the count without destroying the contents of the counter. Show a way to do this for the counters in this exercise.

Use the characteristic equation for the JK flip-flop, $Q(t+1) = J\bar{Q} + \bar{K}Q$, and the order of BCD counting, 0000 to 1001, to derive a state table, similar to Table 4-15 in Chapter 4, for the input requirements for J and K. Simplify the table to find the functions for the Js and Ks.

Components

7472 AND-gated JK flip-flops

SUPPLEMENT FOR LABORATORY EXERCISE 9

A Seven-Segment LED Display

It is easier to examine the contents of a counter or scaler when the contents are displayed as decimal numbers than as binary numbers because the evaluation of a binary number involves adding the powers of two indicated by the binary bits in the particular number. Although the mental arithmetic required to convert a binary number to a decimal number is stimulating, it may be time-consuming and possibly inaccurate.

The counters in Exercise 9 offer a good opportunity to use a seven-segment light-emitting-diode display and a decoder/driver to convert a binary number into a decimal number. A later exercise will show how the displays may be multiplexed, which can reduce the number of components needed to display a number with several decimal digits.

Decoder/Drivers

Decoder/drivers are DIPs that convert the four bits of a binary number into a decimal or hexadecimal number by providing the voltage highs and lows. These, in turn, cause the segments of a seven-segment display to be illuminated or not, as required. The decoder/drivers are usually open-collector elements. The connection is made from the power source through the segment and a series resistor, and the combination is the pullup resistor for the open-collector output circuit. Each segment is connected separately to a pin on the decoder/driver.

The inputs to the decoder/driver (7447) are labeled A, B, C, and D. A is the least significant bit, B is the next higher bit, and so forth. The outputs are labeled a, b, c, d, e, f, and g. These letters correspond to the segments in the display. The 7447 DIP is an "active low" element, which means that the pin on the DIP corresponding to the segment becomes low (0 V) when the segment is to be illuminated. The 7447 is used with common-anode and common-cathode seven-segment displays.

The decoder/driver causes the numbers 0 to 9 to be displayed to correspond to the equivalent binary number. According to the manufacturer's manual, "unique" symbols correspond to the binary equivalents of 10 to 15. A little practice will make them acceptable.

The decoder/driver has additional pins, which we will not use. Pins provide for blanking zeros to the left and right in multidigit numbers, so that only significant figures are seen, and a lamp test.

Seven-Segment LED Displays

Seven-segment light-emitting-diode displays are either common-anode or common-cathode elements. Both kinds are made with different pinout arrangements. The common-anode display has a single connection to +5.0 V, and each segment pin is attached to the decoder/driver segment pin through a current-limiting resistor. A resistor of about 470 Ω is suitable. The resistor must not be less than 100 Ω, but a value of 1 kΩ gives a reasonably bright segment. The segment is likely to be destroyed if no resistor is used.

The common-cathode display has a single connection to ground (0 V), and the decoder/driver allows current through its segment pin to illuminate the segment. A current-limiting resistor is required between the decoder/driver and the segment pin on the display. Either type of display may be chosen to display the decimal digit.

The pinouts for the seven-segment displays are given in the Data Section, Seven-Segment Displays, following these laboratory exercises. You must determine the kind of display you are using and identify the connections to it. Use a resistor in series with the segment when you test the display.

Using the Displays

- After you confirm that the four-bit ripple counter in Laboratory Exercise 9 is operating properly, use the decoder/driver and the seven-segment display to show the decimal and hexadecimal values in the count.
- Without detaching the decoder/driver and the display, rearrange the counter to make it into the BCD counter. Verify that the BCD counter counts through nine and returns to zero.

Components

Seven-segment display
7447 BCD to seven-segment decoder/driver
One-quarter watt resistors

LABORATORY EXERCISE 10

Synchronous Counter and Word Timer

The transfer of binary data must frequently occur in an interval of time limited by a specific number of clock pulses. One example is the serial transfer of the binary code, corresponding to a key struck on a keyboard, into a computer. The synchronous timing circuit prevents the mixup of data and provides a signal to show that the transfer is completed.

In this exercise a synchronous counter and a JK flip-flop are joined together. The JK flip-flop provides the gate signals for the transfer of one byte (eight bits) following a transmit enable (start) pulse.

Synchronous Counter

In the three-bit synchronous counter, each JK flip flop is clocked simultaneously by a master clock signal so that flip-flops toggle together. The logic gates establish the sequence of counting by generating the states for J and K.

The bit length of the synchronous counter can be extended by replicating the first stage. The preset and clear lines are not shown in Figure LE-10. What should be done with them?

- Make the counter and verify its operation. Use a dual-sweep oscilloscope to observe the clock pulse and each output, or place LEDs on the output lines and clock slowly to check the counting sequence.

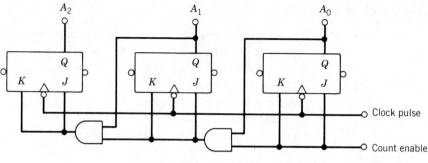

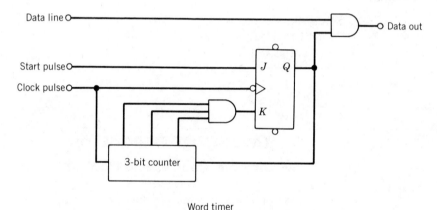

Word timer

FIGURE LE-10
Synchronous Counter and Word Timer

Word Timer

A byte of eight bits is to be transferred during the word time. The simultaneous highs (1) on each output line of the binary counter cause the AND-gate to place a high (1) on K, cause Q to become low (0), and terminate the transmission of the byte. J is low (0) except when the start pulse is given.

- Add the elements to the synchronous counter to make the word timer. After you verify its operation, use a dual-trace oscilloscope and a pulse generator for the clock and for the byte to the data transmission line. Generate a start pulse and observe the number of pulses on the data-out line. Is it necessary to initialize the word timer or is the timer automatically ready after each byte is transmitted?

Experimental Details

In using a pulse generator to provide pulses to flip-flops and other elements, it is important that the pulse lie between 0 and 5 V. Some signal generators have a voltage offset that must be set to ensure that the pulses stay within the voltage ratings of DIPs. The offset voltage is a steady (dc) voltage, which is added to or subtracted from the pulses. When the offset voltage is zero, the pulse may be symmetric about 0 V, for example, -2.5 to $+2.5$ V for a square wave. With the offset voltage set at one-half of the signal amplitude, the pulses can be made to vary from 0 to 5 V. A larger offset voltage, however, may cause a 5-V pulse to vary from $+2.5$ to $+7.5$ V and exceed the allowed signal amplitude for TTL gates. If the offset voltage subtracts from the pulse, the elements may not have large enough signals to respond.

Points at Issue

The three-input AND-gate can be synthesized with NAND-gates and NOR-gates. Show the configurations and the Boolean algebra logic for the syntheses.

Components

7476 Dual JK flip-flops
7400 Quad two-input NAND-gates
LEDs (light-emitting diodes)
Resistors

LABORATORY EXERCISE 11

Shift Registers

The transfer of binary data occurs frequently in computers, and a shift register is often the element involved. A shift register is an array of flip-flops in which each flip-flop transfers its content (bit) to the adjacent flip-flop upon the transfer-enable (clock) pulse. Bits shifted beyond the last flip-flop are lost, and the shift register becomes empty (i.e., all bits are zero) after a number of shifts.

Shift registers are used in many ways; one way will be illustrated in a serial adder made in the next exercise.

Shift Registers

The data are placed in shift registers in series and in parallel. A shift register accepts data placed on its serial input terminal by transferring one bit at a time through the register. Parallel-input shift registers accept the bits on the parallel-input lines in one step when a load-enable signal occurs. You may select shift registers that have capabilities for serial-in, serial-out, parallel-in, parallel-out, or a combination of data transfer schemes. Some shift registers transfer bits in either direction, depending upon the control signals to the DIP. Not all shift registers, however, include all combinations of data transfer and shift direction. In any case, the last pin in the direction of shift in a parallel-out shift register is also a serial-out terminal.

A binary number stored in a shift register is divided by two when it is shifted one step in the direction of the LSB, and multiplied by two if it is shifted toward the MSB. Each additional shift will change the number by an additional power of two.

- Make the shift register shown in Figure LE-11. Load the four-bit A (hex) = 1010 into the register bit by bit by generating a clock pulse when the bit is on the input line. Continue to apply clock pulses; by placing LEDs on the output lines, observe the bits moving through the shift register. Continue to supply clock pulses and observe the shift register as it is emptied.

- Store another four-bit number in the shift register and then connect the serial output to the serial input. Observe the number as it walks repeatedly through the register with a sequence of clock pulses.

- Remove the connection between the output and input lines and store C (hex) = 1100 in the register. Shift the contents one step at a time, and record the numerical value of the contents of the register. Do you observe division by two? Is there a rounding error at some point in the process of repeated division?

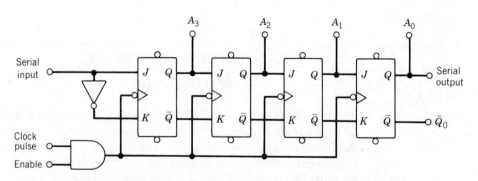

FIGURE LE-11

A Four-Bit Serial-In Parallel-Out Shift Register

Points at issue

Might the preset and clear pins on the JK flip-flops be used to load a binary word in parallel? Show the logic-gate arrangement that will load four bits in parallel. Include a parallel-load enable signal as part of the logic-gate arrangement, so that shifting and parallel loading can take place without conflict, that is, not occur at the same time.

Components

7476 Dual JK flip-flops
7400 Quad two-input NAND-gates
LEDs
Resistors

LABORATORY EXERCISE 12

A Serial Adder

In many cases the addition of two long binary numbers may be accomplished by a serial or sequential process rather than by a large number of full adders in parallel. Serial addition requires only one full adder but takes more time than parallel addition. Speed is sacrificed for less hardware, which is an advantage in many applications.

Shift Registers

Shift registers can be made with JK flip-flops and basic gates, as was done in an earlier exercise, but here we use two four-bit shift registers in DIPs. We illustrate a means of parallel loading shift registers and show that a shift register cannot only accumulate a sum, but also hold one of the numbers to be added. This procedure, like serial addition, is another means of reducing the amount of hardware. In many microprocessors the register that holds one of the two numbers being added is the receptacle for the sum. The principal registers in microprocessors are typically called accumulators because they are used in a manner similar to one of the shift registers in this exercise.

An advantage of shift registers is that they may be connected in series so that a large binary number can be stored and used. A word timer controls the sequence of addition to limit the shifts to the number required to complete the addition exactly. In this exercise the clock pulses are supplied by hand so that the sequence of events can be observed.

Serial Adder

- Make and test a full adder, using a DIP with XOR-gates.
- Configure a *JK* flip-flop as a data flip-flop and verify that it gives the carry-in for each successive pair of bits being added.
- Place the shift registers in the circuit and return the sum bit to the serial input of register *A* so that it accumulates the sum as it is produced.
- Choose several four-bit numbers to load into the shift registers. Confirm that they are stored in the registers. Find the partial sums before adding them with the serial adder so that each step in the adding sequence can be checked.
- After you have confirmed the serial addition for several combinations of initial numbers, connect the output pin of register *B* to its serial input pin. With binary numbers in both registers, clock the registers four times to find the sum. Continue several full cycles of addition without reloading the registers. Report the sequence of sums and show the repeat cycle.
- Explain the reason for the number of cycles required for the sum to repeat.

Experimental Details

The four-bit shift register used in this exercise is the 7495 DIP, a versatile, general-purpose shift register. The pinouts are shown in Figure LE-12. The 7495 has parallel inputs and outputs and a serial input. The LSB output pin is also a serial output pin. Bits may be shifted in either direction.

Data are loaded in parallel as follows. The four bits are placed on the input pins and the mode control (pin 6) is made high (1). One clock pulse on pin 8 transfers the data into the shift register. Serial data input is inhibited when pin 6 is high.

Shift right (toward the LSB) occurs when the mode control is low (0) and the clock pulse is applied to the clock R-shift pin (pin 9). Output pin (O_D) is used for the serial output line.

Points at Issue

Show the arrangement of logic elements that might be used in a hand calculator to write the numbers into the two registers, and add two four-bit numbers when the add button on the calculator is pressed.

Give the sequence of steps in the process. Does your sequence require a clear step to begin?

Components

7495 Four-bit shift register
7400 Quad two-input NAND-gates
7404 Hex inverter
7486 Quad XOR-gate
7476 Dual *JK* flip-flop

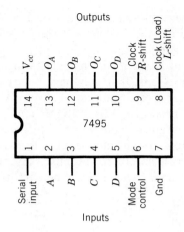

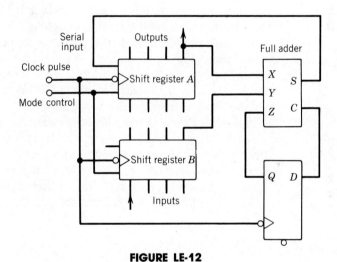

FIGURE LE-12
A Serial Adder

LABORATORY EXERCISE 13

An Elementary Memory Cell

Many cells are required to store the binary data for even the simplest calculator or computer. Hundreds of thousands of bits must be stored in memory arrays so that a computer can operate. Great strides have been made in creating high-density arrays of simple, reliable binary memory cells.

The various kinds of memory elements serve different needs. Shift registers and latches represent two kinds, but here we are considering the cell of a memory that can be addressed in a random fashion. Random access memory cells can be read in any order, rather than in the serial fashion required to read from a magnetic tape.

Random access memories are divided into two classes. Memory cells whose content is set in the manufacturing process are read only memories (ROMs). Memory cells that may have binary values written into or read from them are random access memories (RAMs). (A more descriptive name would be "read and write memories.") Both ROMs and RAMs are addressed by decoders that select the row and column occupied by the memory cell. The binary address is a number of n bits, and the number of cells that n bits can address separately is 2 to the nth power. If $n = 10$, for example, any one of 1024 (1 K) memory locations can be selected. We are interested here in the features of a single RAM cell.

Elementary Memory Cell

The memory element in this exercise is a static RAM cell (Figure LE-13). A static RAM is made with a bistable element (flip-flop) whose state persists until a new bit is written into it. A dynamic RAM serves the same purpose, but it is made with a different technology and requires constant attention to sustain the memory content. The loss of electrical power to a RAM, even for an instant, will destroy the stored binary data. RAMs are volatile but ROMs are not, because the bit pattern established by the manufacturer is permanent.

A control or enable signal, an address decoder, a read line, and a write line are involved in using a RAM. Because writing into a memory cell may change the content of the cell, the write action must be enabled intentionally. Possible change of content, however, is not a concern in reading from a cell because the bit is not changed by reading from the cell. The read/$\overline{\text{write}}$ (R/$\overline{\text{W}}$) line must be made low (0), as indicated by the bar, to write into the cell. Otherwise it remains high (1).

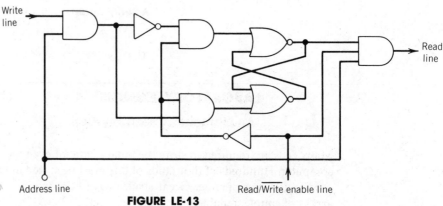

FIGURE LE-13

An Elementary Memory Cell

- Make the RAM cell. Check the operation of the data flip-flop before completing the other parts.
- Confirm the operation of the cell by writing both a zero and a one into the cell and reading the content. Verify that the R/$\overline{\text{W}}$ enable line protects against changing the content.
- Examine the volatile nature of a static RAM cell by removing the electrical power and reading the content after restoring the electrical power. Do this with both a one and a zero in the cell.

Points at Issue

Show the arrangement of a four-by-four array of cells. Use a block or square for each cell but show the address, write, read, and R/$\overline{\text{W}}$ lines that are common to all cells. How many binary bits are needed to address a particular cell in the four-by-four array?

Eight memory arrays are often used in parallel to allow a byte to be addressed in one step. How may eight four-by-four arrays be connected so that 16 bytes (eight bits) may be addressed and read from or written into in a random order?

The RAM cell in Figure 5-9 has different arrangements of logic gates on the address lines and on the R/$\overline{\text{W}}$ lines. The added gates allow that cell to be used in an array of cells, but the RAM in this exercise would not be satisfactory in an array. Why not?

Components

7400 Quad two-input NAND-gate
7402 Quad two-input NOR-gate
7404 Hex inverter
7410 Triple three-input NAND-gate

LABORATORY EXERCISE 14

Random Access Memory

In this exercise we use a 16-bit random access memory DIP (7481) that is addressed, read from, and written into differently from the elementary memory cell in Exercise 13. This exercise illustrates the volatile nature of the static RAM and emphasizes the practical aspects of using the read and write lines.

Binary information is read from a random access memory (RAM) by first addressing the binary cell to be read and then examining the binary state (high or low) on the lines attached to the selected cell. The cell in the memory element used in this exercise is addressed by a high on an X-coordinate line and a high on a Y-coordinate line. An R/$\overline{\text{W}}$ enable is not used in this case because read (sense) lines and write lines are involved separately.

Memory Arrays

Large-size integrated (LSI) circuits include many thousands of identical memory cells, which may be arranged so that a byte (eight bits) or a word (16 bits) can be reached by a single address. An alternate arrangement uses eight (or 16) LSI elements in parallel. In this case all the elements are addressed simultaneously and yield a byte (or a word) with one bit from each memory element.

Is it possible to address the 7481 memory element in this exercise (Figure LE-14) so that a byte can be read from it in one step, or is it necessary to have eight 7481s in parallel to read a byte in one step? Answer this question by sketching the arrangement and listing the addresses.

Sixteen-Bit Random Access Memory

In this exercise a 16-bit static random access memory DIP is used to store and retrieve binary data. The 7481 DIP has 16 cells, arranged in a four-by-four array. A cell is addressed by putting a high on both the X-coordinate line and the Y-coordinate line. All other address lines must be low. In an array of 16 cells, only a 4-bit code is required to address a cell uniquely, but the 7481 has eight address lines.

Draw a demultiplexer made with basic logic gates that makes it possible to use a four-bit number to select a particular cell.

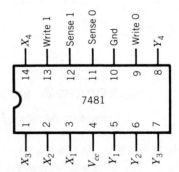

FIGURE LE-14
A 16-Bit Random Access Memory 7481

The write process uses two write lines, a write-1 line and a write-0 line. A one or a zero is written into the cell addressed by the X-Y lines by placing a high (1) on the appropriate write line. That is, a high (1) is placed on the write-0 line to write a zero into the cell. Similarly, a one is written into the cell by placing a high (1) on the write-1 line. The two write lines may be considered as set-reset lines as shown in the RS flip-flops in earlier exercises.

The binary cell is read by addressing it and polling the sense lines. If the cell contains a low (0), the sense-0 line will be low (0), but if the cell contains a high (1) the sense-1 line will be low (0). Because all cells in the 7481 are connected to the sense lines, the memory element is an open-collector (OC) device. Pullup resistors to the supply voltage are necessary on the sense lines. The practice of interconnecting open-collector elements was discussed in Exercise 2, when a common bus was examined.

- Use the 7481 to store a 16-bit word. Use all ones and confirm that the word ($FFFF_{hex}$) is stored. Read a portion of the word several times to confirm that the reading process does not alter the content of memory.
- When you are sure you know the contents of the memory element, remove the supply voltage from the 7481 for an instant and then restore the power. Read the memory contents again. Is the element volatile? Restore all ones, remove the voltage again, and reexamine the contents. Does each cell return to the same default (power on) state each time the power is restored? Why might the default state be the same for the entire RAM each time the power is turned on?

Points at Issue

What value is stored if highs (1s) are simultaneously placed on both write lines? Would an inverter placed between the write-1 line and the write-0 line allow a 1 or a 0 to be read into the cell by placing the desired value on the input of the inverter? Would this be like the input to a D flip-flop?

Show a logic-gate arrangement you would use to establish an independent write-enable signal for the 7481 so that neither write line could change the content of a cell if it were accidentally made high.

Four bits from a scratch-pad memory made with four 7481s in parallel are to be transferred into a four-bit shift register like the 7495 used in Exercise 11. Show the arrangement and list the sequence of steps needed to accomplish this move.

Component

7481 16-Bit random access memory

LABORATORY EXERCISE 15
Diode Characteristics

The simplest electronic element is a diode, a *pn* junction that conducts when it is forward-biased and will not conduct when it is reverse-biased. In fact, there is a very small reverse current, perhaps one-thousandth or less of the forward current, but in nearly all applications of diodes this reverse current is assumed to be zero.

A different kind of diode, the Zener diode, has the same properties as a common diode; that is, a forward current when forward-biased and a trivial reverse current when reverse-biased. The Zener diode, however, also conducts in the reverse direction when the reverse bias exceeds the value established at the time of manufacture.

In this exercise we will measure the current-voltage relations in diodes and examine a simple application.

Forward-Biased Diode

The diode consists of two parts, the anode and the cathode. The symbol for the diode shows the anode as a triangle and the cathode as a bar (Figure LE-15). The cathode is marked with a band or a dot, or the cathode lead is shorter than the anode lead. When the anode is electrically more positive than the cathode, the

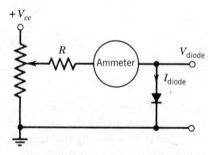

Circuit for diode measurements

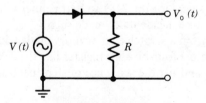

Circuit for rectifier measurements

FIGURE LE-15

Diode Characteristics

diode is forward-biased. The current is very small until the voltage across the diode reaches a value called the threshold or cut in value. The threshold voltage is a property of the junction of the *p*-type and *n*-type semiconductor materials with which the diode is formed. Chapter 8 gives the reason for the *pn*-junction potential (threshold voltage) that must be overcome for forward conduction.

The current through a diode must not exceed its rated value. For the small diodes used in this exercise the rated current is only a few milliamperes; thus a resistor must be placed in series with the diode to limit the current. The resistance value is found by dividing the maximum voltage to be placed across the diode by its rated current. The forward resistance of the diode is $1.0\,\Omega$ or less.

- Measure the current-versus-voltage performance of the diode for both forward-biased and reverse-biased voltages. The voltage may be varied by a resistance voltage divider or by changing the output voltage of a power source. Because the ammeter (current meter) has a very low resistance, it will conduct a large current if a voltage is placed directly across it and will be destroyed. Exercise extreme caution in using the ammeter; it must be in series with the diode and the limiting resistor. Take small steps in voltage, 0.1 V or less, so that the threshold voltage can be readily identified. Plot the data as the points are measured so that you will know whether or not to take readings between the ones you have plotted.

The threshold voltage depends upon the semiconducting material used in making the diode. The cutin voltage is approximately 0.2 V for germanium-based diodes and approximately 0.7 V for silicon-based diodes. Can you tell which kind of diode you are using?

Zener Diode

- Replace the diode used in the previous task with a Zener diode and repeat the measurements. Plot the data to find the cutin voltage and the reverse voltage at which the Zener diode begins to conduct. Measure the points carefully so that you can tell whether the conduction curves are similar. Does the breakdown conduction with reverse bias appear to be the same, or is it probably a different mechanism from forward conduction?

Light-Emitting Diode LED

- Replace the Zener diode with an LED and measure its characteristics. Would an LED be a suitable substitute for a common diode?

Diode as a Rectifier

A common use of a diode is in rectifying an alternating-current voltage source. Rectification is the process of permitting conduction in only one polarity or direction.

- Place the common diode in the rectifier circuit and adjust the function generator to produce a sine wave about 5 V in amplitude and about 1 kHz in frequency. Observe the wave form and measure carefully the intervals of conduction and nonconduction. Be sure that the offset voltage is zero; that is, be sure that the sine wave is symmetrical about 0 V. If the intervals are not equal, explain the reason. Change the function generator to produce a square wave, a sawtooth wave, and other possible wave forms, and observe the action of the diode.
- While rectifying the sine wave, adjust the offset voltage in steps of plus and minus 2 V through its range. Observe and record the action of the diode.
- With a sine wave input carefully adjust the offset so that the intervals of conduction and nonconduction are equal. Find the average voltage for the half-wave rectifier. Can you find the threshold voltage of the diode from the offset voltage required in this adjustment for equal intervals?

Points at Issue

Find the effective resistance of the diodes at a current of 5 mA by measuring $1/R = dI/dV$ from the plots you made of current versus voltage across the diodes. Include both forward and reverse conduction through the Zener diode in finding the Rs for the Zener diode.

Find the power law for the current versus voltage for the common diode. The equation to be applied to the data may be taken as

$$I = G_A V + G_B (V - V_{Th})^N$$

What are the units of G_A and G_B?

Components

1N914 or similar diode
1N4733 or similar Zener diode
Light-emitting diode
Ammeter
Resistors

LABORATORY EXERCISE 16

Common-Emitter Amplifier

The common-emitter amplifier is the most commonly used transistor analog amplifier. This capacitor-coupled voltage amplifier is similar to the vacuum-tube amplifier, which it has supplanted. The transistor amplifier here is very much like the one in the design example of Chapter 10, and it has the characteristic performance of ac amplifiers described in Chapter 9. This exercise illustrates the task of establishing the quiescent operating conditions and shows the frequency dependence of the gain of the amplifier.

Bipolar-Junction Transistors

A bipolar-junction transistor is made by placing a very thin layer of either *p*-type or *n*-type semiconducting material between pieces of the other kind of semiconducting material. The middle element is the base; the others are the emitter and collector. The symbol for the bipolar-junction transistor seen below indicates the arrangement of materials. A *PNP* transistor has negative-type semiconducting material for the base, and the emitter and collector are made of positive semiconducting material. In a *NPN* transistor, the arrangement is just the opposite. The arrow on the emitter of a *PNP* transistor is directed to the base, while the arrow on the emitter of a *NPN* transistor is directed away from the base. It is helpful to remember that the direction of the arrow in the symbol for the transistor is the direction of the current and the voltage polarity, that is, from positive to negative voltage.

The bipolar-junction transistor conducts current from the emitter to the base when that *pn* junction is forward-biased. Most of the current passes through the base into the collector. The essential property of the bipolar-junction transistor is that a very small base current controls the much larger collector current; the collector current may be 100 times the base current. This is the key to the use of a transistor as an amplifier.

In a transistor amplifier the voltages among the three elements—emitter, base, and collector—must fall within certain relative ranges. One step in the design of a transistor analog amplifier is to establish a quiescent operating point about which the amplifier responds to an input signal in a linear manner, so that the amplified signal is an acceptable analog of the input signal.

Common-Emitter Amplifier

The amplifier in this exercise is called a common-emitter amplifier because the emitter is part of both the input circuit and the output circuit. The common-emitter amplifier provides both voltage and current gain. Other transistor amplifier configurations—common-collector and common-base—have different features and advantages, but for small-signal amplification the common-emitter amplifier is most important.

In this exercise we will study the response of the common-emitter amplifier to time-varying small signals. It will be seen that small signals can be amplified faithfully when the quiescent operating point is well chosen. If the input signal is too large, the signal overdrives the amplifier and produces distortion when the transistor operates in a nonlinear, saturated, or cutoff region. The effects of large signals will be seen presently.

The quiescent operating point for the transistor in this exercise is established by the resistors shown in Figure LE-16. The frequency range is governed by the choice

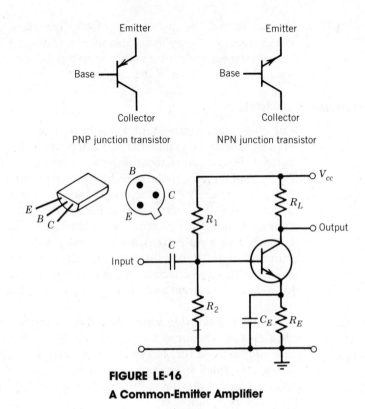

FIGURE LE-16
A Common-Emitter Amplifier

of capacitors and resistors. In the mid-frequency range the gain (amplification) is constant. In the low-frequency and high-frequency regions, however, the gain falls off because of the resistor-capacitor properties of the amplifier. (This behavior is shown in Figures 9-2 and 9-3 and explained in Chapter 9.) We will obtain a similar curve when we measure and plot the ratio of the output voltage to the input voltage versus frequency.

- Construct the amplifier. Omit the capacitors and do not provide an input signal. Meaure the dc voltages from base to ground, E_B, from collector to ground, E_C, and from emitter to ground, E_E. E_C should be about one-half the supply voltage, V_{cc}. Adjust the resistors R_1 and R_2 until E_C has this value.
- Subtract E_E from E_B. This result is the potential across the PN junction formed by the base and the emitter. On the basis of the measurement, can you say whether the transistor is a silicon-based or a germanium-based semiconductor?
- Add the capacitors and apply a small sine wave signal to the input. Use a frequency of 1000 Hz. You may have to make a resistor voltage divider to reduce the input signal enough to produce a good sine wave as output. Find the

gain (ac output/ac input) of the amplifier; this should be the midfrequency gain. Measure the gain from 5 Hz to 100 kHz or higher and plot the gain versus the logarithm of the frequency. Find the half-power points. What is the frequency bandwidth (the frequency range between half-power points) for the single stage amplifier?

- Measure the phase angle between the input and the output signals. Plot the phase angle versus the logarithm of frequency, using the same graph on which you plotted the gain. The phase angle is found by using a dual-trace oscilloscope to measure the difference in position of the peaks of the input and output waves and by calculating the fraction of the 360° angle of a complete cycle, represented by the difference between peaks.

- Gradually increase the amplitude of the input signal; observe and sketch the output waveform. Find the limit of faithful reproduction of the input signal.

Points at Issue

Find the gain in decibels and plot it versus frequency; this is a Bode plot. See Chapter 9 to learn how to make a simplified Bode plot, and show this on your Bode plot.

Components

2N3904, 2N4123, or equivalent *NPN* transistor

Suggested Values

$R_1 = 32\ \text{k}\Omega$ $R_L = 3.2\ \text{k}\Omega$ $C = 0.5\ \mu\text{F}$
$R_2 = 9\ \text{k}\Omega$ $R_E = 1.6\ \text{k}\Omega$ $C_E = 20\ \mu\text{F}$

LABORATORY EXERCISE 17

Multiplexing

Many circuits that involve data transmission and digital displays are simplified by multiplexing, the technique of opening one path or channel after another in a sequence determined by a few control signals. In the case of seven-segment displays a number of digits can be presented by illuminating one digit after another at a rate that exceeds the eye response (about 30 times a second) to give the illusion of continuous display.

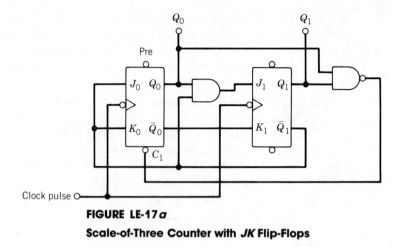

FIGURE LE-17a

Scale-of-Three Counter with JK Flip-Flops

In this exercise we multiplex segments of a seven-segment display to demonstrate the method of multiplexing. The segments, illuminated sequentially in pairs, give the illusion of rotation. Because we use three pairs, a scale-of-three counter is used to generate the sequence (Figure LE-17a).

Scale-of-Three Counter

The binary sequence required to direct the multiplexing is

Line	Q_1	Q_0	$Q_1(t+1)$	$Q_0(t+1)$
1	0	0	0	1
2	0	1	1	0
3	1	0	0	0

The first steps in configuring the scale-of-three counter are the selection of the flip-flop and the identification of the input states. The latter are obtained from the sequence listed above and from the characteristic function of the flip-flop. Chapter 4 presents the general procedure for designing sequential circuits. Here we select JK flip-flops and use their characteristic function $Q(t+1) = J\bar{Q} + \bar{K}Q$.

For instance, line 1 shows that

$$Q_0(t+1) = 1 = J_0\bar{Q}_0 + \bar{K}_0 Q_0 = J_0 1 + \bar{K}_0 0$$

This equation indicates that J_0 must be high for the first transition of Q_0 from 0 to 1. The other Js and Ks can be found in this manner for the three transitions, and a Karnaugh map for each J and K can be used to simplify the configuration. The results are

$$J_0 = \bar{Q}_1, \quad K_0 = \bar{Q}_1, \quad J_1 = Q_0\bar{Q}_1, \quad K_1 = \bar{Q}_0$$

The circuit diagram in Figure LE-17a for the scale-of-three counter satisfies these requirements and enables us to illustrate another feature of sequential circuits. State $Q_1Q_0 = 11$ is not in the scale-of-three sequence, but it may arise when the power is turned on or because of an electrical glitch or noise pulse. Examine the values of the Js and Ks to convince yourself that the count stops with $Q_1Q_0 = 11$. The NAND-gate clears the LSB flip-flop, and the counting sequence is rejoined automatically if the unwanted state arises.

- Make the scale-of-three counter and confirm its operation. Show that it stops when the unwanted state occurs by not connecting the NAND-gate to the clear line and by switching the power on and off to induce the unallowed state. Connect the clear line and show that the problem is no longer present. What should be done with the preset and clear pins on the JK flip-flops?

Multiplexed Display

The multiplexing arrangement we use is a common and practical circuit that can supply the power to illuminate bright displays (Figure LE-17b). The transistors here are not required for the single seven-segment LED display because the JK flip-flops can drive them, but our purpose is to illustrate the more capable multiplexing arrangement.

Seven-segment LED displays are available in two basic types, common anode and common cathode, and in two different pinouts on the elements. The common-anode type is illuminated by placing a low (0) on the segment. A high (1) is required on a segment of a common-cathode LED in order to light it. Both types require

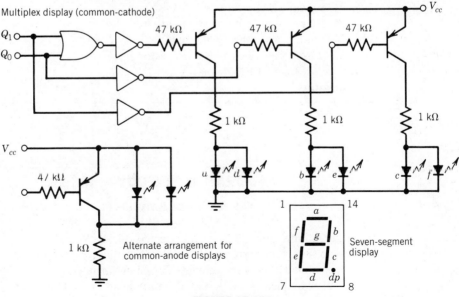

FIGURE LE-17b

Multiplexing

current-limiting resistors to the segments. The pin assignments for the seven-segment LED displays are given in the Data Section, Seven-Segment Displays, following these exercises.

- Construct the multiplexing circuit. A single quad-NOR-gate can satisfy all the gate logic. Use the appropriate transistor connection for the displays you are using. Be sure that your gate logic allows only one pair of segments to be illuminated at a time.
- Drive the multiplexer with a slow clock frequency. You should see the segments appear to rotate around the display. Increase the frequency until you can no longer see the individual pairs. What is the highest frequency at which a flicker in the display is barely observable? With open fingers, pass your hand in front of your eyes and observe the display when the frequency is just above that at which you could see flicker. Can you see flicker through your moving fingers? Why or why not?

Points at Issue

How would the transistor circuit be used with a BCD decoder/driver to illuminate a number on the seven-segment LED display? Show the arrangement for a multiplexed three-digit display with a single BCD decoder and three four-bit latches. Be sure to show the enable signals and the current-limiting resistors.

Components

7476 Dual *JK* flip-flop
7400 Quad two-input NAND-gate
7402 Quad two-input NOR-gate
2N4126 or equivalent *PNP* transistor
Seven-segment LED display
Resistors

LABORATORY EXERCISE 18
Operational Amplifiers

Operational amplifiers were perfected during World War II. They have become increasingly important in analog electronics, especially since integrated circuit operational amplifiers were introduced in the mid-1960s. The operational amplifier is also frequently an essential component in both analog-to-digital converters (ADC) and digital-to-analog converters (DAC).

In this and the following exercises the properties of operational amplifiers and their applications will be studied.

Operational Amplifiers

Operational amplifiers are essentially difference amplifiers with very high open-circuit gain. The result of using negative feedback is an amplifier of good stability, low distortion, wide frequency bandwidth, and great utility. Operational amplifiers are supplied in DIPs and in other forms. Many different operational amplifiers are available with special properties, such as fast response or low drift. In our exercises a widely used generic, general-purpose operational amplifier, the 741, is employed (Figure LE-18).

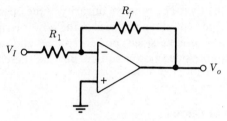

Inverting amplifier

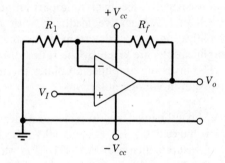

Noninverting amplifier

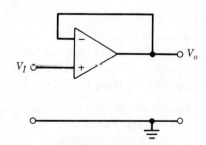

Voltage follower

FIGURE LE-18
Operational Amplifiers

Operational amplifiers have two input terminals. The one indicated by a + sign is the noninverting input and the other, marked with a − sign, is the inverting input. The output is the amplified difference between the two input voltages; voltage connections must be made to both inputs. Operational amplifiers have an open-circuit gain of the order of one million; thus an extremely small voltage across the input terminals produces a large output voltage.

The great utility of operational amplifiers is realized when they are used as feedback amplifiers. With feedback the terminals are maintained at virtually the same voltage; that is, it may be assumed that for most purposes the voltage difference between the terminals is zero.

The power sources for many operational amplifiers are double-sided: there are connections to $+12$ or $+15$ V for one power terminal and to -12 or -15 V for the other. The range of linearity of the operational amplifier output signals is about 80 percent of the voltages to the power terminals, from about $+12$ to -12 V with a 15-V double-sided power source and about $+10$ to -10 V with the 12-V double-sided power source. Beyond these ranges the amplifier output is distorted.

Operational Amplifier Circuits

In this exercise we make three of the simplest operational amplifier arrangements that use negative feedback to impart unique amplification properties. As explained in Chapter 9, negative feedback with a very high-gain amplifier causes the amplification to be determined by the feedback factor. In this exercise the amplifications are given by the ratios of resistor values.

The inverting amplifier amplifies the input signal by the ratio R_f/R_1 with an inversion in voltage polarity.

- Make the inverting amplifier and measure the gain, $A = V_{(out)}/V_{(in)}$ for dc input voltages of 0.5, 1.0, 1.5, and 2.0 V. Use $R_1 = 2.7$ kΩ and $R_f = 10$ kΩ. Repeat the measurements with a 10-kHz sine wave. Plot the amplification versus input-voltage amplitude for both sets of measurements on the same graph. Increase the ac voltage amplitude until the linear range of the amplifier is exceeded, and note the amplified waveform.

- Measure the amplification with a 1-V input as the frequency is increased by factors of 2, 5, and 10 in each decade of frequency. Plot the gain versus frequency on semilogarithmic graph paper. State the frequency at which the gain has dropped by 50 percent of the dc gain.

- Rearrange the circuit to make the noninverting amplifier and repeat the dc measurements above. The gain $A = 1 + R_f/R_1$. Do not repeat the gain versus frequency measurements.

A voltage follower is a noninverting amplifier in which R_f is equal to zero (R_f is replaced by a conductor), and R_1 is infinite (it is an open circuit).

- Make a voltage follower and repeat the gain versus frequency measurements for a 1-V ac input signal as it is increased from 1 kHz until the gain is reduced by 50 percent. Plot these data on the same semilog graph on which you plotted the measurements for the inverting amplifier.

Points at Issue

State the frequency at which the gain has dropped to 70.7 percent of the dc gain for both the inverting amplifier and the voltage follower. This is the frequency that corresponds to the half-power point. What are the frequency bandwidths of the amplifiers?

What is the feedback factor for the noninverting amplifier you made? Does the gain depend upon the feedback factor, as you expect?

Components

741 Operational amplifier
Resistors

LABORATORY EXERCISE 19

Operational Amplifier Characteristics

A real operational amplifier differs from an ideal operational amplifier. In many applications the differences are not important but they do influence other uses. An ideal operational amplifier has a very large open-circuit gain, usually assumed to be infinite, gain stability, zero voltage difference between the input terminals, zero input current, zero output impedance, the ability to ignore a common-mode input signal, and an instantaneous response to a signal.

In this exercise we measure characteristics that are useful in evaluating a real operational amplifier for a specific task. We find the input offset voltage, input bias current, the common-mode rejection ratio (CMRR), and the slew rate.

Input Offset Voltage

A very small voltage, the input offset voltage, is generated internally. It is as if a small battery existed between the input terminals of an operational amplifier, and its voltage is amplified by the large open-circuit gain. The output voltage has the greatly amplified offset voltage added to the desired amplified signal. The first circuit for this exercise shows the offset voltage, V_{os}, as a small battery external to the operational amplifier, but the dashed box enclosing it and the operational amplifier indicates that it is not part of the external circuit.

The offset voltage can be measured by making a high gain inverting amplifier and grounding both input terminals. The voltage found at the output is the offset voltage amplified by the large closed-circuit gain. An analysis of the circuit by Kirchhoff's laws gives

$$A\left[V_{os} - V_0 \frac{R_1}{R_1 + R_f}\right] = V_0$$

Because the real operational amplifier has very large open-circuit gain, A, the equation can be reduced to give the offset voltage approximately as

$$V_{os} \approx V_0 \frac{R_1}{R_1 + R_f}$$

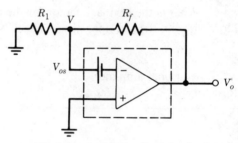

Circuit for measuring input offset voltage

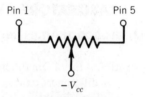

Offset null potentiometer

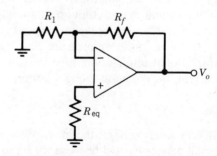

Circuit for measuring input bias current

FIGURE LE-19a

Operational Amplifier Characteristics

Many operational amplifiers have external pins, marked offset null, to allow the offset voltage to be reduced to practically zero. The offset null potentiometer, shown in Figure LE-19a, is placed between the offset null pins and adjusted to minimize the output voltage, that is, to nullify the input offset voltage contribution. Figure 10-5 shows a similar arrangement to equalize the collector currents in a simple difference amplifier so that the quiescent output difference voltage is zero.

- Use $R_1 = 0.1 \text{ k}\Omega$ and $R_f = 10 \text{ k}\Omega$ to measure the offset voltage without the offset null potentiometer connected.
- Connect the offset null potentiometer (10 kΩ is suitable). Minimize and measure the offset voltage. Find the offset voltage when the potentiometer wiper is placed in both extreme positions.

Input Bias Current

The input circuits of operational amplifiers are transistor amplifiers that are biased to operate in the linear region. The input bias currents must exist to enable the amplifiers to work.

Virtually all the current into the summing junction of the inverting amplifier continues through the feedback resistor, but a very small part enters the inverting input as the input bias current. The input bias currents are largest in operational amplifiers made with bipolar-junction transistors, that is, TTL technology. Operational amplifiers with FET (field-effect-transistor) input amplifiers have much smaller input bias currents.

One may think of the input bias current as current that should have gone through the feedback resistor. The input bias current times the resistance of the feedback resistor is the voltage error in the output voltage. For large feedback resistors (1 MΩ) and an input bias current (0.3 μA) an error of 0.3 V arises. Large feedback resistors, R_f, in inverting amplifiers that permit a high resistance input resistor, R_1, to be used, emphasize the error.

An examination of the inverting amplifier circuit shows that the input bias current into the inverting input can come through either R_1 or R_f, that is, they are in parallel in providing the current. Because the input bias current to the noninverting input is of similar magnitude, a resistor equal to the parallel equivalent of R_1 and R_f placed between the noninverting input and ground will generate a voltage to cancel the error in the output voltage.

- Without changing the offset null setting from the optimum position, change the resistors to $R_f = 1 \text{ M}\Omega$ and $R_1 = 10 \text{ k}\Omega$ and measure the output voltage when both input lines are grounded and R_{eq} is not present.
- Add resistors between ground and the noninverting input. Try different values until the best cancellation of the error voltage occurs. Is the parallel combination of R_1 and R_f the optimum resistance? Estimate the magnitude of the input bias currents.

- Measure the voltage at the summing junction with and without R_{eq} in place. Does R_{eq} provide an offset voltage at the summing junction to correct the error in the output voltage? Make diagrams similar to Figure 11-9 to show the voltage relations with and without R_{eq}.

Common-Mode Rejection Ratio

The same voltage signals, ac, dc or both, on each input of an operational amplifier is a common-mode input (Figure LE-19b). An ideal operational amplifier will have a zero output voltage, but a real operational amplifier will have a small output voltage. The common-mode amplification A_{cm} is the ratio of the common-mode output voltage to the common-mode input voltage, $A_{cm} = V_{co}/V_{ci}$.

However, a difference in the voltages between the inverting and the noninverting inputs, $V_{di} = V_{in+} - V_{in-}$, will be amplified by the operational amplifier as desired. We find $A_d = V_{do}/V_{di}$.

The CMRR (common-mode rejection ratio) is the ratio A_d/A_{cm}. CMRR is frequently expressed in decibels. Operational amplifiers typically have CMRRs of 80 to 120 dB. An approximate measurement of the CMRR is suggested as follows.

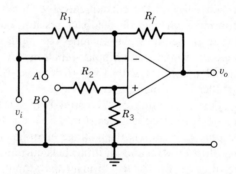

Circuit for measuring the common-mode rejection ratio

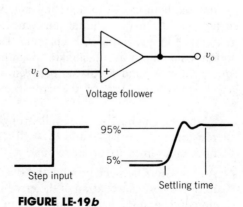

FIGURE LE-19b
Operational Amplifier Characteristics

- Measure the output voltage with R_2 connected to position A. Use a 1-V, 1-kHz input signal, with $R_1 = R_2 = 0.1\,\text{k}\Omega$ and $R_f = R_3 = 100\,\text{k}\Omega$.
- Greatly reduce the voltage of the input signal and then switch the connection to position B. Adjust input voltage to yield a 1-V output signal. Measure the input voltages and calculate the amplification factors from the resistors to find the equivalent input voltages for both connections. Find CMRR and express it as a ratio and in decibels. Is your value reasonable?

Slew Rate

An operational amplifier does not respond to a step change in the input voltage as quickly as a logic gate does. The ratio of the voltage change from 5 to 95 percent following a voltage step, to the time interval in which it occurs, is called the slew rate. The slew rate is given in volts per microsecond.

- Make a voltage follower and use the output of a TTL inverter (7404) driven by a signal generator (clock) as the square wave input. Measure the slew rate with a slow pulse rate, about 1 kHz. Increase the clock frequency and observe and sketch the output waveform at each frequency decade. Compare the waveforms of the inverter and the voltage follower. Are the slew rates the same for both edges of the pulse? Can you also estimate the slew rate of the inverter?
- At the slow pulse rate add (adjust or remove) the offset null potentiometer and examine the slew rate. Does the slew rate depend upon the offset null adjustment?

A careful examination of the transition from low to high shows that the output overshoots and tends to oscillate slightly before becoming steady. The time interval between the start of the transition and the end of the oscillations is called the settling time. What is it in your measurement?

Points at Issue

The 741 operational amplifier is a good general-purpose element but it is not a premium element, that is, it does not have all the best characteristics. Any particular operational amplifier will not have all the desirable features, but one can be selected for the characteristics that are important. Suggest two applications that require much better operational amplifiers than the 741 and the features that are important to the applications.

Components

741 Operational amplifier
Resistors

LABORATORY EXERCISE 20
Voltage Comparators

The operational amplifiers may be employed to compare and measure voltage amplitudes. This exercise illustrates the use of the operational amplifier to make very precise voltage comparisons. Comparisons are the basis for analog-to-digital conversions in many applications and serve to alert a system that it has reached a threshold such as a temperature limit indicated by a voltage from a thermometer circuit.

The magnitude of a voltage or other analog function can be converted to a binary number that represents the magnitude. The process is analog-to-digital conversion (ADC). The voltage comparator is a principal element in many ADC circuits. The fastest means of generating a binary number from an analog voltage is to sample the voltage with a ladder of voltage comparators.

Voltage Comparator

The voltage difference between the inverting and the noninverting inputs of an operational amplifier is amplified by the open-circuit gain of the operational amplifier. Because the open-circuit gain is 100,000 or more, a very small difference between the voltages to the inputs results in a large output voltage. One voltage can be compared with another by placing the reference voltage on the noninverting input and the other (unknown) voltage on the inverting input, or vice versa. While the unknown voltage is less than the reference voltage, the output voltage will be the saturated high voltage (the maximum output). When the unknown voltage is approximately equal to the reference voltage, with a difference of only a few microvolts, the output voltage will fall within the linear range of the operational amplifier, but the output will be the saturated low value when the unknown exceeds the reference voltage by a small amount. The transition between the high and the low output voltages signals the change in relative voltage amplitudes when the reference and the unknown voltages are within a few microvolts of each other. (See upper diagram in Figure LE-20a.)

- Make the voltage comparator and adjust a voltage divider to provide a reference voltage of 1 V. Use a sawtooth or ramp voltage as the unknown voltage and observe the transition when the relative amplitudes of the reference voltage and of the unknown voltage are reversed. Estimate the accuracy of the comparison. Is there hysteresis in the transitions? That is, does the switch in the output voltage occur at the same level as the unknown voltage changes upward and downward?

Comparator Ladder

A comparator ladder is formed by many voltage comparators with sequential reference voltages that correspond to steps at which the analog voltage is to be quantized. (See lower diagram in Figure LE-20a.) The comparators whose

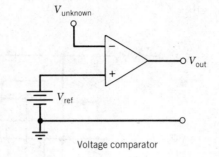
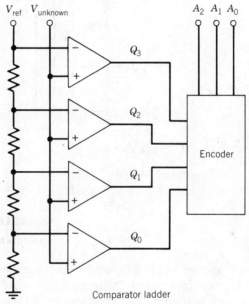

FIGURE LE-20a
Voltage Comparators

reference voltages are less than the applied voltage will be in the state of saturated low output, and the other comparators will have saturated high output. As the applied voltage increases or decreases, the comparators change their output state. The number of comparators below the applied voltage yields the fraction of the full-scale voltage. (Full scale is the voltage at which all comparators will have switched.) The number may be encoded into a binary or a digital number to be read or used.

- Make a comparator ladder with four operational amplifiers. Use a resistor string to give the sequential reference voltages, spaced equally from 0 to +4.0 V.
- Make a logic-gate circuit (Figure LE-20b) to encode the operational amplifier outputs into a binary number. Use LEDs to display the bit pattern as the

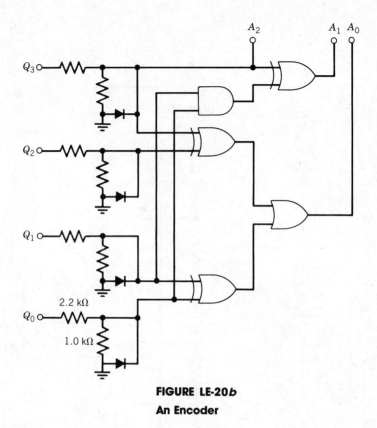

FIGURE LE-20b
An Encoder

applied voltage is varied. Because the output voltage of the operational amplifiers swings between approximately $+14$ V and -14 V, the outputs cannot be attached directly to the TTL gates in your encoder. Design and make the circuit necessary to ensure that the voltage to the TTL gates lies between 0 and $+5.0$ V.

- Place a variable voltage on the input of the comparator ladder; observe and measure the transition voltages along the ladder. Use a sawtooth voltage as the input to the comparator ladder. Increase the frequency of the sawtooth voltage while observing the rate at which the input signal is digitized. Does the comparator ladder provide fast analog-to-digital conversion? Estimate the conversion time. What limits the rate of conversion?

Points at Issue

Four-bit analog-to-digital conversion is quite imprecise. How many comparators are required to give one percent accuracy? What precautions are required to make the resistor chain adequate for a one percent conversion?

How many comparators are required if the comparator ladder and associated circuits are to yield 10-bit conversion accuracy? Is a comparator ladder a reasonable means of achieving the 10-bit accuracy? An ADC with 256 comparators for digitizing video signals is commerically available. What is an alternative to the long comparator ladder for 10-bit conversion?

Components

741 (or 747 dual) operational amplifiers
Resistors
LEDs
7400 series logic gates
1N941 or similar diodes

LABORATORY EXERCISE 21
Summing Amplifiers

The sum of several voltages may be obtained with the versatile operational amplifier. The signals may be either dc, ac, or any combination. The algebraic sum, inverted by the amplifier, is produced faithfully by the inverting amplifier. The sum is limited only by the voltage range of the operational amplifier and its frequency range.

The summing amplifier is an essential component of many digital-to-analog converter (DAC) circuits. It is used in analog computers and in many instrument circuits.

Summing Amplifier

In the summing amplifier each voltage is connected to the inverting input of the operational amplifier through its own resistor. The feedback current of the inverting amplifier is the sum of the input currents, and each is given by its input voltage divided by its input resistance. Because the input currents depend on voltage and resistance, it is possible to give different weights to input signals. This will be illustrated below when we make a summing amplifier whose output is proportional to the numerical value of a binary number.

The magnitude of the output signal depends upon the input voltages, the input resistances, and the feedback resistance. The equation that relates the quantities is found by applying Kirchhoff's laws to the first circuit diagram in Figure LE-21. The virtual ground at the inverting input of the operational amplifier and the essentially zero voltage between the inverting and noninverting inputs are used in

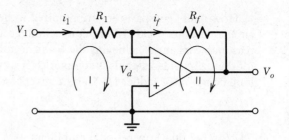

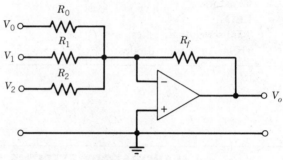

Bit-weighted summing amplifier

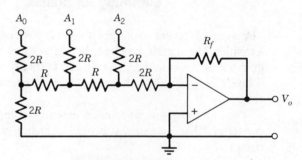

R-2R summing amplifier

FIGURE LE-21
Summing Amplifiers

writing the Kirchhoff loop and node equations, as in Chapter 9. The result is the equation

$$\frac{-V_{\text{out}}}{R_f} = \frac{V_0}{R_0} + \frac{V_1}{R_1} + \frac{V_2}{R_2}$$

No restriction on the nature of the voltages is implied in the equation. The algebraic addition of voltages is restricted only by the voltage range of the operational amplifier and its frequency limitations. The resistance values may be chosen to ensure that the voltage range limitations are met.

In the special case when all resistors are identical we find that the voltages are summed with equal weight, that is,

$$-V_{out} = V_0 + V_1 + V_2$$

- Make a summing amplifier with identical resistance values (approximately $10\,k\Omega$ is suitable) to add the following sets of three voltages.

	V_0	V_1	V_2
Set 1	$V_0 = 1.0$ dc	$V_1 = 0.5$ dc	$V_2 = 0.25$ dc
Set 2	3.0 dc	-1.0 dc	-1.5 dc
Set 3	2.0 1-kHz	1.0 dc	0.50 dc

- Confirm that in each set the voltage sum is proportional to the input voltages. Vary the amplitude of the ac voltage in Set 3, and observe the sum. Is the summing amplifier suitable for mixing audio signals? Try adding two different ac signals.

Bit-Weighted Sum

The output of a summing amplifier can be made proportional to the numerical value of a binary number by using bit-weighted input resistors. Each voltage input is either 0 V for a zero bit or $+5$ V for a one bit. The input resistor values establish the bit weights. If we rewrite the first equation, we have

$$-V_{out} = \frac{V_0 R_f}{R_0} + \frac{V_1 R_f}{R_1} + \frac{V_2 R_f}{R_2}$$

The quantities V_0 and R_0 apply to the least significant bit (LSB), V_1 and R_1 apply to the next more significant bit, and so forth. Because each bit in a binary number has twice the value of the one to its right, a factor of two must exist between the currents from adjacent inputs. Because all the input voltages, V_0, V_1, and V_2, are either 0 or 5 V, the factors of two are governed by the input resistances. If the ratio R_f/R_0 is unity, then the ratio R_f/R_1 must be two and R_f/R_2 must equal four. With these values the currents are weighted by the ratios one, two, and four. More significant bits may be summed by continuing the power-of-two ratios between R_f and R_n for the nth bit.

- Change the summing amplifier into a bit-weighted summing amplifier. Use 0 and 5 V as inputs for several binary numbers to measure the bit-weighted sum.

The bit-weighted sum requires that precise ratios of resistances be provided, a difficult task for large binary numbers. (Another problem, the settling time, which arises when large binary numbers are converted to a proportional voltage, is discussed in Chapter 12.) The next circuit to be discussed is a practical solution to these problems.

R-2R Ladder Summing Amplifier

The *R-2R* resistor ladder is a clever arrangement that splits the current to the summing junction (the inverting input) of the operational amplifier. The current is divided into equal parts at each node (junction) in the ladder. The analysis of the *R-2R* ladder is done best by using the superposition of currents. The current from the +5.0 V placed on the input for the most significant bit (the one closest to the summing junction) is found by assuming that the input of each other bit is grounded. Similarly, the current contribution of any other bit may be found by assuming that the input of interest carries +5.0 V and that all others are grounded.

- For the binary number 101 calculate the current from each node and verify that the net current to the summing junction is proportional to the magnitude of the binary number.

- Make the ladder network and use all eight combinations of inputs to measure the output voltages. For inputs use the power source for +5.0 V and the ground for 0 V.

- Repeat the instructions just mentioned with inverters as the source of the inputs. That is, put an inverter between +5.0 V and the input corresponding to a zero bit and another inverter between ground and the input corresponding to a one bit. How well does the sum voltage match the magnitude of the binary number when TTL elements are used as the signal sources? Measure the inverter voltage outputs. How much do they differ from 0 and +5.0 V? Would you choose a DAC with TTL logic gates attached directly to the resistors in the *R-2R* ladder?

Points at Issue

If you can read a meter needle position to 3 percent of full scale, how many binary bits should be used in the DAC to match the accuracy of the meter reading?

Assume that an ammeter with the full-scale reading of 1.0 mA attached to the output of the summing amplifier is to be used to read the magnitude of any four-bit binary number. Show the circuit and the resistance values to give a full-scale reading that corresponds to the binary number 1111. The inputs are 0.0 and 5.0 V. Does your answer depend upon the *R* value used in making the resistance ladder?

Components

741 Operational amplifier
7404 Hex inverter
Assorted resistors

LABORATORY EXERCISE 22
Peak-Voltage Detection

An operational amplifier circuit can measure a voltage independently of its waveform. In this exercise we will make a peak-voltage detector, which yields the peak value for evaluation by a dc voltmeter (Figure LE-22).

The usual voltage measurement of a sine wave signal yields the rms (root-mean-square) value. The peak voltage is easily obtained as the square root of two times the rms value. A nonsinusoidal voltage wave, however, does not have this simple relation between an average measured by a voltmeter and its peak voltage. The relation between average and peak voltage of a known waveform, such as a sawtooth wave, can be found by Fourier analysis, but it may not be possible to analyze an irregular waveform without copying it in detail.

Peak-Voltage Detector

The noninverting amplifier has ideal properties for detecting a voltage. The advantages arise from the high input impedance and the low output impedance. The first allows a voltage to be placed on the input of the operational amplifier

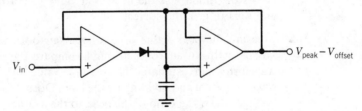

Peak-voltage detector

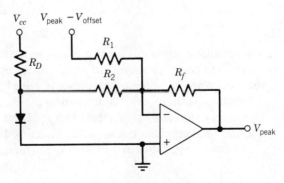

Summing amplifier

FIGURE LE-22
Peak-Voltage Detection

without loading (drawing significant current from) the voltage source. The second allows a high rate of charging a capacitor to the peak voltage, which can then be measured. You will see that these two advantageous properties also make it possible to detect the voltage of the charged capacitor and to give the peak voltage that can be measured by a common voltmeter.

The peak voltage is found because the capacitor on the output of the first operational amplifier is charged through the diode whenever a part of the voltage being measured is greater than the voltage of the capacitor. The low output impedance of the operational amplifier causes the capacitor to be charged quickly. An error exists in the measured voltage, however, because of the voltage drop across the diode. The voltage reached by the capacitor is less than the peak voltage by a constant value. Can you suggest what the magnitude of the voltage difference might be?

The capacitor charged to a peak value tends to hold that voltage because there is no discharge path. If the voltage waveform changes so that there is lower peak voltage, the peak-voltage detector will not quickly indicate the new value. What do you suggest as a remedy for this limitation?

- Make the peak-voltage detector. Measure the peak voltage of a sine wave and confirm the relation between the rms value and the peak value. Adjust the signal generator so that there is no dc-offset voltage. Measure the voltage drop across the diode; use the measured value to correct the indicated peak value and thus to give the correct value.

- Measure the peak value of a voltage sawtooth wave. Measure the sawtooth voltage with an ac voltmeter and compare the values. Sketch the sawtooth waveform and estimate the average value by measuring the area of the sawtooth wave and dividing by the base. Does the ac voltmeter give the average voltage? State the ratio of the peak to the average that you measure and to the average that you found from your plot of the waveform. Do they agree?

Diode Bias

A versatile operational amplifier may be used to make an easy correction of the measured peak voltage for the offset voltage caused by the voltage drop across the diode. When a summing amplifier is added to the output of the peak-voltage detector, the diode drop is restored to the voltage of the peak-voltage detector to yield an accurate measurement. This application shows the usefulness of the basic operational amplifier circuits—the inverting amplifier, the noninverting amplifier, and the voltage follower—in solving instrumentation difficulties.

- Add the summing amplifier to the peak-voltage detector. Adjust the current through the diode to give identical voltage drops across both diodes. Measure the peak voltages for the sine wave and the sawtooth wave. Estimate the error between the actual peak values and those measured with the diode-offset voltage correction. Is the current magnitude through the diode in the summing amplifier critical in achieving accurate measurements?

Points at Issue

Frequently we have found that precise resistance values are not required for suitable circuit performance. Is this true in the offset-voltage correction? What is the required precision in the resistor value?

The peak-voltage detector with the offset-voltage correction is to be made available as an instrument for general use in your laboratory. What provisions for adjustments and calibration should be included in the instrument?

Components

741 Operational amplifier
1N941 or similar diodes
Capacitor
Resistors

Suggested Values

$$C = 1000 \, \text{pF} \qquad R_1 = R_2 = R_3 = 10 \, \text{k}\Omega$$

LABORATORY EXERCISE 23

An Integrator

The arithmetic uses of operational amplifiers were illustrated in earlier exercises, where they were used to multiply by a constant (inverting and noninverting amplifiers and a voltage follower) and to add voltages (summing amplifier). This exercise illustrates the integration of a voltage. Integration is one of the two basic processes (the other is comparison) used in converting an analog signal into its binary equivalent (ADC).

The virtual ground is a unique property of an inverting amplifier made with an operational amplifier. In this arrangement the current to the summing junction continues through the feedback impedance to the output terminal. This property is the basis of arithmetic circuits made with operational amplifiers.

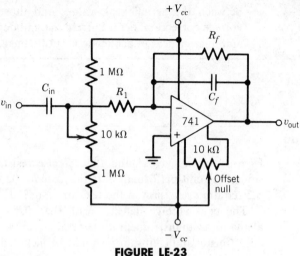

FIGURE LE-23
An Integrator

Integrator

The feedback resistor R_f in the inverting amplifier is replaced with a capacitor C_f to make the integrating circuit (Figure LE-23). The current to the summing junction is determined by the input voltage and the input resistor, as before. The operational amplifier keeps the summing junction at virtual ground by providing an output voltage that sinks the current; that is, it ensures that the current to the summing junction continues through the feedback impedance. The charge (q) on the capacitor must be the integral of the current to the summing junction to sink the current. The charge, voltage, and capacitance are related by

$$q = Cv_{out} = -\int i_f \, dt + Q$$

Because of the virtual ground, the input current i_{in} is identically equal to the feedback current i_f. We use $i_{in} = v_{in}/R_1$ to rewrite the equation as

$$v_{out} = -\frac{1}{R_1 C_f} \int v_{in} \, dt + V_{out}$$

Q and V_{out} are the constants of integration established by the initial conditions. Lowercase letters are used for the voltages and currents to indicate that they are not dc values. That is, any voltage waveform, including a constant voltage, may be integrated as long as the voltage amplitudes are within the range of the operational amplifier (plus or minus 80 percent of the supply voltage). The factor before the integral ($1/R_1 C_f$) is a scaling factor with units of inverse seconds. Real-time integration occurs when the magnitude of the scaling factor is unity.

Experimental Details

The integrator circuit responds to the input voltage in typical fashion; that is, it adds the input components with respect to time. In addition to the input-signal voltage, whose integral is desired, other voltages contribute to the integral. The other voltages include an unwanted dc voltage offset of the input signal and the intrinsic input offset voltage of the operational amplifier itself. The latter two voltages must be dealt with in order to obtain a true integral of the input signal.

The integral of a constant (dc) voltage is a ramp voltage that grows linearly in time at the rate proportional to the dc voltage amplitude and inversely proportional to the scaling factor, $R_1 C_f$. The negative sign in the equation above indicates that the outramp is opposite in sign to the dc input voltage. Unless the dc components are eliminated, the output of the operational amplifier integrator output reaches saturation, and the input signal is no longer integrated faithfully. An input signal with a dc component is integrated as a ramp plus the integral of its time-varying portion, and saturation will be reached because of the dc portion.

The true integral will be displayed if care in making an integrating operational amplifier circuit reduces or nulls the offset voltages. One means of reducing the contribution of an offset voltage is to place a high resistance, R_f, in parallel with the capacitance, C_f. A long-time constant of the feedback impedance, $R_f C_f$, will allow a time-varying signal to be integrated while it returns the dc portion as a feedback signal. As a result, a constant voltage is added to the integrated time-varying (ac) portion. As long as the sum of the constant voltage and the ac portion does not reach the saturation limit, the ac part is integrated properly, but the output signal is offset by the constant voltage, whose magnitude is a function of the input-offset voltage. The offset in the output is not generally known in advance.

Integrator Performance

The use of an operational amplifier as an integrator is less straightforward than the simple theory suggests. Some practical aspects are shown in the measurements to be made here. The operational amplifier, 741, is useful for the integrator, but an advanced system would employ an operational amplifier with features that allow it to serve better than the 741.

- Make the operational amplifier integrator. Use a square wave input voltage as the signal to be integrated. Predict the waveform of the integral of the square wave. Examine the output and calculate the output amplitude with

$$-\frac{1}{R_1 C_f} \int v_{in}\, dt = -\frac{1}{R_1 C_f} V_{in}\left(\frac{T}{2}\right)$$

where T is the period of the square wave and $V_{(in)}$ is the amplitude of a square wave symmetrical around 0 V, that is, with zero offset voltage. Compare the measured and the calculated amplitudes.

- Change the period (frequency) of the square wave. Predict and verify the new amplitude.
- Change the input to a sawtooth voltage. Predict the output waveform and compare it with the measured output.
- Ground the input through R_1 and measure the dc output voltage. Adjust the offset-null potentiometer to bring the dc output as close to 0 V as you can. Remove the feedback resistor, R_f, and measure the ramp rate. Reset the integrator; that is, replace R_f and then readjust the offset null to a different value (at the output), and repeat the measurement of the ramp rate. Compare the ramp rates to the magnitude of the offset voltages. Is the integration formula correct?
- Choose a slowly rising ramp and measure the linearity of the ramp by recording the time versus voltage. With care 30 sec or more are required for the ramp voltage to reach saturation. Is the integration of a constant truly proportional to time?
- With a square wave as the input, allow a slow ramp to exist. Observe that the output signal is the sum of the input and that the offset is integrated. What happens when the ramp has grown to a saturation level? Was the square wave integrated properly before it reached the saturation level?
- Replace the resistor, R_f, and use a square wave with a slight offset as the input. Does the integrated square wave ride on an output-offset voltage whose magnitude is proportional to the input offset and the magnitude of R_f? Change R_f to confirm the dependence on it. Vary the input-offset voltage to measure the output-offset voltage. Find the relation between the output-offset voltage and the time constant $R_f C_f$; or is it dependent upon R_f/R_1?

Points at Issue

A dc voltage is measured by integrating it. The ramp voltage is compared with a reference voltage by using a voltage comparator to stop the integration and to reset the measurement to start again. While the ramp is growing, a counter is accumulating the counts from a clock, and the count total is used to evaluate the dc voltage amplitude. Show the configuration of elements to measure the voltage. What calibration is required and how may it be done?

Components

741 Operational amplifier
Metalized polyester capacitors
Potentiometers
Assorted resistors

Suggested Values

$$C_{in} = 0.2\,\mu\text{F} \qquad C_f = 0.15\,\mu\text{F}$$
$$R_1 = 100\,\text{k}\Omega \qquad R_f = 1\,\text{M}\Omega$$

LABORATORY EXERCISE 24
The Schmitt Trigger

The Schmitt trigger is a useful circuit because it has the property of turning on at one signal level and turning off at a different signal level. This property is called hysteresis. Hysteresis is useful in a control circuit that must start a device at one level and stop it at a different level. An application is the control of a furnace, which should not turn on and off repeatedly as the temperature waivers about a set level. It should come on and stay on until the temperature has increased sufficiently and then go off.

A number of logic gates have Schmitt trigger inputs for the same purpose, that is, to require a higher level signal to cause a logic gate to change state and a lower level signal to return to the initial state. Noise arising in transmitting digital signals may be rejected by the Schmitt trigger inputs. The typical hysteresis in logic gates with Schmitt trigger inputs is 0.8 V. In this exercise, however, we examine the properties of a Schmitt trigger made with an operational amplifier. Chapter 13 discusses a transistor circuit, Figure 13-2, and an operational amplifier circuit, Figure 13-3, that are Schmitt triggers.

Operational Amplifier with Hysteresis

The circuit in Figure LE-24 differs from the circuit in Figure 13-3 because of the addition of a resistor voltage divider on the input line. The voltage divider allows a dc offset voltage to be added to an input signal. If the signal generator you use in the exercise has an offset-voltage control, the resistance voltage divider on the input line is not required and may be omitted. The capacitor on the input line is removed for slowly varying signals.

The operational amplifier with hysteresis is essentially a voltage comparator in which the reference voltage changes as the output changes from the saturated high to the saturated low state. The saturation voltages, $+V_s$ and $-V_s$, are approximately 80 percent of the supply voltages, but they are not necessarily equal in magnitude; thus the reference voltages may not be symmetrical about 0 V. A resistance voltage divider from the output of the operational amplifier to ground establishes the voltage interval, that is, the magnitude of the hysteresis between the positive and negative reference voltages. The reference is positive when the

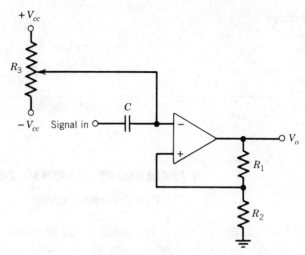

FIGURE LE-24
The Schmitt Trigger

operational amplifier is in the positive saturated state, $+V_s$, and negative in the other state, $-V_s$. The magnitude of the reference voltages is given by

$$V_{ref} = V_s \frac{R_2}{\langle R_1 + R_2 \rangle}$$

A narrow range of hysteresis may be chosen by letting R_2 be much smaller than R_1.

- Make the Schmitt trigger circuit. Measure the saturated voltages and compute the reference voltages for each state.
- Use a sine wave, a square wave, and a triangular wave as input signals to the Schmitt trigger. Measure the voltages that cause the changes in state. Do the measured values correspond to the calculated values? Shift the dc offset of the input signals to find the levels at which the Schmitt trigger does not respond. Why does the Schmitt trigger fail to respond?
- Change the relative sizes of R_1 and R_2 to expand or contract the difference between triggering levels. Is there a limit in the minimum difference between triggering levels?
- Increase the frequency of the input signal and observe the slope of the transition between saturated states. Is the slope governed by the slew rate? Calculate the slew rate. Increase the frequency to find if the change of states of the operational amplifier becomes inhibited.

Points at Issue

The output voltage of the Schmitt trigger made with an operational amplifier exceeds the 0 to 5-V range allowed for the inputs to TTL logic gates. Design a circuit that will accept the Schmitt trigger output and drive TTL gates.

A device is being controlled by a circuit involving a Schmitt trigger, and you wish to know how many times it is cycled on and off in a day. Show the circuit that will provide the record.

Explain how the Schmitt trigger may be used to reject electrical noise on the input signal.

Components

741 Operational amplifier
Assorted resistors
Capacitor

Suggested Values

$R_1 = 100\,k\Omega \qquad R_2 = 10\,k\Omega$
$R_3 = 1\,M\Omega$ potentiometer
$C = 0.2\,\mu F$

LABORATORY EXERCISE 25
Active Filters

Circuits formed only with resistors, capacitors, and inductances are passive circuits because they can merely respond to the input signals. Active circuits also contain amplifiers or other elements that may draw energy from power sources as they respond to the input signals. Active circuits can behave in a superior manner relative to passive circuits, as this exercise will show.

To illustrate an advantage of an active circuit, we compare an active low-pass filter with a passive low-pass filter. A low-pass filter propagates low-frequency signals and attenuates high-frequency signals. Bandpass filters pass signals that lie between certain frequencies, and high-pass filters attenuate low frequencies and pass high-frequency signals.

Filters are characterized by the frequencies at which the signals are attenuated to one-half of the maximum transmitted power. In a bandpass filter the frequency interval between half-power points is the frequency bandwidth of the filter. A low-pass filter has a frequency bandwidth from 0 Hz to the half-power frequency. Both passive and active filters may be placed in series to sharpen the transition from transmitting to attenuating signals.

Passive Low-Pass Filter

A passive low-pass filter is the resistor-capacitor circuit shown in Figure LE-25. By applying Kirchhoff's laws to the ac circuit we find that the output voltage, v_o, is related to the input signal, v_i, by the frequency-dependent equation

$$v_o = \frac{1}{1 + j2\pi f RC} v_i = \frac{1}{1 + \dfrac{R}{-jX_C}} v_i$$

At low frequencies, where the imaginary term in the denominator is smaller than unity, the output voltage is approximately equal to the input voltage. The magnitude of the denominator increases as the frequency of the input signal increases and the output voltage is reduced from its zero-frequency value. The electrical phase, the time relation between the ac output voltage and the ac input voltage, changes as the imaginary term in the denominator grows.

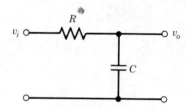

Passive low-pass filter

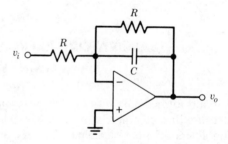

Active low-pass filter

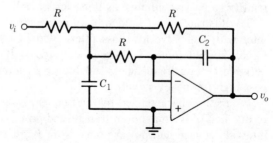

Second-order low-pass filter

FIGURE LE-25

Active Filters

The half-power point is reached as the frequency increases to the point where the imaginary term in the denominator becomes equal in magnitude to the real term, that is, when $1 = 2\pi f RC$. The equation states that the filter response, v_o/v_i versus frequency, is 70.7 percent of the zero-frequency value at the half-power point. Because the electrical power is proportional to the square of the voltage, the power is half (0.707×0.707) of the zero-frequency power, which is the term to designate this particular circumstance. The phase difference between the input voltage and the output voltage is 45° at this point.

The attenuation of the filter (the reduction in the transmitted power) may be expressed in decibels. At the half-power point

$$p_{1/2} = 10 \log \frac{p_o}{p_i} = 20 \log \frac{v_o}{v_i} = -3 \text{ dB}$$

Active Filters Versus Passive Filters

The importance of an active filter as a circuit element is seen from the effect of the load (the circuit attached to the output terminals) on the passive low-pass filter. When the load, Z_L, is in place, the output voltage is given by

$$v_o = \frac{1}{1 + \frac{R}{Z_O}} v_i$$

where

$$Z_O = \frac{-jX_C Z_L}{Z_L - jX_C}$$

Z_O is the impedance formed by the capacitor of the filter in parallel with the load impedance, Z_L. The load impedance modifies the frequency response and shifts the half-power point; that is, Z_O replaces $-jX_C$ in the equation given in the previous section.

The advantage and importance of the active low-pass filter is that the frequency response of the active filter is decoupled from the load impedance. The half-power point is not shifted. The load impedance need not be known to predict the filter response, and a variety of loads may be attached to the filter without changing its response.

Filter Responses

- Construct the passive low-pass filter and measure the ratio of the output voltage to the input voltage as a function of frequency. Use approximately a 1-V signal as the input. Make measurements at the beginning of each frequency

decade and at two intermediate frequencies. Choose the resistor and the capacitor to give the half-power point at about 1 kHz. Plot the data on semilog graph paper.
- Add a similar capacitor as the load on the filter and repeat the measurement. Plot the data on the same graph.
- Construct the active filter and repeat both of the steps described above.
- Construct the second-order active low-pass filter and repeat the first step described above.

Compare the responses of the filters you constructed.

Points at Issue

Use Kirchhoff's laws to find the equation that gives the response of a passive low-pass filter if its load is an identical low-pass filter.

What configuration of a resistor and a capacitor creates a high-pass filter? Find the equation to show that it blocks dc signals and passes high-frequency signals.

Find the rate of attenuation in decibels per octave and decibels per decade at high frequencies for all of the first-order and second-order low-pass filters you measured.

Components

741 Operational amplifier
Assorted resistors
Assorted capacitors

Suggested Values

$$R = 1.6 \text{ k}\Omega, \quad C_1 = 0.15 \text{ } \mu\text{F}, \quad C_2 = 0.067 \text{ } \mu\text{F}$$

LABORATORY EXERCISE 26

MOSFET Electronic Switches

MOSFET (metal-oxide-semiconductor field-effect transistor) electronic switches may be opened and closed by digital signals at rates greater than 1 MHz. In this exercise we examine some of the unusual, and until recently impossible, circuit phenomena that these electronic switches can achieve, as in Figure LE-26. The next exercise shows the utility of the electronic switches in sampling analog signals.

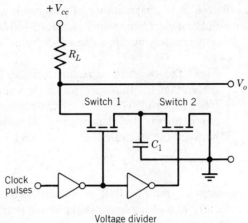

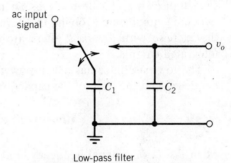

FIGURE LE-26
MOSFET Electronic Switches

Electronic switches are incorporated in LSIC (large-scale integrated circuits) as analog and digital electronics are made part of a single element for analog-to-digital or digital-to-analog conversion. These switches can be used to circumvent the limitations in forming resistors in high-density integrated circuits.

Switched-Capacitor "Resistor"

A switched-capacitor resistor whose resistance depends upon the switching rate will be made with a small capacitor and a MOSFET switch (C4066). The switched-capacitor resistor in a resistor-divider circuit is a variable resistance, across which a portion of the voltage appears. A change in frequency of the clock driving the switch is a means of changing the voltage output.

The resistance of the switched-capacitor resistor is inversely proportional to the switch frequency; that is,

$$R \propto \frac{1}{Cf}$$

where C is the capacitance in picofarads and f is in hertz. The resistance of the conducting electronic switches, about 100 Ω, is usually smaller than the resistance being generated, but its value contributes to the actual resistance.

- Make the switched-capacitor resistor. Determine its resistance by measuring the voltage fraction across it in the voltage-divider circuit. Vary the frequency from 1 to 100 kHz, and plot the resistance versus switch frequency. Does the curve fit the $1/f$ dependence? Determine the constant of proportionality between the measured resistance and $1/Cf$.

Low-Pass Filter

Among the reasons that make switched-capacitor resistances important is the difficulty of making integrated circuits that combine analog and digital functions and operate in the audio frequency range. The area on a silicon chip restricts the capacitance that can be made as an integral part of an integrated circuit, but switching circuits are evolving to alleviate the limitations. The capacitors used in the exercise with low-pass filters are as large as the packages containing the integrated circuits.

The low-pass filter has a half-power frequency proportional to the switching frequency and the ratio of the capacitors. The relation is

$$F(-3\,dB) \propto f\left(\frac{C_1}{C_1 + C_2}\right)$$

when $f \gg F$. F is signal frequency, and f is the switching frequency. You should note that the half-power frequency is proportional to the ratio of capacitances; thus large capacitors are not required for audio frequency signals.

- Make the switched-capacitor low-pass filter. (The switched-capacitor resistor in the low-pass filter is the same as that in the voltage divider, but a simpler symbol is used.) Set f equal to 50 kHz and vary F from 10 Hz until the half-power point is passed. Plot the filter response. What is the constant that makes an equation out of the proportionality? Change the switch frequency by about 25 percent and confirm that the half-power point is changed similarly.

Points at Issue

The switched-capacitor resistor in a voltage-dividing circuit can become an analog frequency meter. Show a circuit that can act as a frequency meter without loading the switched-capacitor resistor. Estimate the sensitivity of the frequency meter; that is, the change in voltage versus the change in switch frequency, over the useful range of your frequency meter. What do you think could be the useful ranges of such a setup?

Components

C4066 Quad bilateral switch
Assorted capacitors
Resistor

Suggested Values

$$C_1 = 33\,\text{pF}, \qquad C_2 = 1000\,\text{pF}$$

LABORATORY EXERCISE 27
Signal Sampling

This exercise demonstrates the features of an electronic switch by using it to sample a voltage signal. Electronic switches may be opened or closed at a high rate (more than 1 MHz), far beyond the rate possible with any mechanical switch. The switches are used with analog signals for signal gating, signal chopping, modulating and demodulating, and in analog-to-digital conversion (ADC). Digital signals may be multiplexed, and digital control may be implemented by the switches. In the previous exercise an electronic switch was used to simulate resistance in a circuit without a resistor.

Voltage sampling is used to provide a momentary signal for measurement or analysis. Analog-to-digital conversion, for example, requires a voltage amplitude at each conversion point. This amplitude is frequently provided by a sample-and-hold circuit that keeps the voltage constant while its amplitude is being digitized.

Signal chopping is used to transform a nearly steady voltage into an alternating voltage so that it may be amplified by a capacitor-coupled amplifier. The long-time drift, a problem in dc-amplifying circuits, can be circumvented and the significant changes in the slowly changing voltage amplitude can be assessed more reliably. Signal chopping is demonstrated in the exercise.

Bilateral Switch

The quad bilateral switch chosen for this exercise, the C4066, contains four enhancement MOSFET (metal oxide-semiconductor field-effect transistor) switch circuits. Figure LE-27 shows the arrangement of each switch. (Chapter 8 discusses enhancement transistors, and Figure 8-20 shows the semiconductor arrangement of an enhancement MOSFET.)

Two features are important for operation as switches. One is the channel between the source and drain, formed when the enhancement MOSFET is conducting. The term "bilateral" indicates that the conduction can be in either direction; that is, either end of the channel may be considered the source and the other end the drain.

478 Laboratory Exercises

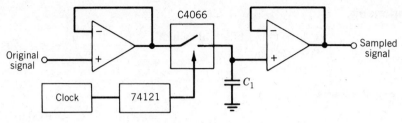

Signal sampler

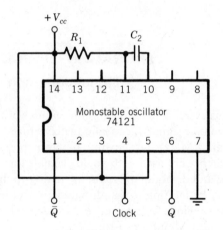

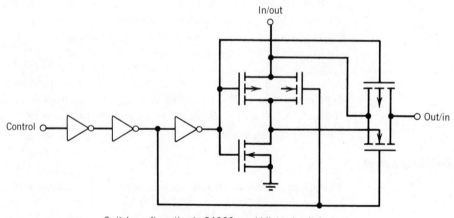

Switch configuration in C4066 quad bilateral switch

FIGURE LE-27
Signal Sampling

The second feature of the enhancement MOSFET is the absence of a channel between the source and the drain unless the control voltage creates (enhances) the channel. At each end a *pn* junction is formed by the input and output regions (the source and drain regions) and the substrate. Depending upon the signal-voltage polarity, one *pn* junction is reverse-biased and no conduction can occur. The lack of conduction is equivalent to a very large resistance. The open switch has a resistance of about 10 GΩ (10^{10} Ω.) The resistance of the closed switch is about 100 Ω.

Caution is required in handling the MOSFET because a static electrical discharge between the ground point and DIP being held by a person may destroy the element by piercing the oxide layer of the transistor. Ground yourself before picking up a MOSFET by holding on to the ground of the circuit board into which the DIP is to be inserted. Insert the DIP carefully. MOSFET elements should be stored in conducting foam or conducting plastic tubes.

Voltage Sampling

The circuit for sampling a voltage is made with operational-amplifier voltage followers and a bilateral switch. Two advantages of the voltage follower, high input impedance and low output impedance, enable the voltage source to be sampled without loading it and allow the capacitor to be charged quickly to the magnitude of the voltage at the sample point. The switch isolates the capacitor between sampling points. The output voltage follower may be considered an impedance-matching circuit element.

The capacitor can be charged to the voltage of the sample point in the time available only if the time constant, RC, is smaller than the sampling interval. R is approximately the resistance of the conducting switch; thus a very small capacitor must be used for small sampling intervals suitable for rapidly changing voltages. The output voltage follower draws a slight current from the capacitor in the interval between sampling and diminishes the voltage stored on the capacitor.

In this exercise the voltage followers are made with 741 operational amplifiers, which do not have extremely high input resistances. As a result, the capacitor discharges significantly in the period between sampling. (You will note this in your measurements.) Operational amplifiers with FET input transistors are selected for commercial voltage sample-and-hold circuits to minimize the drain of charge from the capacitor.

The small sampling interval during which the switch is closed is made by a short pulse on the switch-control line. Here we use a monostable oscillator as the control voltage source. The monostable oscillator is gated on by a clock pulse. The pulse lasts only a short time, until the monostable oscillator returns to its stable state and stays off until the next clock pulse induces a subsequent pulse. The 74121 monostable oscillator generates approximately a 35-nsec pulse, but the external resistor-capacitor circuit can extend the pulse duration from 40 nsec to 28 sec. In this exercise we use a sampling pulse interval of 100 μsec or more.

- Make the voltage-sampling circuit. Place a sine wave on the input voltage follower. Observe the switch action by placing a low (0) on the switch-control line. Vary the amplitude of the sine wave voltage and observe the output. Does the switch conduct when the input signal is above a certain magnitude? Sketch the output waveform and explain what you observe.

- Place a high (1) on the switch-control line and observe the output. Vary the amplitude of the input signal, and sketch the output signal when the input signal is larger and smaller than the voltage on the switch-control line.

- Make the monostable oscillator. Add the external resistor and capacitor and verify that a suitable pulse width is provided. Drive the circuit with a square wave clock pulse. The on-time of the 74121 should be only about 5 percent of the clock period.

- Attach the on-pulse to the switch-control line. Observe the output signal and sketch the waveform. Use input signal amplitudes larger and smaller than the on-pulse. Observe the decay of the voltage on the capacitor.

- Use a sine wave input signal whose frequency is smaller than the clock signal. Observe the sine wave signal to determine if it is chopped. That is, is the output signal a modulated series of square waves that follows the amplitude of the input signal? Use a sawtooth wave. Is it chopped properly?

Point at Issue

The manufacturer of the C4066 states that the crosstalk between switches in the DIP is -50 dB. What does this mean?

Explain the operation of a switch circuit in the C4066.

Components

C4066 Quad bilateral switch
74121 Monostable oscillator
741 Operational amplifiers
Capacitors
Resistors

Suggested Values

$$R_1 = 220 \text{ k}\Omega, \quad C_1 = 100 \text{ pF}, \quad C_2 = 56 \text{ pF}$$

LABORATORY EXERCISE 28
A Rate Meter

It is useful to convert the rate of pulses into an analog signal to present data for quick evaluation. The sequential presentation of the rate itself as a numerical display gives the same information, but it requires the observer to recall the earlier numerical values in order to recognize that the rate is increasing or decreasing. For pulses that occur at random the subsequent readings may vary widely, and the average rate may be difficult to evaluate.

The conversion of a sequence of pulses into an analog signal is properly called digital-to-analog conversion (DAC). In this exercise the process is analogous to the rectification of the pulses with a capacitor filter. The same system is used in power supplies to produce a steady voltage from an alternating voltage.

Rate Meter

Pulses originate from a clock or from sources such as a radioactive detector or an event counter. In the latter cases the pulses may vary in amplitude and time interval. In order to present the rate rather than the character of the pulses, it is useful to generate identical pulses to be converted into the analog rate. This permits each pulse to be weighed equally in the conversion into a rate. A monostable oscillator is used for this purpose.

At the slowest event rate the pulses from the monostable oscillator occupy less than one-tenth of the average interval between events. A small analog rate is shown. As the event rate increases, the monostable oscillator generates more separate identical pulses which are converted into a larger analog signal. A change of the rate meter scale requires only a change in the pulse width of the monostable oscillator.

At high rates, randomly spaced pulses may occasionally occur very close to each other. The pulse that follows closely may be ignored because the monostable oscillator pulse overlaps it. This situation is a source of "dead-time" error, which can be reduced by using a scale that allows a higher counting rate.

For infrequent clock or random pulses the capacitor filter smooths the signal so that the analog meter holds a relatively steady position. This condition occurs when the time constant, $R_L C$, is made long with respect to the period between pulses. R_L is the load resistance in parallel with the capacitor through which it discharges. In the circuit in Figure LE-28, R_L is input resistor, R_1, to the operational amplifier. The operational amplifier serves two functions. First, it allows a low-resistance analog meter, the ammeter, to be used with a suitably high load resistance across the capacitor filter C and, second, it permits the meter to be initialized to zero.

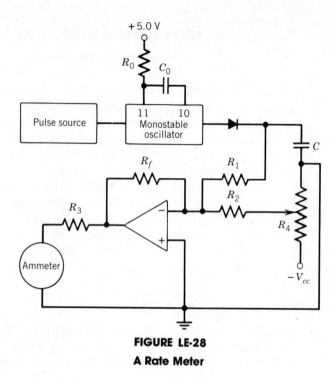

FIGURE LE-28
A Rate Meter

- Make the monostable oscillator operate and observe the pulse duration. An offset (dc) component in the clock output may prevent the monostable oscillator from generating a pulse.
- Complete the rate meter. Use a clock rate from 10 to about 1000 pulses/sec, and record and plot the output reading. A voltmeter may be substituted for the resistor, R_3, and the ammeter.
- Double the resistance, R_0, external to the monostable oscillator and remeasure the pulse duration. Then repeat the measurements above. What scale change has occurred?

Points at Issue

Each of three widely different pulse rates is to be read on a rate meter. Show the arrangement for such a meter.

Components

74121 Monostable oscillator
741 Operational amplifier

1N941 or similar diode
0–100 Microammeter or voltmeter
Assorted capacitors
Assorted resistors

Suggested Values

$$R_0 = 10\ \text{k}\Omega \quad C_0 = 0.9\ \mu\text{F} \quad C = 33\ \text{pF}$$
$$R_1 = R_2 = R_3 = 100\ \text{k}\Omega \quad R_f = 320\ \text{k}\Omega$$
$$R_4 = 10\ \text{k}\Omega$$

LABORATORY EXERCISE 29

Interfacing

Logic gates and analog elements are fabricated with several different technologies. The ones encountered in this work are TTL (transistor-transistor logic) and MOS (metal-oxide-semiconductor) devices. Difficulties may arise in driving one kind by another. Other driving difficulties arise when the voltage swing, the power available, or an offset (dc) voltage is wrong for the subsequent circuit elements. A signal reflected from an electrical discontinuity in a circuit is another problem. An unwanted second pulse, the reflected pulse, may be injected into the circuit. We examine this phenomenon in this exercise.

Several techniques may provide an interface between elements. A transistor used as an emitter follower or as a common-emitter amplifier is frequently a solution. The common-emitter amplifier is the subject of another exercise, and both it and the emitter follower are discussed in Chapter 10. Another possibility is an operational amplifier used as a voltage follower, an inverting amplifier, or a noninverting amplifier. Other times a high-pass filter or the combination of a capacitor and an offset-voltage generator is sufficient. In still other cases a TTL element of one kind, a 74xx, can replace or be replaced with a similar element, a 74LSxx or 74Hxx, to drive or to be driven by a MOS gate. Table 10-4 lists some differences between the TTL gates.

Voltage-Controlled Oscillator

Interfacing a voltage-controlled oscillator (VCO) will show some of the options available. The LM566 is an integrated circuit that generates a square wave and a triangular wave whose frequency is varied by a control-voltage input. The frequency can be changed by about a factor of 10 around a central value fixed by an external capacitor-resistor circuit. That is, for example, from 100 Hz to 100 kHz with a central frequency of about 1 kHz. The VCO is a useful signal generator for testing circuit developments.

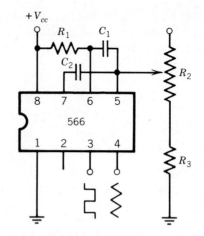

Voltage-controlled oscillator

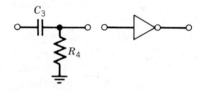

High-pass filter

FIGURE LE-29a
Interfacing

- Make the voltage-controlled oscillator circuit. Measure the frequency range and the voltage pattern of the square wave and the triangular wave. Measure the voltage offset at the outputs of the LM566.
- Attach the output of the LM566 to the input of an inverter (7404) and find out if the inverter is driven by the square wave. Is the waveform acceptable for the TTL gate?
- Add a capacitor-resistor high-pass filter between the square wave output of the LM566 and the inverter. Find the range of time constants, $R_4 C_3$, that allow the inverter to operate. Measure the dc voltage component on the input to the inverter that barely allows the inverter to operate. (See Figure LE-29a.)

Transistor Interface

Two transistor interfaces are shown in Figure LE-29b. One is suited for pulses or ac signals and the other is used for slowly or infrequently changing inputs. The function of both circuits is to provide sufficient current and a large voltage swing and to delete an offset (dc) voltage.

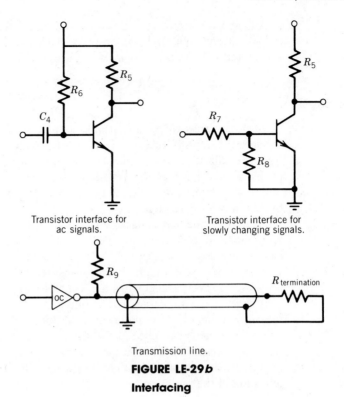

Transistor interface for ac signals.

Transistor interface for slowly changing signals.

Transmission line.

FIGURE LE-29b

Interfacing

- Make the circuit with the input capacitor. Use the output of the LM566 through the transistor to drive the inverter. Measure the voltage swing on both the input and the output of the transistor. How would you adjust the voltage swings to different limits?
- Replace the capacitor with a resistor and make the second transistor circuit. Measure the performance of this circuit.

Interconnected Equipment

Many components are designed to work together, but new problems arise in connecting different pieces of equipment together. Frequently, tristate buffers are used to interface components connected by a bus or cable. A problem arises, however, when the current from a gate or buffer is inadequate to charge the input capacitance of the cable or bus leading to another circuit board or piece of equipment. The transmitted signal takes too long to reach the high state (approximately 5 V). In high-speed circuits the signals are not propagated as expected. A pullup resistor of $1\,k\Omega$ on the output of a TTL gate or element may solve the problem even though the TTL gate is not an open-collector element. Interconnecting equipment is a challenging task for an experienced worker.

A long conductor carrying multiplexed signals from remote sensors may require a high current buffer to serve properly. Although the problem is not illustrated in this exercise, it can be recognized in unreliable transmission of signals.

Line Termination

An electrical pulse on a coaxial cable is propagated at approximately 80 percent of the speed of light. If the line is infinitely long, a pulse continues to travel from the source indefinitely. A mismatch or a discontinuity in the line, however, causes some of the electrical energy to be reflected back to the source. The time between the initial pulse and its reflection is twice the distance to the discontinuity divided by the speed of propagation.

A cable or line has a characteristic resistance that is determined by its physical dimensions—the diameter of the conductors and the separation between them. Commonly used coaxial cables have characteristic resistances of 50 to 100 Ω or more; but that is not the resistance measured by an ohmmeter. The characteristic resistance of two parallel wires may be several hundred ohms.

A shorted coaxial cable, that is, one that has the center conductor attached to the outside sheath, is mismatched at that place. A positive voltage pulse traveling down the cable is cancelled at the short—no voltage difference can exist between the center conductor and the sheath—so a voltage of opposite polarity must arise to add to the initial pulse. The negative pulse propagates to the source as a reflection of the initial pulse. Since the pulses are attenuated as they propagate in either direction, the successive reflected pulses are diminished in amplitude. Similarly, an open end of a cable is a discontinuity. If a resistance equal to the characteristic resistance is placed between the center conductor and the sheath of a coaxial cable, the energy of the pulse is absorbed and there is no reflection. Reflections are avoided by terminating the cable with the characteristic resistance to make the cable appear unending.

The discontinuity is usually not a short or an open but the mismatch that arises when a cable is attached to an instrument or a component, and a reflected wave occurs. The proper termination is a resistance that, in parallel with the input resistance of the instrument or component, matches the characteristic resistance of the cable or line.

- Obtain a coaxial cable 100 or more meters long. Perhaps a drum of cable from the electronics shop may be available. Use an open-collector inverter as the signal source. The distant end of the cable should be open because the *OC* inverter may not provide enough power to give a suitably large reflected pulse with a shorted cable. Try a shorted cable after you have adjusted the system. Trigger the inverter with a period that is several times the estimated reflection time and observe the pulse and the reflections.
- Place a large resistor across the open end of the cable and observe the pulses. Repeat this step with smaller resistors until you select a resistance that makes

the reflected pulses disappear. What happens if still smaller resistors are used? What is the characteristic resistance of the cable? Sketch the waveforms.

Points at Issue

The propagation speed of a pulse in a cable or line may be used to measure the distance to a mismatch. Conversely, the propagation speed may be measured with a cable of known length. What is the propagation speed of the pulse in the cable that you used?

Components

LM566 Voltage-controlled oscillator
7404 Hex inverter
7405 *oc* hex inverter
2N4124 or equivalent *NPN* transistor
RG-662/U or other coaxial cable
Assorted fixed and variable resistors
Assorted capacitors

Suggested Values

$C_1 = 0.5\ \mu F \qquad C_2 = 820\ pF \qquad C_3 = 0.1\ \mu F$
$C_4 = 0.1\ \mu F \qquad R_1 = 120\ k\Omega \qquad R_2 = 10\ k\Omega$
$R_3 = 5\ k\Omega \qquad R_4 = 10\ k\Omega \qquad R_5 = 2\ k\Omega$
$R_6 = 10\ k\Omega \qquad R_7 = 10\ k\Omega \qquad R_8 = 2\ k\Omega$
$R_9 = 470\ \Omega$

LABORATORY EXERCISE 30

Voltage and Current Regulation

Controlled and steady voltages and currents are required for computers and instruments and are needed in many research and development activities. A regulated voltage source maintains the voltage very close to the set value as the current drawn varies within the current ratings of the regulator. Similarly, a constant current is maintained by a current regulator independently of the change in the load resistance. Voltage and current regulation with operational amplifiers and three-terminal regulators is discussed in Chapter 14, and their circuits are

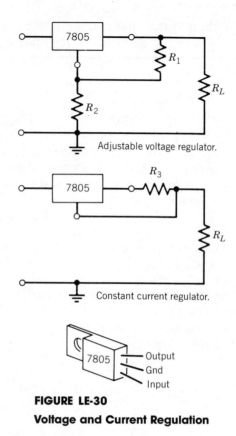

FIGURE LE-30
Voltage and Current Regulation

shown in Figures 14-14 to 14-17. Three-terminal regulators are basically the same as the operational amplifier circuits, but internal amplifiers and overload protection circuits improve their performance.

Three-terminal regulators are available with specific values of voltage (± 5.0, ± 12.0, or ± 15.0 V) that they will maintain. Each voltage regulator has a maximum current rating (100 mA, 1 A, or 5 A) that must not be exceeded. The performance of a power supply with a three-terminal regulator is described in Chapter 14. Regulators are packaged in several different forms that are usually related to the power dissipation expected in the application. Low-power regulators may be DIPs while higher-power regulators are in a form that may be attached to a heat sink as a means of keeping their temperatures within limits. In this exercise we use a positive 5-V, 1-A regulator (7805) in a TO 220 package. This package is frequently used on computer circuit boards. (See Figure LE-30.)

Adjustable Voltage Regulation

A voltage different from the fixed value of the voltage regulator may be provided by a simple circuit that uses the regulated output voltage as the source to generate an offset for the internal reference voltage in the regulator. The intrinsic amplification of the regulator contributes to the stability of the output voltage, but in all applications an error voltage must exist to be used as the control signal. Because of the high gain in the control circuit a good regulator requires a very small difference between the output voltage and the reference voltage.

- Make the adjustable voltage regulator. Determine the range of the regulated voltage. Use a load resistor that draws one-tenth of the rated current of the three-terminal regulator at the voltage midway in the adjustable range. (The suggested current is small so that the power source of the solderless breadboard will not be overloaded. If a more powerful power supply is available, currents closer to the rated current of the regulator should be used.) Measure the output voltage as the current is doubled and halved by changing the load resistance. What limits the range of adjustable voltages? How are the limits related to the current into the load?

Constant-Current Regulator

Constant currents are important in experiments and instruments. A constant current ensures a steady magnetic field in a magnet whose coils change resistance with temperature. Constant illumination from a light source in an optical instrument requires current regulation.

- Make the constant-current regulator. Measure the range of current regulation by doubling and halving the series resistance that provides the offset voltage for the regulator. Use a load resistor that draws about one-tenth of the rated current.

- Replace the load resistor by a small light bulb that requires about one-tenth of the current capacity to burn brightly. A combination of a bulb and a resistor in parallel with it may allow an available bulb to be used. As part of the load, place a resistor that has approximately one-quarter of the resistance of the lighted bulb in series with the light bulb (or combination). Adjust the current for a brightly lighted bulb. Place a shunt (a wire bypass) on the series load resistor, and observe the change in light intensity when the shunt is added and removed. Measure the voltage across the light bulb for both conditions. Adjust the current to dim the bulb so that a change in intensity may be more obvious, and repeat the measurements with the shunt. Do you consider the constant-current regulator to be adequate for the light source of an optical instrument? If not, what would you suggest that might improve the current regulator for this application?

Points at Issue

An inverting amplifier made with an operational amplifier becomes a constant-current source when a constant voltage on the input resistor causes a constant current to the summing junction. The current in the feedback resistor (the element whose current is being maintained constant) is steady even if the feedback resistance varies. Can you estimate the range of current that a 741 operational amplifier can give in this simple arrangement? What is the power into the element whose current is controlled? How does the power change when the resistance of the element changes?

Can you suggest some other applications of a constant-current source?

A standard calorimeter and an identical calorimeter holding a radioactive sample are held at the same temperature by heating the standard calorimeter at the same rate that the radioactive decay is heating the second calorimeter. The activity of an encapsulated radioactive sample is measured in this way. Can you suggest an arrangement so that the power to an element heating the standard calorimeter can be maintained constant?

Components

7805 Voltage regulator
Small light bulb
Assorted 10-W resistors
Assorted 0.25-W resistors

Suggested Values

$$R_1 = R_2 = 5\,k\Omega, R_3 = 10\,\Omega, 10\,W$$

STUDENT'S GUIDE TO EXERCISES

Each laboratory exercise begins with a discussion of the purpose of the exercise and the principles that are to be learned. Read each exercise completely through before you start to use the electronic elements. There are points and questions to be addressed as you do the exercise. It is important that you become aware of them before you start making and using the circuits.

Each portion of the exercise that requires you to do something with the electronic equipment is marked with a "•." The portions not marked can be done outside of the laboratory provided you obtain the data and information while you are constructing the circuits and making the measurements.

You will work most efficiently if you practice two simple procedures in making the circuits on the solderless experimental boards. First, connect the circuit elements in a systematic and orderly manner and use short wires. Second, try each element as it is added to the circuit to make certain that it has electrical power to it and that it works as it should before it is connected to the other elements. All exercises can be completed within a two and one-half hour laboratory session.

The exercises require only a few pieces of laboratory equipment and use a limited number of integrated circuits. The integrated circuits are the least complicated ones and represent the elements that form all integrated electronic circuits. You will find that working with the simplest elements will enable you to learn at the best rate. After you have become acquainted with electronics and begin to apply the things that you learned, you will frequently choose more sophisticated elements to satisfy your needs.

Each exercise is built upon the ones that proceed it. It is expected that the exercises will be done in the sequence in which they appear. The text and the exercises introduce material in the same order so that you may relate the information in the text to your laboratory experiences. Enjoy the insight into modern electronics that the laboratory experience gives you.

SEVEN-SEGMENT DISPLAYS

Seven-segment LED displays are available with either common-anode or common-cathode connections and with different pinouts. The pinouts are given in Table B-6. They operate with 5-V sources and must have current-limiting resistors. Resistor values from 220 Ω to a kilohm or more are appropriate. A typical value is 470 Ω.

The pinout is easily determined by examining the element. The common-anode LED requires a low (0) through the resistor to a segment to illuminate it when its anode is high (1). The common cathode is tied to ground (0), and a segment on it is illuminated when the segment lead through the current-limiting resistor is made high (1).

TABLE B-6
Pin Assignments for Seven-Segment Light-Emitting Diodes

Pin	Common Anode		Common Cathode	
1	Cathode a	No pin	Anode a	No pin
2	Cathode f	Anode	Anode f	Cathode
3	Anode	Cathode f	Cathode	Anode f
4	No pin	Cathode g	No pin	Anode g
5	No pin	Cathode e	No pin	Anode e
6	Cathode dp	Cathode d	Anode dp	Anode d
7	Cathode e	No pin	Anode e	No pin
8	Cathode d	No pin	Anode d	No pin
9	No connection	Anode	No connection	Cathode
10	Cathode c	Cathode dp	Anode c	Anode dp
11	Cathode g	Cathode c	Anode g	Anode c
12	No pin	Cathode b	No pin	Anode b
13	Cathode b	Cathode a	Anode b	Anode a
14	Anode	No pin	Cathode	No pin

RESISTOR COLOR CODE

The resistance value of a resistor is indicated by the color of the bands painted on the resistor body. The three, starting closest to the end of the body of the resistor, indicate the resistance value. The fourth band states the precision of the resistance.

The first (end) band gives the first number of the resistance value.

The second band gives the second number.

The third band indicates the power of 10 by which the other numbers are multiplied to give the resistance in ohms.

The order of the colors in the spectrum of white light corresponds to the order of the digits in the code. However, the counting starts with black and then brown before the spectrum colors are used and ends with gray and white. The numerical values are

Color	Value
Black	0
Brown	1
Red	2
Orange	3
Yellow	4
Green	5
Blue	6
Violet	7
Gray	8
White	9

The following examples illustrate the meaning of the color bands.

First Band	Second Band	Third Band	Value
Brown	Black	Black	$10E0 = 10\ \Omega$
Brown	Black	Red	$10E2 = 1000 = 1\ k\Omega$
Red	Black	Red	$20E2 = 2000 = 2\ k\Omega$
Red	Red	Red	$22E2 = 2200 = 2.2\ k\Omega$
Yellow	Violet	Brown	$47E1 = 470 = 0.47\ k\Omega$
Orange	Black	Green	$30E5 = 3{,}000{,}000 = 3\ M\Omega$

The resistor tolerance indicated by the fourth band is

No fourth band	± 20 percent
Silver	± 10 percent
Gold	± 5 percent
Red	± 2 percent

BIBLIOGRAPHY

The books in the bibliography provide an alternate presentation of much of the material in this text and other topics not presented here. The general electronics texts are annotated to suggest their usefulness. The technical journals are good sources of up-to-date information on advances in digital and analog electronics and computers. The lists are not comprehensive.

Electrical and Electronic Circuits

1945   Bode, Hendrik W., *Network Analysis and Feedback Amplifier Design*, D. Van Nostrand, New York. *An important classic.*

1947   Clifford, Harry E., and Wing, Alexander H., Editors, *Electronic Circuits and Tubes*, Cruft Electronics Staff, McGraw-Hill, New York. *A compilation of the state of electronics immediately after World War II and before transistors.*

General Electronics Text Books

1975   Havill, R. L., and Walton, A. K., *Elements of Electronics for Physical Scientists*, Macmillan, London. *Transistor circuits and their solid-state nature.*

1976   Lee, Samuel C., *Digital Circuits and Logic Design*, Prentice-Hall, Englewood Cliffs, N.J. *Combinational and sequential logic design and fault detection at an advanced level.*

1977   Taub, Herbert, and Schilling, Donald, *Digital Integrated Electronics*, McGraw-Hill, New York. *A respected text covering digital electronics and analog-to-digital conversion.*

1978   Henry, Richard W., *Electronic Systems and Instrumentation*, John Wiley, New York. *Mathematical techniques for studying electronic circuits with limited attention to operational amplifier and digital circuits.*

1979   Blakeslee, Thomas R., *Digital Design with Standard MSI and LSI*, Second Edition, John Wiley, New York. *An important book that emphasizes the use of ROMs, PLAs, and other MSI and LSI elements and the problems of timing, noise, etc., that prevent proper circuit operation.*

1979   Diefenderfer, A. James, *Principles of Electronic Instrumentation*, Second Edition, Saunders College Publishing, Philadelphia. *A useful presentation of electronics, circuits, transducers, and instrumentation.*

1979   Millman, Jacob, *Microelectronics*, McGraw-Hill, New York. *Semiconductor devices, digital and analog circuits and systems and the techniques of bipolar-junction transistor amplifier design.*

1979   Mano, M. Morris, *Digital Logic and Computer Design*, Prentice-Hall, Englewood Cliffs, N.J. *An excellent account of digital design foundations and techniques.*

1980   Horowitz, Paul, and Hill, Winfield, *The Art of Electronics*, Cambridge University Press, Cambridge, Eng. *An excellent account of digital and analog electronics including practical aspects of making circuits perform.*

1981   Malmstadt, Howard V., Enke, Christie G., and Crouch, Stanley R., *Electronics and Instrumentation for Scientists*, The Benjamin/Cummings Publishing Co., Menlo Park, Calif. *A well-accepted text on analog and digital instrumentation that covers many practical aspects.*

1981   Sprott, J. C., *Introduction to Modern Electronics*, John Wiley, New York. *Emphasis on analog circuits and elements with brief attention to digital circuits and communication electronics.*

1982   Barnaal, Dennis, *Analog and Digital Electronics for Scientific Application*, Breton Publishers, North Scituate, Mass. *Two separate introductory texts, one analog, one digital, bound together.*

1983   Brophy, James J., *Basic Electronics for Scientists*, Fourth Edition, McGraw-Hill, New York. *Introductory electrical, analog, and digital circuits.*

1983   Higgins, Richard, J., *Electronics with Digital and Analog Integrated Circuits*, Prentice-Hall, Englewood Cliffs, N.J. *A good presentation of digital and analog circuits and their applications.*

Specialized Books

1978   Ennes, Harold E., *Boolean Algebra for Computer Logic*, Howard W. Sams, Indianapolis, Ind.

1978   McWhorter, Gene, *Understanding Digital Electronics*, Texas Instruments, Dallas, Tex. (Distributed by Radio Shack.)

1979   Camp, R. C., Smay, T. A., and Triska, C. J., *Microprocessor Systems Engineering*, Matrix Publishers, Portland, Oreg.

1979   Cannon, Don L., and Luecke, Gerald, *Understanding Microprocessors*, Texas Instruments, Dallas, Tex. (Distributed by Radio Shack.)

1979   Zuck, Eugene L., Editor, *Data Acquisition and Conversion Handbook*, Data Intersil, Mansfield, Mass.

1980   Artwick, Bruce A., *Microcomputer Interfacing*, Prentice-Hall, Englewood Cliffs, N.J.

1980   Heiserman, David L., *Handbook of Digital IC Applications*, Prentice-Hall, Englewood Cliffs, N.J.

1980   Tokheim, Roger L., *Digital Principles*, Schaum's Outline Series, McGraw-Hill, New York.

1980   Young, George, *Digital Electronics: A Hands-On Learning Approach*, Hayden, Rochelle Park, N.J.

1982 Kaufman, Milton, and Wilson, J. A., *Electronics Technology*, Schaum's Outline Series, McGraw-Hill, New York.

Operational Amplifier Books

1973 Graeme, Jerald G., *Applications of Operational Amplifiers*, Burr-Brown Research Corporation, McGraw-Hill, New York.

1980 Jung, Walter G., *IC OP-AMP Cookbook*, Second Edition, Howard W. Sams, Indianapolis, Ind.

1981 Irvine, Robert G., *Operational Amplifier Characteristics and Applications*, Prentice-Hall, Englewood Cliffs, N.J.

1983 Gayakwad, Ramakant A., *Op-Amps and Linear Integrated Circuit Technology*, Prentice-Hall, Englewood Cliffs, N.J.

Technical Journals

Byte, McGraw-Hill, Peterborough, New Hampshire. *A monthly covering techniques and developments in small computer systems.*

Computer, IEEE Computer Society, Institute of Electrical and Electronics Engineers, New York. *A monthly presenting developments in computers, software, architecture, and technical challenges.*

IEEE Spectrum, Institute of Electrical and Electronics Engineers, New York. *A monthly presenting developments in electrical and electronic engineering for all members of the IEEE.*

IEEE Transactions on Computers, Institute of Electrical and Electronics Engineers, New York. *A monthly with technical articles on computer architecture and networks, hardware, software, reliability, artificial intelligence, and other state-of-the-art topics.*

IEEE Transactions on Electron Devices, Institute of Electrical and Electronics Engineers, New York. *A monthly reporting the advancements in the science and technology of semiconductor devices.*

Proceedings of the IEEE, Institute of Electrical and Electronics Engineers, New York. *A monthly with technical articles on electrical and electronic engineering.*

Technical Data Sources for Integrated Circuits

The TTL Data Book for Design Engineers, Second Edition, Texas Instruments, 1976 and Supplements, Dallas, Tex.

Signatics Logic-TTL Data Manual, 1978, Signatics, Sunnyvale, Calif.

Logic Data Book, National Semiconductor, 1981. Santa Clara, Calif.

CMOS Data Book, National Semiconductor, 1981. Santa Clara, Calif.

The Linear Control Circuits Data Book for Design Engineers, Second Edition, 1980, Texas Instruments, Dallas, Tex.

Data-Acquisition Data Book, 1982, Analog Devices, Norwood, Mass.

TM990 Introduction to Microprocessors, 1979, Texas Instruments, Dallas, Tex.

Motorola Microprocessor Data Manual, 1981, Motorola, Austin, Tex.

Microcomputer Components-Z80 Designer's Guide, 1982, United Technologies-Mosteck, Carrollton, Tex.

IAPX 86,88 User's Manual, 1981, Intel, Santa Clara, Calif.

INDEX

Acceptors, 208. *See also* Semiconductors
Active circuit elements, 173, 235
Adder, 46, 51, 412, 416
   serial, 108, 433
   sum and carry equations, 51, 416
   summing operatonal amplifier, 307, 338, 459
Advantage in pulse transmission, 343
Aliasing, 336
Alpha, 219. *See also* Bipolar junction transistors
Amplification, 236, 247, 265
   practical magnitudes, 247
Amplifiers, 235
   analog, 236, 258, 291, 442
   bode plot, 249
   characteristics, 236
   depletion MOSFET, 277. *See also* Transistor circuits
   design example, 258. *See also* Transistor circuits
   difference, 264, 302
   distortion, 244
   gain, 236
   half power points, 238, 301
   input and output impedance, 246
   noise, 245
   stability, 250
   summing, 306, 338, 459
   universal curve, 238
   see also Feedback
Analog-to-digital conversion (ADC), 317
   basic techniques, 319, 345
   comparator ladder, 320, 456
   compression and expansion, 324
   delta modulation, 335
   dual slope, 329
   monitoring voltages, 329, 332
   pulse modulation, 342
   quantization uncertainty, 320
   rate, 319
   restoration of analog signals, 337
   servo-tracking, 328
   sine wave sampling, 318, 334
   single slope, 331
   staircase, 332
   subranging, 323
   successive approximation, 322
   voltage comparator, 292, 320, 456
   voltage-controlled oscillators, 333

AND-gates, 6, 405
  diode equivalent, 41, 407
  see also Logic gates
Arithmetic, 391
  with operational amplifiers, 305
  with shift registers, 108
  see also Computer arithmetic
Arithmetic logic unit(ALU), 67, 131, 134, 143
ASCII(American standard code for information interchange), 355
Avalanche, 216. See also Semiconductor

Bandwidth, 238, 442
  with feedback, 247
  of operational amplifiers, 300
Bar codes, 355
Basic electrical circuit theory, 149
BCD (binary coded decimal) numbers, 56, 391
  decimal arithmetic, 399
  packed, 400
  R-2R digital-to-analog conversion, 340
  scaler, 91, 99, 425
BCD-to-seven-segment display decoder, 62, 427
Beta, 242. See also Feedback
Beta h(fe), 220, 257. See also Bipolar junction transistors
Bilateral switch (MOSFET), 353, 474, 477
Binary logic elements, 4. See also Logic gates
  electrical equivalent, 4
Binary numbers, 46, 54
  addition, 46, 51, 108, 412, 416
  codes, 391
  decoder, 55
  encoder, 320, 456
  subtraction, 80, 391
Bipolar junction transistors, 217
  alpha, 219
  amplifier, 256, 442. See also Transistor Circuits
  beta h(fe), 220, 257
  generic characteristics, 221
  quiescent operating point, 220, 256
Bistable elements, 69. See also Flip flops
Bits-positional value, 47, 54, 58, 307
Bit-weighted sum, 307, 338, 459. See also Digital-to-analog conversion
Bode:
  plot, 249
  simplified plot, 250, 301, 442
  stability criterion, 252
    gain margin, 252
    phase margin, 252
Boole, George, 6, 18

Boolean algebra for digital systems, 16
  chart simplification, 27
  DeMorgan's theorems, 21
  graphical representation, 23
    Karnaugh maps, 29
    Venn diagrams, 24
  summary of rules, 22
Bridge circuit, 160. See also Circuits
Bridge rectifier, 375. See also Electrical power
Bubble memory, 106
Buffers, 43
  amplifiers, 261, 300, 313
  controllable, 354
  tri-state, 61
Byte, 56, 130

Calculation of mathematical functions, 119
Capacitance, 167
  blocking DC signals, 182, 236
  coupled amplifier, 236, 442
  filter, 368. See also Electrical power
  intrinsic, 212, 237, 248, 281
  pn-junction, 121
  property of, 170. See also Circuits
  reactance, 181
  stray, 237, 248
  transient response, 167
Channel in FET, 222, 273. See also Transistors
Characteristic functions of flip flops, 85
  JK flip flop, 83
  RS flip flop, 76
  toggle flip flop, 81
Check sum, 356
Choke (inductance), 378
Circuit elements, active and passive, 173, 235
Circuits, 149
  AC (alternating current), 174
    phase, 239
    reactances, 181, 237
    resistance-capacitance, 182, 187
    resistance-inductance, 189
    simplification, 190
  bridge, 160
  current divider, 159
  DC (direct current), 149
    resistance, 150, 158
    simplification of resistor circuits, 156
      example, 158
  Norton's theorem, 165
  Q (quality), 194
  resonance, 192
    parallel circuit, 195
      three frequencies, 196
    series circuit, 192

Index

Thevenin's theorem, 162
time-dependent circuits, 166
transistor, 255. *See also* Transistor circuits
voltage divider, 4, 152
Clock (oscillators), 69, 101
CMOSFET(complementary MOSFET), 230, 278
    4000 DIP series, 38
CMRR(common mode rejection ratio), 265
    measuring, 454
    operational amplifier, 304
Coaxial cable termination, 483
Collector, 217. *See also* Bipolar junction transistors
Color code (resistor), 492
Combinational digital electronics, 37
    adder, 46, 53
    decoders, 55
        BCD-to-seven-segments, 62
        binary-to-octal, 58
    demultiplexer, 61
    digital comparator, 65
    encoder, 59
    multiplexer, 60
Common data bus, 407
Common emitter amplifier, 256, 442. *See also* Transistor circuits
Common mode rejection ratio, 265. *See also* CMRR
Comparator, 65, 320
    binary magnitude, 65
    Ladder, 320, 456. *See also* Analog-to-digital conversion
Comparison test in algorithm, 126
Complement of logic function, 8, 18
Complex numbers, 180, 192
Compression-expansion, 324. *See also* Analog-to-digital conversion
Computer arithmetic, 391
    conversion between number systems, 56, 392
    decimal arithmetic, 399
    double precision, 400
    signed numbers, 396
    Wallace algorithms, 401
Computer program, 127
    multiplication algorithm, 131
    operational code, 129
    square root algorithm, 125, 136
Conduction band, 205. *See also* Semiconducters
Contact bounce, 347, 425
    defeat by single-shot multivibrator, 348
    defeat by software, 349
Controllable buffer-inverter, 354
Co-processors, 147, 401

Counters, 90. *See also* Scalers
CPU(central processing unit), 124. *See also* Microprocessor
Current, 149, 168
    amplifier, 384
    continuity in inverting amplifier, 305
    coupled feedback, 260
    divider, 159
    limitation in inverting amplifier, 297
Current regulation, 262, 385, 487. *See also* Electrical power
Current sink, 306
Current-voltage relation, 149. *See also* Circuits
Cutin Voltage, 214, 440. *See also* Diodes

DAC, 2. *See also* Digital-to-analog conversion
Data flip-flop, 73. *See also* Flip flops
Data transmission, 344
dB (decibel), 240
    sound reference, 241
Dead time, 103, 332
    error, 481
Decoder/driver for seven-segment displays, 62, 427
Degenerative feedback, 241. *See also* Feedback
Delta modulation, 335. *See also* Analog-to-digital conversion
DeMorgan's theorems, 21
    demonstration by Venn diagrams, 24
    logic gate equivalents, 22
Demultiplexer, 61
Depletion, MOSFET, 226. *See also* Transistors
Depletion layer, 211. *See also* Semiconductors
Difference amplifier, 264, 302
Differentiating circuit, 170
Digital-to-analog conversion, 2, 306, 317, 337, 459, 481
    basic techniques, 345
    bit-weighted sum, 306, 338, 459
    glitch, 342
    pulse modulation-demodulation, 342
    rate meter, 343, 481
    R-2R current ladder, 340
    settling time, 339
    summing amplifier, 306, 338, 459
Digital display, 62
    4-1/2 digits, 331
    3-1/2 digits, 331
Diodes, 40, 212, 407, 440
    characteristics, 214, 440
    LED, 62, 427, 440, 445
    practical, 215
    rectifier, 367. *See also* Electrical power
    zener, 216, 440

DIP (dual inline package), 38
  power connections, 6, 405
  7400 series, 39
  voltages, 38, 405
Discrete element circuit theory, 151
Divide-by-N counter, 91. See also Scalers
Division by Wallace algorithm, 401
  by two with shift register, 108, 431
Donors, 210. See also Semiconductors
Don't-care values, 93
Doping, 210. See also Semiconductors
Double Precision Arithmetic, 400
DRAM, 282. See also Dynamic random access memory
Dual-slope, 329. See also Analog-to-digital conversion
Dynamic elements, 80
Dynamic random access memory(DRAM), 114, 282
  refreshing, 283
Dynamic shift registers, 110, 281
  transfer-refresh, 281

Edge triggering flip flops, 80
EEPROM(electrical erasable programmable ROM), 121, 288
Electrical circuit equivalent of binary logic, 4
Electrical circuit theory, 149
Electrical power, 198, 363
  AC-DC equivalent, 199
    root-mean-square voltage, 199, 364
  bridge rectifier, 375
  capacitance filter, 368
    additional filtering, 376
  design example, 374
  DIP, 6, 405
  efficiency, 380, 389
  full wave, 371
  half wave, 367
  performance curves, 372
  performance example, 381
  power factor, 198
  rectifier, 367, 440
  regulator, 263, 379
    current, 263, 487
    operational amplifiers, 383
    shifting voltage, 384
    three-terminal, 379, 487
  ripple, 369, 377
  switching, 386
  transformer, 366
  two-sided, 293, 378
  voltage doubler, 378
  voltage polarity, 375, 378

Electromotive force(EMF), 149, 198
Electronic switches, 353, 474. See also MOSFET
Emitter, 217. See also Bipolar junction transistor
Emitter-follower, 261, 313. See also Transistor circuits
Enable, 61, 113, 115, 272
Encoder, 59, 456
Energy Bands-Gap, 205. See also Semiconductors
Enhancement MOSFET, 228. See also MOSFET
EPROM (Erasable Programmable ROM), 121, 287
  erasing, 287
  programming, 287
Equipment interconnections, 485
Equivalence gate (XNOR-gate), 14, 412. See also Logic gate
Error correction (parity), 356
Euler's relation, 180
Excess-three code, 391. See also Computer arithmetic
Exclusive OR-gate, 14, 412. See also Logic gates
Exclusive NOR-gate, 15, 412. See also Logic gates
External regenerative feedback, 251
Extrinsic semiconductor, 208. See also Semiconductors
Eye response, 445

Fan-out, 43, 266
Feedback, 235, 241
  advantages of negative feedback, 243, 293, 298
  band width expansion, 247, 301
  beta (factor), 242
  operational amplifiers, 293
Field effect transistors(JFET), 222. See also Transistors
Field ionization, 216
Filters, 368, 376
  active filters, 471
  low-pass, 187
  see also Electrical power
Flip flops, 69
  characteristic equations summary, 85
  data, 73, 77, 113, 116, 418
  edge triggered toggle, 80
  indeterminate state, 70, 418
  initialize, 70
  JK, 78, 81, 422
    state diagrams, 84
  master-slave, 79, 81, 421
  quiescent inputs, 73
  RS, 71, 74, 418
  toggle, 80, 422
Floating gate read only memory(EPROM), 287

Index

FPLA(field programmable logic array), 358
Frequency:
  band width, 238, 300
  half power point, 238
  parallel resonant, 195
    maximum f(max), 196
    purely resistive f(0), 196
  power line, 363
  series resonance f(r), 192
Full adder, 51, 416

Gate, 6. *See also* Logic gate
Gated data latch, 78, 418
Gain, 236. *See also* Amplifiers
Gain margin, 252. *See also* Bode
Glitch, 342. *See also* Digital-to-analog conversion
Gray code, 392. *See also* Computer arithmetic

Half adder, 46, 412
Half power point, 238, 301
  in Decibels dB, 240
  see also Amplifiers
Halt subtractor, 48, 412
Hamming code (parity check), 356
Hexadecimal numbers, 56
h(fe) forward-current transfer function, 257. *See also* Bipolar junction transistor
High pass filter, 190
Holes, 207. *See also* Semiconductor
Hysteresis, 350, 469

Impedance, 183
  improvement, 352
Indeterminate state, 70, 418. *See also* Flip flops
Inductance, 171, 173, 179, 378
  energy storage in switching power supplies, 388
  property of, 173
Inexpensive voltmeter, 352
Injection-gate bulk switching device, 230
Input impedance, *see specific devices*
Instability in amplifiers, 235. *See also* Amplifiers
Insulated-gate FET, 226. *See also* MOSFET
Insulators, 205. *See also* Semiconductors
Integrator, 308, 465
  analog-to-digital conversion, 329
  resistance-capacitor circuit, 170
  see also Operational amplifier
Intel Corp, 123
Interconnecting logic gates, 43
Interfacing equipment, 483
Intrinsic capacitance, 248, 281. *See also* Capacitance
Intrinsic semiconductors, 206. *See also* Semiconductors
Inverter, 10, 405, 407. *See also* Logic gates
Inverting Amplifier, 293, 448. *See also* Operational amplifiers

J (rotation operator), 180
JFET(junction field effect transistor), 222, 273
  characteristics, 225
  equivalent resistor, 225, 273
  input resistance, 276
JK flip flop, 78, 422. *See also* Flip flops

Karnaugh map, 29
Kilby, Jack, 125
Kirchhoff's laws, 149
  conventions, 151
  summary, 155

Latch, 112
LC filter section, 376. *See also* Electrical power
LED, 62, 440. *See also* Diodes
Line termination, 483
Logarithmic, 325. *See also* Analog-to-digital conversion
Logarithmic voltage-current relation, 310
Logic functions, 45, 410
Logic gates, 4, 6, 405
  AND-gates, 6, 41, 405, 407
  configurations, 255. *See also* Transistor circuits
  equivalence (XNOR-gate), 15, 412
  exclusive OR-gate (XOR), 14, 412
    from elementary gates, 14, 412
    programmable buffer-inverter, 354
  inverter, 10, 405
  NAND-gate, 8, 405
    electrical circuit equivalent, 10
  NOR-gate, 12, 405
    electrical circuit equivalent, 13
  open-collector, 44, 119, 271, 407
  OR-gate, 11, 42, 405, 407
  synthesis, 40, 407
  tri-state, 272
  wired-OR-gate, 44, 407
Long word (Binary), 130
Lookup tables, 358
Low-pass filter, 187, 471
LSB(least signficant bit), 47

Magnetic bubble memory, 106
Magnitude:
  binary positional numbers, 55
  comparator, 65
  comparison by ALU, 146
  complex quantity, 184
Majority carriers, 211. *See also* Semiconductors

Master-slave flip flops, 81. *See also* Flip flops
Memory, 105
  cell, 70, 435
  DRAM, 282. *See also* Dynamic random access memory
  EPROM(erasable programmable ROM), 121, 287
  PROM, 120. *See also* Programmable read only memory
  RAM, 105. *See also* Random access memory
  registers, 105
  ROM, 105. *See also* Read only memory
  volatile, 106, 114, 432
Microprocessor, 123
  architecture, 139
  CPU (central processing unit), 124
  operational code, 129
  size definition, 130
Minority carriers, 211
  role in thermal runaway, 260
  see also Semiconductors
Minterm, 59
Missing pulse detection, 349
Mnemonics, 126. *See also* Computer program
Modulation, 343
Modulus, 99, 425
Moll's relation, 310
Monitoring voltages, 329, 332
Monostable elements, 69, 102, 477, 481
MOSFET(metal oxide semiconductor field effect transistor), 38, 110, 226, 276
  complementary, 230, 278
  depletion, 226
  enhancement, 228
  handling, 229, 474
  inverter, 278
  load transistors, 278
  NAND-gate, 278
  NOR-gate, 278
  switches, 353, 474
Multiplexer, 60, 110
  display, 445
Music and speech sampling rate, 334

NAND-gate, 8. *See also* Logic gate
Negative Feedback, 241. *See also* Feedback
Noise, 245, 310. *See also* Amplifiers
Noninverting amplifiers, 298. *See also* Operational amplifiers
NOR-gate, 12, 405. *See also* Logic gates
Norton's theorem, 165
NPN, 217. *See also* Bipolar junction transistors
n-type, 209. *See also* Semiconductors
Number systems, 54, 391

Numerical data processor, 147, 401
Nyquist sampling rate, 334
Nyquist stability criterion, 251

Octal number system, 56
Offset voltage, 312. *See also* Operational amplifiers
Ohm heating, 181
Ohmic junction, 215. *See also* Semiconductors
Ohm's law, 50
One's complement, 50, 397. *See also* Computer arithmetic
One-shot multivibrator, 103, 477, 481
Open-circuit gain, 242, 292
Open-circuit voltage, 162
Open-collector gates, 44, 119, 271, 407. *See also* Logic gates
Operational amplifiers, 291, 448, 451
  arithmetic uses, 305, 338, 459
  characteristics, 451
  common mode rejection ratio(CMRR), 304, 454
  emitter follower, 261, 313, 448
  frequency dependence, 300, 450
  input-bias current, 312, 451
  input-offset voltage, 312, 451
  inverting amplifier, 293, 448
    sinking current, 306
    summing node, 306
    virtual ground, 293
  limitations, 293
  matching output current requirements, 313
  noninverting amplifier, 298, 448
  open-circuit gain, 292
  slew rate, 301, 451
  voltage follower, 299
  voltage supply, 293
Operational code, 127
  for hypothetical microprocessor, 129
  see also Computer program
OR-gate, 11, 405. *See also* Logic gates
Oscillators or clocks, 69, 101
  astable, 344
  LC resonance, 192, 195
  relaxation, 102

Packed BCD, 400. *See also* Computer arithmetic
Parity, 355
Passive circuit elements, 173, 235, 471
Peak voltage detection, 463
Phase angle, 235. *See also* Amplifiers
Phase margin, 252. *See also* Bode
Pinouts, 39
  Seven-Segment Displays, 491
  see also specific elements

Index **503**

PLA(programmable logic array), 358
pn-junction, 211. *See also* Semiconductors
PNP, 217. *See also* Bipolar junction transistors
Point-contact transistor, 217. *See also* Transistors
Positive feedback, 235. *See also* Feedback
Positive logic, 4
Programmable read only memory(PROM), 120
    configurations, 275. *See also* Transistor circuits
    electrical erasable (EEPROM), 121
    erasable (EPROM), 128, 287
    programming, 120, 287
PROM, 120. *See also* Programmable read only memory
Propagation delay, 266
p-Type, 210. *See also* Semiconductors
Pull-up resistor, 44, 427, 483
Pulse height analysis, 331

Q (quality factor), 194. *See also* Circuits
Q (state of flip flop), 70
Q(t + 1) (sequential state of flip flop), 70
Quantization uncertainty, 320. *See also* Analog-to-digital conversion
Quiescent inputs, 73
Quiescent operating points, 220, 258

Radix point, 392
RAM, 105. *See also* Random access memory
Random access memory (RAM), 105, 118, 435, 437
    dynamic, 114, 282. *See also* Dynamic random access memory
Rate meter, 342, 481
RC (resistance-capacitance), 167. *See also* Circuits
Reactances, 181. *See also* Circuits
Read only memory(ROM), 105, 119
    applications, 119
    configurations, 274. *See also* Transistor circuits
Rectification, 367. *See also* Electrical power
Regenerative feedback, 241. *See also* Feedback
Registers, 105
    data, 112
    shift, 107. *See also* Shift registers
    see also Memory
Regulation, 262, 379, 487
    simple circuits, 262. *See also* Transistor circuits
    three-terminal regulators, 379, 487
    using operational amplifiers, 383
    see also Electrical power
Relaxation oscillators, 102
Resistors, 150, 177
    color code, 492
    switched-capacity, 353, 474
    see also Circuits

Resonance, 149. *See also* Circuits
Ripple counters, 90. *See also* Sealers
Ripple voltage, 369. *See also* Electrical power
RL (resistance-inductance), 171. *See also* Circuits
RMS (root-mean square), 199, 364
ROM, 105. *See also* Read only memory
Rotational operator j, 180
Rounding error, 431
RS flip flops, 71. *See also* Flip flops
RTL (resistor-transistor logic), 266
R-2R current ladder, 340, 459. *See also* Digital-to-analog conversion

Sample analog signals, 317, 447
    rate for music and speech, 334
Sample-and-hold, 323, 477
Satellite transmissions, 344
Scalers, 69
    decade, 91, 99, 425 *ie. BCD counter*
    design of sequential circuits, 92
    divide-by-N, 91
    rejoining sequence, 97, 445
    ripple, 90, 425
    scale-of-five, 92
    scale-of-three, 445
    start-stop, 425, 429
    synchronous, 98, 425, 429
    up-down counter, 100
Scaling by toggle flip flop, 81
Schmitt trigger, 350, 469
    inputs on DIPs, 350, 469
Schottky junction, 215. *See also* Semiconductors
Security systems, 350
Semiconductors, 205
    avalanche, 216
    n-type, 209
    Ohmic junction, 215
    pn-junction, 211
        depletion layer, 211, 222, 273
        diode, 212
        electrical properties, 211
            dependence on host atoms, 215
        Moll's relation, 310
        uncovered charges, 211
    p-type, 210
    thermister, 207
Sequential digit electronics, 69
    analysis, 85
    design, 92
    state diagrams, 84, 88
Serial adder, 108, 433
Series resistors, 152. *See also* Circuits
Series resonance, 192. *See also* Circuits

504 Index

Series voltage regulator, 262, 383. *See also* Electrical power
Servo ADC, 329. *See also* Analog-to-digital conversion
Settling time, 339. *See also* Digital-to-analog conversion
Seven-segment display, 62, 427, 447
   multiplexed, 445
   pinouts, 491
Shift registers, 98, 107, 431, 433
   dynamic, 106. *See also* Dynamic shift registers
   universal, 110
Shunt, 263, 384
   current, 165
   resistance (bypass), 165
   see also Voltage regulator
Signal mixing, 306, 459
Signal sampling, 477
Signed numbers, 396. *See also* Computer arithmetic
Silicon, 205. *See also* Semiconductors
Simplified bode plot, 249. *See also* Bode
Sine wave sampling, 318, 334. *See also* Analog-to-digital conversion
Single-shot multivibrator, 102, 348, 477
Single-slope ADC, 331. *See also* Analog-to-digital conversion
Sinusodial voltages, 174. *See also* Circuits
Slew rate, 301, 451. *See also* Operational amplifier
Solderless experimental board, 405
Solid state electronic devices, 205. *See also* Semiconductors
Speech digitizing, 334
Square root algorithm, 125, 136
   Wallace, 401. *See also* Computer arithmetic
   see also Computer program
Square wave fourier components, 174
Stability criteria, 250. *See also* Amplifiers
Stack, 112, 135, 141. *See also* Shift register
Staircase ADC, 332. *See also* Analog-to-digital conversion
State diagram, 84, 88
State of logic element Q, 70
Static RAM, 115. *See also* Random access memory
Static shift register, 281. *See also* Shift registers
Stray capacitance, 237, 248. *See also* Capacitance
Student's guide to exercises, 490
Subtraction, 49, 145, 397. *See also* Computer arithmetic
Summing amplifier, 306, 338, 459. *See also* Digital-to-analog conversion
Switched-capacitor resistor, 353, 474

Switching power supplies, 386. *See also* Electrical power
Synchronous circuits, 98

Termination of coaxial cable, 483
Thermal runaway, 260
Thermistors, 207. *See also* Semiconductors
Thevenin's theorem, 162, 246
Time Constant, 166. *See also* circuits
Transfer-refresh process, 281. *See also* Dynamic shift registers
Transformer, 366. *See also* Electrical power
Transient response, 166. *See also* Circuits
Transient suppression, 347
Transistor, 217
   CMOSFET, 230
   interface, 483
   JFET (junction field effect transistor), 222, 273
   MOSFET, 226, 276
      depletion, 226
      enhancement, 228
   see also Bipolar junction transistor
Transistor circuits, 255
   common emitter amplifier, 256, 442
      design example, 258
      thermal runaway, 260
   current gain, 262
   depletion MOSFET analog amplifier, 277
   difference amplifier, 264
   dynamic random access memory, 282
   dynamic shift register, 281
   emitter-follower amplifier, 261, 313
   EPROM, 287
   monostable oscillator, 102, 348
   MOSFET logic gates, 278
   open-collector gate, 271
   programmable read only memory, 275
   read only memory, 274
   Schmitt trigger, 350
   simple regulating circuits, 262
   transistor-transistor logic gates, 38, 266
   tri-state gate, 272
Transistor-transistor logic(TTL), 38, 266
   unconnected pins, 115, 425
Trignometric functions, 358
Tri-state devices, 61, 110, 272
Truth table, 6, 405. *See also* specific gate
TTL (transistor-transistor logic), 38. *See also* Transistor circuits
Two's complement, 50, 145, 397. *See also* Computer arithmetic
Two-sided power supply, 293. *See also* Electrical power
Two-terminal linear circuit, 162

UART(universal asynchronous receiver transmitter), 108, 140
Unallowed (indeterminate) states, 71, 418. *See also* Flip flops
Uncovered charges, 211. *See also* Semiconductors
Universal amplification curve, 238. *See also* Amplifiers
Universal shift registers, 110. *See also* Shift registers
Up-down counter, 100. *See also* Scalers

Valence band, 205. *See also* Semiconductors
Venn diagrams, 24
Virtual ground, 293. *See also* Operational amplifiers
Voltage comparator, 292. *See also* Analog-to-digital conversion
Voltage controlled oscillators, 333, 483
Voltage divider, 4, 152
Voltage follower, 299
  as input element, 304, 352
  see also Operational amplifier
Voltage measurements, 160, 329, 332

Voltage regulation, 263, 487. *See also* Electrical power
Voltage sampling, 317, 447

Wallace algorithms, 401. *See also* Computer arithmetic
Weighted sum, 306, 338, 459. *See also* Digital-to-analog conversion
Wire connections in exercises, 405
Wired OR-gate, 44, 407
Word (binary), 130
Word-time generator, 109, 429
Write command, 115

XNOR-gate (equivalence gate), 15, 412
XOR-gate (exclusive OR-gate), 14
  adding circuits, 46, 53, 412, 416
  from basic gates, 14, 412
  programmable buffer-inverter, 354

Zener diode, 216, 440
  breakdown mechanisms, 216
  reference voltage, 217, 262, 384

7481

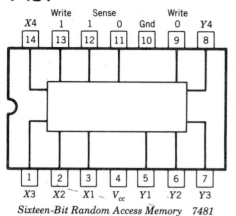

Sixteen-Bit Random Access Memory 7481

7486

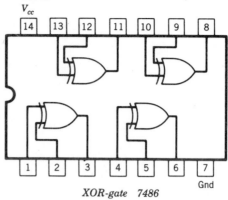

XOR-gate 7486

7483

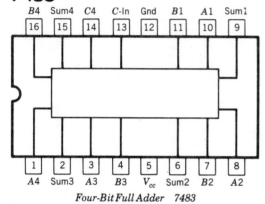

Four-Bit Full Adder 7483

7495

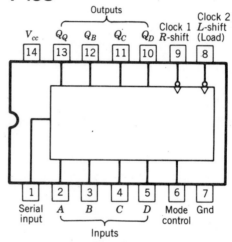

Four-Bit Shift Register 7495

7485

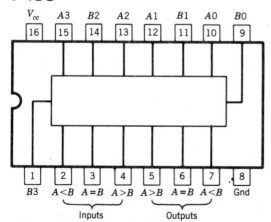

Four-Bit Magnitude Comparator 7485

74121

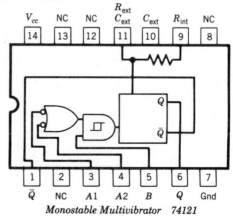

Monostable Multivibrator 74121

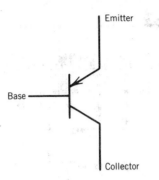

PNP Bipolar Junction Transistor   *NPN* Bipolar Junction Transistor

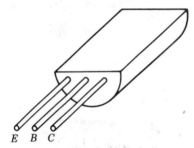

Transistor lead arrangement

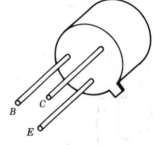

Transistor lead arrangement

78XX

Three-Terminal Positive Voltage Regulator

79XX

Three-Terminal Negative Voltage Regulator

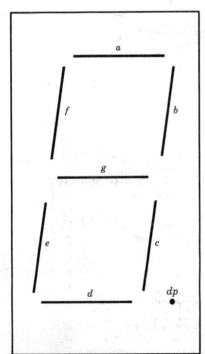

Seven-Segment Display